Sustainable Development Goals Series

The **Sustainable Development Goals Series** is Springer Nature's inaugural cross-imprint book series that addresses and supports the United Nations' seventeen Sustainable Development Goals. The series fosters comprehensive research focused on these global targets and endeavours to address some of society's greatest grand challenges. The SDGs are inherently multidisciplinary, and they bring people working across different fields together and working towards a common goal. In this spirit, the Sustainable Development Goals series is the first at Springer Nature to publish books under both the Springer and Palgrave Macmillan imprints, bringing the strengths of our imprints together.

The Sustainable Development Goals Series is organized into eighteen subseries: one subseries based around each of the seventeen respective Sustainable Development Goals, and an eighteenth subseries, "Connecting the Goals," which serves as a home for volumes addressing multiple goals or studying the SDGs as a whole. Each subseries is guided by an expert Subseries Advisor with years or decades of experience studying and addressing core components of their respective Goal.

The SDG Series has a remit as broad as the SDGs themselves, and contributions are welcome from scientists, academics, policymakers, and researchers working in fields related to any of the seventeen goals. If you are interested in contributing a monograph or curated volume to the series, please contact the Publishers: Zachary Romano [Springer; zachary.romano@springer.com] and Rachael Ballard [Palgrave Macmillan; rachael.ballard@palgrave.com].

Himachalam Dasaraju · Tulus T.H. Tambunan
Editors

Role of Micro, Small and Medium Enterprises in Achieving SDGs

Perspectives from Emerging Economies

palgrave
macmillan

Editors
Himachalam Dasaraju
School of Commerce
and Management
Sri Venkateswara University
Tirupati, India

Tulus T.H. Tambunan
Center for Industry, SME
and Business Competition Studies
Universitas Trisakti
Jakarta Barat, Indonesia

ISSN 2523-3084 ISSN 2523-3092 (electronic)
Sustainable Development Goals Series
ISBN 978-981-99-4828-4 ISBN 978-981-99-4829-1 (eBook)
https://doi.org/10.1007/978-981-99-4829-1

Cover illustration: Barbara Boensch/Alamy Stock Photo

This Palgrave Macmillan imprint is published by the registered company Springer Nature Singapore Pte Ltd.
The registered company address is: 152 Beach Road, #21-01/04 Gateway East, Singapore 189721, Singapore

Paper in this product is recyclable.

Preface

The book purports to explore the various phases of transformation in the domain of micro, small, and medium enterprises (MSMEs), which are a crucial part of UN Sustainable Development Goals (SDGs) in the context of developing or emerging economies. MSMEs are the key elements and driving force to achieve progress in all areas of the economy across the world. The book reviews the importance, progress, and role of MSMEs in achieving the SDGs in emerging economies. The vital contribution of MSMEs to broader social-economic objectives, including job creation, poverty alleviation, women empowerment, and gender equity makes them a key priority area for achieving the 17 UN SDGs. It gained a lot of importance in achieving SDGs in developed economies in general and in developing economies in particular, owing to its inherent characteristics of inclusive growth and sustainable development.

MSMEs play an important role in most developing economies contributing up to 45% of total employment and up to 33% of gross domestic product (GDP). These numbers would be significantly higher when enterprises in informal MSMEs are included and are major creators of jobs mostly in developing and less developed countries. In emerging markets, 4 out of 5 new positions in the formal sector were created by MSMEs, which is about 90% of total employment. As reported by the World Bank, around 600 million jobs will be needed in the next 15 years to absorb the ever-growing global workforce, mainly in Asia and sub-Saharan Africa.

MSMEs are one of the strongest drivers of economic development, innovation, and employment generation. According to Asian Development Bank (ADB) statistics, in the Asian region specifically, MSMEs account for 96% of total enterprises, employ 62% of the labor workforce, and contribute an average of 42% of GDP. The MSMEs Economic Indicators Database 2019 indicated the region-wise number of formally registered MSMEs across 176 economies. The East Asia and Pacific region is having a higher number of MSMEs followed by South Asia, Sub-Saharan Africa, Europe, Central Asia, North America, etc. Employment creation through MSMEs will often directly benefit the poor and vulnerable, particularly women and youth, thereby directly reducing poverty, increasing economic status, and so on. Entrepreneurs, in the MSMEs sector, are the lifeline of the economy in the process of effective utilization of untapped local resources such as men, money, and material resources.

Therefore, MSMEs are vital for poverty eradication, mostly among women and other socially underprivileged groups. Keeping in view its critical contribution to the economy, the UN has assigned a great role to MSMEs for taking a lead in achieving most of the economic-related SDGs, including promoting inclusive and sustainable economic growth, increasing employment opportunities and decent work, especially for the poor, advancing sustainable industrialization and innovation, and creating a positive push for a higher quality of life, better education, and good health for all. MSME development has the potential for wide-reaching impacts on the SDGs globally, including SDG 1 (end poverty), SDG 2 (zero hunger), SDG 3 (good health and well-being), SDG 5 (gender equality), SDG 8 (promote inclusive and sustainable economic growth, employment and decent work), and SDG 9 (improve sustainable industrialization and fostering innovation).

According to the Business and Sustainable Development Commission, sustainable business models could open economic opportunities worth $12 trillion and create 380 million jobs by 2030, with more than 50 percent located in developing countries. Such opportunities for MSMEs have been identified under each goal. MSMEs possess immense potential for the achievement of all SDGs directly or indirectly.

The purpose of this book which consists of 15 chapters is to discuss the role of MSMEs in achieving the SDGs in various emerging countries, especially in Asia. Excerpt Chapter 1 (Introduction), the remaining 14 chapters are structured into three sections to present a distinction between those that develop a country panorama of the contribution of MSMEs to the achievement of the SDGs (Part I with Chapters 2, 3, 4 and 5), those whose approach is more focused on a particular area of the SDGs (Part II with Chapters 6, 7, 8, 9, 10, and 11), and those interested in the operational management or development of MSMEs (Part III with Chapters 12, 13, 14).

Chapter 1 the Introduction, discusses MSMEs' significance, furtherance, and key role in bringing off SDGs in developing countries. It analyzed theoretically the MSMEs' contributions toward achieving SDGs. The cardinal contribution of MSMEs to broader social-economic objectives, including job creation, poverty alleviation, empowerment of people, women empowerment, and gender equity, etc., makes them a key priority area for achieving the SDGs. It assumes great importance in achieving SDGs in the context of developed economies in general and developing economies in particular, owing to its inherent characteristics of inclusive growth and sustainable development. The generation of huge employment and inclusive growth through MSMEs will often directly benefit the poor and vulnerable sections of the people, particularly women and youth, thereby directly reducing poverty, the severity of hungry, increasing economic status, and empowerment. Besides, it leads to nurturing entrepreneurship, which is an engine of inclusive growth and stimulus for economic prosperity and empowering people. All the developing economies have taken several measures to promote SMEs by initiating various flagship programs and initiatives to stimulate further them. Besides various progressive measures of governments initiated globally, the MSMEs sector is not free from myriad problems, and it requires more stimuli to strengthen it further. In addition, the COVID-19 pandemic lockdown has badly disrupted the functioning of MSMEs worldwide. It is a dire need of the day to nurture and promote this sector and make it viable to access emerging technologies, global markets, markets, and competitiveness, and to play a vital role in reaching UN targets of SDGs by 2030.

Chapter 2 entitled "Role of MSMEs in the Achievement of SDGs in India in the Current Scenario," describes SDGs in India and how MSMEs in the country can attain these goals in the current scenario. The MSMEs in India produce more than 8000 products. They have generated employment for more than 11 crore people second only to agriculture. Further, they contributed 30 percent to GDP and 40 percent to exports. The chapter states that the MSMEs in India can play a vital role in the achievement of SDGs by 2030. Thus, these enterprises can contribute to the economic as well as social development of the country.

Chapter 3 titled "The Potential Role of MSMEs in Achieving SDGs in Indonesia" examines the contribution of MSMEs in achieving SDGs in Indonesia with some secondary data. It analyzes the direct or indirect relationships between the existence or growth of MSMEs and the achievement of SDGs such as reducing poverty and eliminating hunger. By creating job and business opportunities and the generation of GDP, MSMEs indirectly participate in reducing poverty, which also means eliminating hunger, and *ceteris paribus*. Also, by providing business opportunities to women, low-skilled/unskilled youth, and disabled persons, MSMEs contribute to reducing inequality.

Chapter 4 with the title "MSMEs, The Growth Engine of India to Achieve Sustainable Development Goals and Self Reliance," shows that MSMEs are a significant pillar of the Indian economy contributing considerably to its growth and development and achieving the SDGs. These enterprises produce more than 6000 varied products providing employment opportunities to around 70 million people and contributing to 45% of manufacturing production and 40% of exports. The sector has the potential to strengthen the socio-economic growth of the country by resolving the unemployment problem and reducing regional inequalities and economic imbalance. This is in consensus with the direct or indirect attainment of many of the 17 SDGs. The supporting data presented in this chapter clearly depicts the role of MSMEs in the reduction of disparities based on region and gender in varied sectors of the country where Uttar Pradesh and West Bengal tend to be the exemplary states benefitting from the initiative. The role of MSMEs in the economic growth of the country through the achievement of SDGs along with the various challenges faced by the sector and their possible remedies are suggested in the chapter. The vibrant demography of the MSMEs needs to be analyzed to understand the prevailing conditions, its strength and gaps to facilitate suitable policy interventions to uplift the sector. Schemes should be devised to support those streams which are yet to be nurtured. Financial support from the government and favorable policies can push India to become a high-growth manufacturing-based economy thus, realizing its goal of being self-reliant and contributing to transforming the world by the attainment of UN-laid goals through SDGs.

Chapter 5 entitled "MSMEs and SDGs: Evidence from Bangladesh" begins with a statement that SDG achievement necessitates appropriate development priorities and programming, with local governments best prepared to adopt policies and programs that alleviate poverty, decrease inequality, promote climate vulnerability, and promote gender equality. While the big corporates and listed firms have progressed in their journey to achieve the SDGs, there is limited evidence from the SME sector. The attainment of SDGs requires a multi-stakeholder approach to rural sustainable development through SME expansion. The relationship between SMEs and sustainability is mutually dependent, and the success of the sustainability agenda is heavily influenced by SMEs. SMEs cannot expand without incorporating sustainability concepts into their business strategy. This chapter aims to explore the state of SME adoption of SDGs practices. A quantitative research approach using a survey instrument is adopted to achieve the aims of this chapter. Respondents for this survey are SME owners and employees from the agricultural, fishing, manufacturing, wholesale, and retail sectors in Bangladesh. A sample of 243 respondents provides the data for analysis. Overall, this chapter provides an understanding of where the SME sector stands concerning the adoption of sustainable development goals and what needs to be done to initiate policy guidelines for progressing SMEs' adoption of SDGs in Bangladesh and ensure the sustainability of the SME sector.

Chapter 6 is about the role of MSMEs in Africa's performance with regard to SDGs with its title "Africa's Performance with Regard to SDGs and the Important Role of MSMEs." African countries participate in many global programs, especially those of the United Nations Organisation and its affiliates. In this chapter, the continents' participation in the global efforts to ensure human activities are carried out with due regard to future generations' welfare is discussed. As most activities involve MSMEs, their role is discussed. It also discusses the evolution of the sustainability concept. Africa's performance with regard to SDGs, and the important role of MSMEs in the achievement of SDGs in Africa.

Chapter 7, "The Role Palyed by Research and Development in Uganda's MSME Toward Promoting of SDGs" is a case from Uganda. The chapter starts with a brief discussion of the global efforts at improving the quality of life sustainably. In particular, the authors report on findings of research in which they sought to establish the effect of research and development (R&D) on the growth of MSMEs in the greater Kampala metropolitan area (GKMA), Uganda. Quantitative research methods were adopted. The descriptive, correlational, survey, and cross-sectional research designs were used on a sample of 226 MSMEs where data was collected using a survey research questionnaire. Findings revealed that 18.7% of the growth of MSMEs was accounted for by R&D in the region.

Chapter 8, entitled "MSMEs as a Driving Force for Achieving SDGs in Developing Economies: Special Focus on Gender Equality in India," reviews the importance, progress, and contributions of MSMEs in accomplishing the SDGs in developing countries. MSMEs are having significant contributions to the comprehensive social and economic objectives, making them a crucial right-of-way priority for attaining the 17 SDGs. It was tasked with great importance in attaining SDGs in the developed and developing economies as well, due to its immanent attributes of inclusive growth and sustainable development all over the world. It aims to explore the various phases of transformation in the domain of gender equality, which is a crucial part of the UN SDGs in the context of emerging economies. Gender equality and empowerment of women are the key elements to achieving progress in all areas of the economy. It is one of the eight-millennium goals and seventeen sustainable development goals as well set for the well-being of people across the world. Further, this chapter provides a brief review of women's empowerment, gender equality, the global gender gap index, strategies and paradigms toward women's empowerment, and economic empowerment in women's development. It provides insights into the relationship between MSMEs and gender equality in the Indian context as a part of UN SDGs. The conceptual study on this topic is validated by the secondary data as far as possible to have better insight.

Chapter 9 with its title "Role of MSMEs for Achieving SDGs in Developing Economies with Special Reference to India and Bangladesh" argues that enterprises are crucial to achieving the SDGs, especially Goal 1 Ending Poverty, Goals 8 Decent Work and Goal 9 Economic Growth which is based on infrastructure, industry innovation. MSMEs' contributions go beyond these three objectives; they also directly or indirectly aid in the accomplishment of other objectives. In this chapter, an effort has been made to examine how MSMEs contribute to the achievement of the SDGs. The function of MSMEs has been divided in achieving various objectives in various activity aspects. The chapter has also included some suggested actions that can help MSMEs become more sustainable for developing economies, especially in Bangladesh and India, because there is much similarity in both countries, which will ultimately aid in the achievement of the SDGs.

Chapter 10 with the title "The Role of MSMEs in Achieving the SDGs, Especially on Poverty Reduction in 5 ASEAN Countries" focuses on the first goal of the 17 SDGs, namely the importance of MSMEs in poverty alleviation. It analyzes the effect of MSME growth in poverty alleviation in 5 ASEAN countries during the research period 2010–2019, namely Indonesia, Myanmar, Philippines, Thailand, and Vietnam. The dependent variable in this study is poverty as measured by the proportion of the population living below the international poverty line, in terms of the number of monetary units to meet the minimum living needs. This study uses the panel data method and the results show that there is a significant relationship between MSME growth, the Trade Openness Index, per capita income growth, and the Human Development Index on poverty alleviation in 5 ASEAN countries. The results of this research give a strong message that the sustainable production growth of MSMEs is one or even the main prerequisite for the ability of these enterprises to play their role in achieving the SDGs.

Chapter 11 entitled "Contribution of MSMEs to SDGs Through Corporate Social Responsibility" is about the importance of corporate social responsibility (CSR) through which MSMEs can contribute largely to SDG achievement in India. The author argues that the SDGs are ambitious and require revolutionary changes in business activities nationally. This transformation may demand doing business differently like adapting new business models, bringing in new innovation, and adopting new technology. This situation opens up new business opportunities for the business world in general and MSMEs in particular. It is also expected that businesses should operate more sustainably and ethically. CSR is the way through which businesses contribute toward societal development. A country like India has aligned its CSR guidelines or laws with SDGs to achieve these global goals at the national level. Seventeen SDGs can be roughly divided into 3 categories, Environment- SDG 6, 13, 14, 15, Society SDG 1, 2, 3, 4, 5, 7, 11, 16, and Economy- SDG 8, 9, 10, 12, SDG 17th Can be a part of all three categories. CSR activities can address most of these SDGs. CSR activities of all MSMEs together can contribute largely to SDG achievement in India.

Chapter 12 is a story from Nigeria with the title "Increasing the Role of Small Businesses in Achieving the SDGs Through Strengthening Managerial Capabilities: The Case of Nigeria." The chapter argues that MSMEs are vital and they contribute to broader social economic objectives, including employment creation making this sector a significant priority area for achieving the SDGs. However, their survival has generated concern, and academic attention nationwide. These businesses have faced dwindling performance and challenges attributed to the ambiguity of managerial capabilities and entrepreneurial self-efficacy. With this argument, the chapter examined the mediating role of self-efficacy in the relationship between managerial capabilities and the performance of MSMEs in achieving SDGs in Nigeria. For this chapter, the author has conducted a survey research approach, using 460 owner-managers operating in Lagos State. Based on the results, this chapter concludes that entrepreneurial self-efficacy improved the relationship between managerial capabilities and small business performancse, with theoretical and practical consequences for entrepreneurial theory and practice. The study recommended that MSMEs should prevent managerial ineptitude to increase performance and in achieving SDGs in Nigeria.

Chapter 13 is about MSMEs in Malaysia and Botswana. With its title "Configuring Ecosystems for MSMEs to Achieve SDGs: A Co-Design Study in Botswana and Malaysia," the chapter starts by stating that while these enterprises in the two countries are contributing immensely to their nations' economic development, the challenges of high unemployment, poverty, and inequality still exist. These challenges (and others) defeat the efforts tailored to achieving the most sustainable development goals. Therefore, this chapter argues that much work needs to be focused on exploring local SME ecosystems, where entrepreneurs are embedded. Developing entrepreneurship in emerging economies requires more than just giving out money to SMEs, but supporting them with tools to understand, nurture, and manage local ecosystems. This chapter draws from case studies between two emerging economies, i.e., Botswana and Malaysia. The chapter discusses key findings drawn from co-design activities with SMEs and their stakeholders in both cases. This chapter argues that co-design approaches are significant in helping SMEs collaboratively understand and access resources embedded in local ecosystems, thus bringing them closer to achieving sustainable development goals. The discussions add to the literature on how SMEs and policymakers can use co-design to reveal important factors that affect framework conditions at the local level and how policies developed at the national level can be tailored to local conditions to support SMEs ecosystems and, by large, UN sustainable development goals.

Finally, Chapter 14 entitled "MSMEs in Achieving SDGs with Reference to Nepalese Education Consultancies Through COVID-19 and Their Status with SDGs for Nepal" begins with the story of the negative impact of the COVID-19 pandemic on MSMEs in Nepal. The impacts of lockdowns on Nepalese Education Consultancies, part of MSMEs have been extremely negative. These enterprises have evolved larger in number in Nepal till 2019. Among others, the increasing craze of Nepalese students opting to fly abroad to study is assumed to have been a major causative. Nevertheless, shut down of businesses due to lockdowns that brought economic recession forced many to close down amid the outbreak of the COVID pandemic from 2019 to 2021, few consultancies managed to survive even through COVID lockdowns and afterward, as they somehow invented new strategies and techniques of sustainability. This assumption lent this chapter an objective to explore the strategies or techniques which the surviving education consultancies could have invented that helped them keep their services moving on. As the education consultancies equally as other MSMEs make financial and educational contributions, it was hence assumed that the consultancies that survived through the COVID lockdowns also managed to pay income tax having its role to the revenue which contributed to GDP in addition to the educational enhancement on youth empowerment as the 4th indicator of SDGs for Nepal.

Guided by the assumptions, this study was framed on qualitative design and it was to better understand the encountered problems and the invented techniques of Nepalese education consultancies through COVID-19 lockdowns and their contribution to achieving financial growth and educational empowerment as two major indicators of achieving SDGs for Nepal as one of the developing countries.

The book is designed to stimulate academic interest and make it useful to the academies, researchers, students, the general public, and policymakers in the domain of MSMEs and the SDG of the UN across the world.

Tirupati, India — Himachalam Dasaraju
Jakarta Barat, Indonesia — Tulus T.H. Tambunan

Contents

Notes on Contributors

Dr. Shweta Bapat management faculty by profession is a doctorate in Corporate Social Responsibility from renowned University of Pune (Now known as SPPU) and has a postgraduate diploma and master's degree in management. She is President of Business Ethics Foundation, Pune, and Chief mentor to CSRNexus LLP. She is also Head of the Department of Human Resource Management at Kaveri College of Arts, Science and Commerce Pune.

In her 20+ years of rich experience in academics and industry, she has successfully carried out different responsibilities and played various roles like secretary of Business Ethics Foundation Pune, A member of CSR Experts committee of ICSI, New Delhi, An External Expert for an International conference on Business Economics and Sustainable Development organized by NISM, Mumbai and TERI-SAS New Delhi, A Mentor at CRB tech solutions, guest speaker and panellist at various seminars/ webinars and conferences.

She has published various research papers in renowned national and international journals. She has published 5 textbooks and 1 reference book as co-author, and various articles on national and local platforms at her credit. She has won Best Research Paper Award, Best Achievement Awards, Best Teacher Award, and an Award for Excellence in Human Resource Management, Nation Builder from different renowned institutes.

Dr. Tukundane Benson B.B.A., M.B.A., Ph.D., is an academician with over 14 years of work experience in University education. Currently, he is the Deputy Director for the Higher Degrees Directorate at Kampala International University, having spanned through the ranks ranging from lower administrative levels. He has taught, researched, and supervised students at all levels in the areas of management science where he specialized.

Dr. Himachalam Dasaraju has served as a Professor, Head, Chairman & Dean of Commerce at Sri Venkateswara University, Tirupati for more than 32 years. He was honored with **Best Teacher Award in 2011** for his notable contribution to the students and teaching community in the Commerce and Management discipline, Bharat Jyothi Award 2013 by IIFS, Award for Teaching Excellence, 2013 by Indus Foundation, USA, Distinguished TeacherAward 2015 by MTC Global. Around 25 researchers got their Doctoral degrees (Ph.Ds) under his guidance and completed 12 major research projects. He is a panel member of experts for various Indian and global universities. He extensively delivered special lectures at various universities across India. He is the recipient of the Social Scientist Award under Indo-French Cultural Exchange Programme, France in 2003. He also served as Visiting Professor at the University of Montesquieu, Bordeaux, for 2 years and visited Maison De Sciences De'L Homme, Paris, France, five times. He got **Commonwealth Visiting Fellowship** and visited Essex Business School, University of Essex, UK in 2008–09 and then he was invited as **Visiting Professor** at **University of Essex, UK**. He visited Bordeaux, France (1999), (2007), Paris, France (2003), (2005), (2006), and (2007), Brussels, Belgium (2003), Singapore (2004), Bali, Indonesia (2005), Kuala Lumpur, Malaysia (2005), Bangkok, Thailand (2006), Netherlands (2006), Colchester, UK (2008), (2009), (2011), Naples & Rome, Italy (2011), Southampton, UK (2013) for various academic and research purposes. He was awarded **Emeritus Professorship** by the UGC and **Senior Fellowship** by the ICSSR, and completed successfully. He was honored with the **Lifetime Achievement Award** in 2017 and 2018 by SVURTA & IIPA and VIF Global respectively. He is a **NAAC Peer Team** member, founder member of "Global Association for MSMEs and SDGs Research in Developing Countries", Indonesia and Advisory member for Global College, Kathmandu, Nepal. At present, he is the **Collegium member of ICSSR**. A book on "Role of MSMEs in Achieving SDGs in Developing Countries" with Professor Tulus Tambunan of Trishakti University, Indonesia by Palgrave Macmillan, London is in progress. Right now he is having Collaborative research on "Carbon Accounting & Auditing: Case of India and France" with Prof. Stephane Trebucq, University of Bordeaux, France.

Dr. Sudarsan Kasa has more than two decades of teaching and research experience and published 6 books and 35 research papers in reputed National and International journals, 14 Research Papers Published in Edited Books, 10 Research Papers Presented in International Conferences/Seminars, 35 Research Papers Presented in National Conferences/Seminars. He delivered talks at National and International Conferences. He organized International and National Conferences/ Seminars/workshops. His specialization areas are Accounting, Finance and Entrepreneurship.

Prof. Dr. Dhanonjoy Kumar has contributed in a number of ways to the growth of leadership, management, and entrepreneurs. He is currently a professor at the department of management, Islamic University, Kushtia, Bangladesh. He has published his qualitative work in numerous prestigious journals. In addition to having over 50 publications (in journals, books, book chapters, conference proceedings, etc.), he has delivered over 30 technical talks in a variety of settings (both international and national). He received invitations to give speeches at various international gatherings in Turkey, Malaysia, and India. His study focuses on management, women's empowerment, SMEs, MSMEs, leadership, and the SDGs.

Ernestus Florian Lamaile is an alumnus of the Development Economics Study Program, Faculty of Economics and Business, Atma Jaya Catholic University of Indonesia.

Prof. Emuron Lydia B.Sc., B.B.A., M.M.S., Ph.D., is a renowned researcher, author, educator, and mentor. She has both national and international exposure in the administration of University education, of both private and public sectors at the top management level. She currently works at Kampala International University, a leading private University, and doubles as a Deputy Vice-Chancellor in the Ugandan public Sector.

Sazir Mayanja Nsubuga is a graduate of Makerere University, London School of Insurance, and the Cyprus Institute of Marketing. His doctorate is in finance, and he holds qualifications in banking, insurance, accounting, and microfinance. He was awarded a Certificate of Achievement by the Association of Chartered Certified Accounts (ACCA) for excelling in their examinations in 2008. He was a founding member of the Network for Educational and Multidisciplinary Research Network Africa (NEMRA) in 2017. He is an extensively published researcher and a member of both the American Finance Association and the International Finance Management Association. He coordinates activities of Researcheasilearn, a training platform.

Dr. Badziili Nthubu is a Lecturer at the Botswana International University of Science and Technology (BIUST). He teaches courses in Industrial and manufacturing, Energy, and Mechanical engineering.

Before being a Lecturer in 2021, Dr. Nthubu was working for Imagination Lancaster's £13.2 million projects, funded by Research England as a Post-Doctoral Research Associate. He was working for the factory and workplace cluster exploring future factories and workplace ecosystems through design research. Dr. Nthubu made huge accomplishments during his tenure as a researcher for Imagination Lancaster by forming significant networks both in the UK and Botswana. Through his important networks, He published work in top referred journals and conference venues such as the Design Journal, Journal of Design, Business and Society, Association for Computing Machinery (ACM), Springer and others. His excellent research output helped Imagination Lancaster Design Centre to reach its research target.

Prior to joining Lancaster University in the UK, Dr. Nthubu was a Teaching Instructor at BIUST teaching courses in the Department of Mechanical, Energy and Industrial Engineering. Dr. Nthubu also spent 5 years with Botswana Power Corporation, Morupule Power Plant A&B as a Training and Development Officer in Mechanical. He also spent 3 years at Botswana Defence Force where he was working as a Design and Manufacturing Engineer.

Dr. Nthubu earned a Doctor of Philosophy (Ph.D.) in Design from Lancaster University, where he was a Commonwealth Scholar. His Ph.D. research focused on design visualization methods and techniques to explore manufacturing and innovation ecosystems. He also earned a Master of Mechanical Design and Theory from the North-eastern University in China.

Dr. Timilehin Olasoji Olubiyi is a seasoned professional, adviser, and scholar with over twenty years of hands-on continuous experience in investment banking, public sector investment advisory, fund management, consulting, training, business administration, research, and entrepreneurial development. Dr. TimiOlubiyi, as he is affectionately known, is a seasoned professional and a scholar of high repute with good working with government officials on advice and policy formulation on various matters. He has held key roles in top stockbroking and financial institutions in Nigeria. A well-known investment banker and Securities & Exchange Commission (SEC) registered capital market operator, author, researcher, and business expert who has previously served in several government and policy roles.. An affiliate of the Department of Business Administration and Marketing Babcock University Nigeria, an affiliate of Babcock Centre for Executive Development Ogun State, Nigeria and facilitator and project supervisor at the Department of Business Administration National Open University of Nigeria. Dr. TimiOlubiyi has several papers and opinions published in high-impact academic journals and several renowned newspapers worldwide on small business, entrepreneurship, knowledge management, enterprise continuity, succession, strategy, policy, sustainability, and leadership. He is also a reviewer and member of the editorial board of many indexed journals.

Dev Raj Paneru Ph.D. has contributed transnationally and at the international level a researcher in the area of foreign language teacher education under the European Doctorate in Teacher Education (EDiTE) as part of the European Commission's Horizon 2020 Initiative, affiliated with the Department of Educational Sciences of the Faculty of Arts at Masaryk University, Brno, Czech Republic.

His primary research engagements include investigating and exploring innovative pedagogical dimensions of teaching and technology with a special focus on foreign language teaching some effective means to enhancing twenty-first-century teacher skills that align with a broad-based vision articulated as transformative teacher learning for better student learning in the emerging Global, European, and most recently Asian and local contexts.

The additional research areas of interest and contributions entail exploring and examining to determine the major dynamics of financial, economic, and corporate leadership management also linking a number of research studies focused on development and changes in the entrepreneurship learning and process.

Dr. Shibaprasad Parhi is a faculty member in the Department of Business Management at Adani Institute of Infrastructure Management, Ahmedabad, India.

Prof. Ravinder Rena is Professor of Economics and Honorary Professor in Durban University of Technology. He has more than 30 years of teaching research experience in several world-famous universities at Monarch, NSW, Durban University. He has published more than 150 research papers. He is an expert of BRIC financial team, he is also in the top 2% social scientist list. He has served as founder editor in chief for Interscience journal.

Dr. Meenu Singla has a rich experience of 17 years in academics and industry-related assignments. Her academic interest is in the areas of Human Resource Management relating to job satisfaction, Employee Retention, E-HRM, Supply Chain Management, Operations Research, Entrepreneurship and Leadership, Women Entrepreneurship, Corporate Social Responsibility, Teaching Pedagogy, etc. She has authored two books on Service Marketing and Training and Development. She is taking up various industrial interface projects with numerous industrial units that have equipped her with the connection of theoretical and practical applications of diverse aspects. The other projects included various aspects of HR, Production and Operations Management, and Supply Chain Management in Industrial units like VSGM Ltd, Trident Industries Limited, Vardhman Corporate Office, Vardhman Special Steels, Ludhiana, and Auro Textiles, Baddi (H.P).

She has varied research papers and articles indexed in Scopus, Emerald and other reputed journals. She is an active researcher and has knowledge about the software like SPSS, NVIVO, Mendeley, STATA, etc. She has participated and presented her research work at various National and International Conferences and workshops including IITs and IIM. She is an active social worker and works for women's empowerment. She has been awarded for her social work and initiatives. She is acting as a Resource Person for Punjab Police Academy Phillaur for the training of various Probationary Officers Class-I and other educational institutes of high repute.

Prof. Dr. Susela Devi K. Suppiah has worked and contributed in various capacities for the development of accounting and the accounting profession. Currently, she is the President of the Malaysian Accounting Association. She is a member of the Malaysian Institute of Accountants (MIA) and the Malaysian Institute of Certified Public Accountants. She served as Technical Advisor to the Malaysian delegation to the International Accounting Standards Committee in 1998. In 2005, she was an Advisor to the Confederation of Asia Pacific Accountants (CAPA) and represented CAPA at the International Accounting Standards Board's Roundtable Discussions of Financial Reporting Standards for SMEs.

She was appointed by the Asian Development Bank in 2002 to conduct 2 "Train the Trainers" programs on International Accounting Standards for participants from Bangladesh, Nepal and Pakistan (in cohort 1) and Vietnam, Cambodia and Mongolia (cohort 2). She served as a project manager for the Malaysian Accounting Standards Board for projects such as the Conceptual Framework for Financial Reporting and Operating Segments.

Her research expertise is in qualitative research and has published her qualitative work in many high-ranked journals. She has more than 100 publications (in Journals, Books, Book chapters, Conference proceedings, etc.) and presented more than 80 technical lectures in various forums (International & National). She was invited to deliver talks in various International forums in Japan, Korea, Bangladesh, Sri Lanka, and New Zealand. She has collaborated with the United Nation's International Standards on Accounting and Reporting (ISAR) on a case study on capacity building for the accounting profession.

She has served as Executive Director of the Malaysian Accounting Standards Board and Chief Editor of the Asian Journal of Business and Accounting (Scopus indexed). Her current focus is to enhance the ecosystem for capacity building of professional accountants to empower Micro and Small Medium Enterprises and improve accounting education in institutions of higher learning with a view to nation-building.

Her research includes Sustainability Reporting and integrated Reporting. She believes that the accounting profession contributes significantly to the nation's economic development and sustainability. The latest quest is to establish a sustainable business research community.

Dr. Sivakumari Supramaniam is a senior lecturer in the Department of Marketing Strategy & Innovation, Sunway University Business School.

She holds a Ph.D. in Marketing (Waikato University, New Zealand) and a Master in Marketing (Northumbria University UK). Her main research approach is qualitative, drawing on social constructionism, and relationality. Applying her qualitative expertise, she currently leads a sustainable project for SMEs mainly addressing the application of Culture of Health in Malaysia. She also works in collaboration with other local and abroad universities in the areas of sentiment analysis, gender, and workplace and influencer role in Marketing. Dr. Sivakumari has published in Management Decision Journal, and Journal of Consumer Marketing, and recipient of the best paper award from the American Marketing Association (2017).

Tulus T.H. Tambunan is professor of Economics, Universitas Trisakti, Indonesia. Currently, he is also the director of the Center for Industry, Small and Medium Enterprises, and Business Competition Studies at the same university. He is also the director of the Ph.D. program, the initiator, and the chief editor of *The International Journal of Small and Medium Enterprises and Business Sustainability* at the same university. He holds a Ph.D. in economics from Erasmus University in Rotterdam, the Netherlands. He has done many studies on various issues related to micro, small, and medium enterprises, economic crises, regional trade, and inclusive development. Some of his publications are *The Readiness of MSMEs in Facing ASEAN Economic Community 2015* (New York: Nova Science Publisher, Inc, 2014); *Economic Crisis and Vulnerability: The Story from Southeast Asia* (New York: Nova Science Publishers, Inc, 2011), *Global Economic Crisis and ASEAN Economy. Theory and Empirical Findings* (Lambert Academic Publishing (LAP), Saarbrücken, 2010), *Trade Liberalization and SMEs in ASEAN* (New York: Nova Science Publishers, Inc., 2010), *SME in Asian Developing Countries* (Palgrave Macmillan Publisher, 2009), and *Fostering Resilience through Micro, Small and Medium Enterprises Perspectives from Indonesia* (Palgrave Macmillan, 2022). Other than that, he is also the editor of some international books, including *Development of MSMEs in Developing Countries Stories from Asia, Africa, and Latin America* (New Delhi: AkiNik Publications, 2019), *COVID-19 and Its Reflection on SMEs in Developing Countries* (as the first editor; Nova Scientific Publisher, New York, 2022), and *Economic Crises: A Review and Directions for Research* (as the first editor; Nova Scientific Publisher, New York, 2022).

Dr. H.M. Thakar has the following Educational Qualifications: B. Com. (Hons.),1984, First Class, Banaras Hindu University, Varanasi; M. COM.,1986, First Class, Banaras Hindu University, Varanasi; Ph.D., 1989, Banaras Hindu University; M.B.A., 2019, First Class, IGNOU. He holds the following positions:

He has completed 2 minor research projects. He has also worked as judge in the Avishkar competition organized in Shivaji University every year.

Thakar has the following Educational Qualifications: B. Com. (Hons.), 1984, First Class, Banaras Hindu University, Varanasi; M. Com., 1986, First Class, Banaras Hindu University, Varanasi; Ph.D., 1989, Banaras Hindu University; M.B.A., 2019, First Class, IGNOU. He holds the following positions:

- Lecturer, Department of Management Studies, North Maharashtra University, Jalgaon, from 1991–1997.
- Reader, Department of Management Studies, North Maharashtra University, Jalgaon, 1997–2006.
- Head of the Department from 1994–1997 & 2000–2006.
- Director, Department of Commerce and Management, Shivaji University, Kolhapur, from 2006 to Jan. 2022 till retirement.

His Administrative Experiences are as follows:

- Member of Board of Studies in Management, Academic Council, Research & Recognition Committee, Purchase Committee, Board of Examinations, Kaviyitri Bahinabai Choudhari North Maharashtra University, Jalgaon.
- Chairman, Board of Studies in Management, Kaviyitri Bahinabai Choudhari North Maharashtra University, Jalgaon.
- Member, Board of Studies in Management, Shivaji University, Kolhapur.
- Member, Board of Studies in Accounting, Rajasthan University, Jaipur.
- Member, Research & Recognition Committee, Bombay University, Mumbai.
- Member, Research & Recognition Committee, Savitribai Phule Pune University, Pune.

- Member, Research & Recognition Committee, SantGadge Baba University, Amravati.
- Member, Board of Studies in Management, CSIBER, Kolhapur.
- Member, Board of Studies in Management, Rajarambapu Institute of Technology, Sakhrale.
- Member, Board of Studies in Management, DKTE, Ichalkaranji.
- Member of Scrutiny Committee of Major Research Projects in UGC in 2008 and 2010.
- Coordinator, M.B.A. CET Admission Process from 2003 to 2012.

Member of selection committee of teachers in various universities and affiliated colleges in Kolhapur, Jalgaon, Mumbai, Pune and Aurangabad. He has guided over 12 Ph.D. candidates, 1 M.Phil. and more than 300 M.B.A. projects.
He has delivered lecture on:

- "Accounting" in All India Radio, Jalgaon on Management Department of North Maharashtra University, Jalgaon in September 1995.
- "International Management" in Refresher course in Commerce organized by North Maharashtra University, Jalgaon in July 1996.
- "Business Communication" in Nahata College of Commerce, Bhusawal in July 1996.
- "Motivation" in Supreme Industries Ltd., Jalgaon in October 2000.
- "Strategic Management" in Dhanaji Nana Choudhari College of Commerce, Faizpur in October 2002.
- "M.B.A. Admission" in Baheti College, Jalgaon in October 2003.
- "Communication Skills" at Rotary Club, Jalgaon in March 2006.
- "Career in Management" at Institute of Management and Research, Jalgaon in August 2007.
- "Group Discussion and Personal Interview in M.B.A. Admission Process" at Arts, Commerce and Science College, Palus in January 2008.
- "Employee Motivation" in GOSHIMA, Kolhapur in February 2008.
- "Organizational Behaviour" at Distance Education Centre, Shivaji University, Kolhapur in 20th September 2008.
- "Capital Market and Derivative Market" at Mahatma Gandhi Mission's Institute of Management, Aurangabad in 27th September 2008.

He has completed 2 minor research projects. He has also worked as judge in the Avishkar competition organized in Shivaji University every year.

Nuning Trihadmini is a lecturer and researcher at the Master of Applied Economics Study Program, Faculty of Economics and Business, Atma Jaya Catholic University of Indonesia. Completed a master's program at the University of Indonesia in monetary economics and banking and a doctorate from the same university in international finance. Currently, he is an associate professor who is actively researching in the fields of development economics, SME, international finance, macroeconomics, banking, and monetary economics.

Dr. Ravindra Tripathi is Associate Professor and HoD in the Department of Humanities and Social Sciences, Motilal Nehru National Institute of Technology Allahabad. He has more than 40 research papers in Scopus-indexed and ABDC-listed journals to his credit. His research area includes Financial Management, Taxation and Economics.

Dr. Nithya US is a faculty member in the Department of Commerce at Sree Narayana College for Women, Kottayam, Kerala, India.

Mr. Uma Shankar Yadav is currently working in the Department of Humanities and Social Sciences, Motilal Nehru National Institute of Technology Allahabad. His research area is Welfare Economics and Entrepreneurship and sustainable development. He has published more than 25 research papers in Scopus Indexed and ABDC-listed journals .he has published more than 6 chapters in an edited book in different famous publications, and he is serving as an editorial board member and reviewer in several SSCI, Scopus, ABDC listed journal.

List of Figures

List of Tables

CHAPTER 1

Introduction: Role of MSMEs in Achieving SDGs—Perspectives from Emerging Economies

Himachalam Dasaraju and Tulus T.H. Tambunan

Introduction

Micro, small, and medium-sized enterprises (MSMEs) are the backbone of many economies, representing 95% of all companies worldwide and accounting for 60% of employment. In many countries, more than 90% of all enterprises can be classed as MSMEs, and a large share of those can be classed as micro firms, with lesser than ten employees. While they may be small individually, new ILO data show that micro and small enterprises,

H. Dasaraju
School of Commerce and Management, Sri Venkateswara University, Tirupati, India

T.T.H. Tambunan (✉)
Center for Industry, SME and Business Competition Studies, Universitas Triskti, Jakarta Barat, Indonesia
e-mail: ttambunan56@yahoo.com

H. Dasaraju and T.T.H. Tambunan (eds.), *Role of Micro, Small and Medium Enterprises in Achieving SDGs*, Sustainable Development Goals Series, https://doi.org/10.1007/978-981-99-4829-1_1

together with their workers, account for a staggering 70% of employment worldwide (ILO, 2019). MSMEs play an important role in most developing economies in general. Formal MSMEs contribute up to 45% of total employment and up to 33% of gross domestic product (GDP) in emerging market economies. These numbers would be significantly higher when enterprises in informal MSMEs are included and are major creators of jobs mostly in developing and less developed countries. In emerging markets, 4 out of 5 new positions in the formal sector were created by MSMEs, which is about 90% of total employment.

The World Bank reports that around 600 million jobs will be needed in the next 15 years to absorb the ever-growing global workforce, mainly in Asia and sub-Saharan Africa (World Bank, 2018). The MSMEs have a dominant share in the number of firms and employment in all G20 countries, and so can play an important role in economic development and job creation in each country. In Japan, for example, MSMEs (firms with either less than 300 employees or 300 million yen in the capital in general) account for 99.7% of firms, employing approximately 70% of the workforce. In the manufacturing sector, they produce about half of the value-added. The legal or statistical definitions of MSMEs differ across countries and also across sectors (ICSB, 2019). Many MSMEs depend on international trade for their activities, either because they export their products through direct or indirect channels or import inputs to manufacture the products they sell domestically. They are major employers of women and young people, and a key driver of innovation (WTO, 2020).

Status of MSMEs

MSMEs are one of the strongest drivers of economic development, innovation, and employment generation. According to Asian Development Bank (ADB) statistics, in the Asian region specifically, MSMEs account for 96% of total enterprises, employ 62% of the labor workforce, and contribute an average of 42% of gross domestic product (GDP) (OECD, 2017). The MSMEs Economic Indicators Database 2019 indicated the region-wise number of formally registered MSMEs across 176 economies, as mentioned in Table 1.1. The numbers in the parenthesis indicate the rank in terms of the number of MSMEs in various regions.

Table 1.1 indicates that the East Asia and Pacific region is having a higher number of MSMEs followed by South Asia, Sub-Saharan Africa, Europe, Central Asia, North America, etc., employment creation through

Table 1.1 Region-wise MSMEs (numbers)

No.	*Region*	*MSMEs per region*	*Rank*
1	East Asia and the Pacific	102,289,112	(1)
2	Europe and Central Asia	35,027,833	(4)
3	Latin America and the Caribbean	17,217,911	(6)
4	The Middle East and North Africa	10,066,801	(7)
5	North America	33,729,168	(5)
6	South Asia	75,924,258	(2)
7	Sub-Saharan Africa	48,185,712	(3)

Source https://www.smefinanceforum.org/data-sites/msme-country-indicators (MSME-EI, 2019)

MSMEs will often directly benefit the poor and vulnerable, particularly women and youth, thereby directly reducing poverty, increasing economic status, and so on. Entrepreneurs, in the MSMEs sector, are the lifeline of the economy in the process of effective utilization of untapped local resources such as men, money, and material resources.

According to SME Finance Forum, the MSMEs are at the heart of job creation, accounting for 9/10 of businesses and 2/3 of jobs worldwide. By the end of 2030, 600 million jobs will be needed to absorb the ever-growing global workforce, both skilled and unskilled. But meeting the financial needs is a big challenge and hinders the growth of MSMEs in all regions of the world. MSMEs need more financial support to expand and flourish. Around 131 million or 41% of formal MSMEs, more so in developing countries, have unmet financing needs. The unmet demand for credit is estimated to be around $4.5 trillion. The women-owned businesses account for 23% of MSMEs and 32% of the financing gap (IFC, 2019).

The MSMEs constitute over 90% of total enterprises in most developing economies and are credited with generating the highest rates of employment growth and account for a major share of industrial production and exports. The MSMEs continue to be on the move and recognized globally for their contributions to innovation and improving economic conditions and employment generators.

The WTO Secretariat has published an information note looking at how MSMEs are being affected by the COVID-19 pandemic. It indicated the impact of supply chain disruptions on MSMEs and the extent to which smaller businesses are represented in the economic sectors hardest hit by the crisis. The report notes that supply chain disruptions can have a particularly severe impact on MSMEs because sourcing from new suppliers or absorbing price increases is more challenging for a smaller firm with limited supply options and capital. The report looks into a wide range of measures taken by governments to support MSMEs. These include measures to address cash flow issues, expand trade opportunities for MSMEs, and make them more resilient.

MSMEs and Sustainable Development Goals (SDGs)

MSMEs play a major role in most economies, particularly in developing countries. MSMEs account for the majority of businesses worldwide and are important contributors to job creation and global economic development. As said before, they represent about 90% of businesses and more than 50% of employment worldwide. Formal MSMEs contribute up to 40% of national income (GDP) in emerging economies. These numbers are significantly higher when informal MSMEs are included. However, access to finance is a key constraint to MSME growth, it is the second most cited obstacle facing MSMEs to grow their businesses in emerging markets and developing countries (https://www.worldbank.org/).

The MSMEs are thus the key contributor, at least potentially to economic development in both developed and developing economies across the world. The MSMEs have been assigned an important role in developed and developing countries as well, due to their huge potential for solving socio-economic challenges. Creating opportunities for MSMEs in emerging markets is a key way to advance economic development and reduce poverty (IFC, 2020). The vital contribution of MSMEs to the broader socio-economic objectives, including employment generation and poverty alleviation makes them a key priority area for achieving the SDGs. It indicates the pivotal role to be played by MSMEs in emerging economies where poverty is persistent and lack of opportunity to work for their livelihood is a serious problem, hindering the social status of the people.

Therefore, the MSMEs are vital for poverty eradication, mostly among women and other socially underprivileged groups. Keeping in view its critical contribution to the economy, the UN has assigned a great role to MSMEs for taking a lead in achieving most of the economic-related SDGs, including promoting inclusive and sustainable economic growth, increasing employment opportunities and decent work especially for the poor, advancing sustainable industrialization and innovation, and creating a positive push for a higher quality of life, better education, and good health for all (OECD, 2017). Job creation through MSMEs will often directly benefit the poor and vulnerable, particularly women and youth, thereby directly reducing poverty, increasing income, and positively impacting household investments in education and health over time. MSME development has the potential for wide-reaching impacts on the SDGs globally, including SDG 1 (end poverty), SDG 2 (zero hunger), SDG 3 (good health and well-being), SDG 5 (gender equality), SDG 8 (promote inclusive and sustainable economic growth, employment and decent work), and SDG 9 (improve sustainable industrialization and fostering innovation).

There are 17 SDGs ratified by the United Nations General Assembly in September 2015, which were presented as "a universal call to action to end poverty, protect the planet and ensure that all people enjoy peace and prosperity" which are targeted to be achieved by 2030. Not surprisingly, strengthening the capacity of MSMEs—typically, though not exclusively, through greater access to finance—is identified as a key target in specific SDGs. As shown in Table 1.2, MSMEs involved in specific sectors could help achieve many objectives under different goals (Blue Orchard, 2017).

The SDGs targets are ambitious and require the transformation of public and private activities toward the prosperity of people and the planet. This transformation is associated with adopting new business models, bringing in innovation and technology, and doing business differently—more sustainably and ethically. This process opens up new business opportunities for the private sector in general and particularly for MSMEs (Lessidrenska, 2019). According to the Business and Sustainable Development Commission, sustainable business models could open economic opportunities worth \$12 trillion and create 380 million jobs by 2030, with more than 50% being located in developing countries. Such opportunities for MSMEs have been identified under each goal (Table 1.3). MSMEs possess immense potential for the achievement of all SDGs directly or indirectly. Individual MSMEs have the potential to adopt actions in their business practice to contribute to the goal. New business models/solutions of MSMEs that will help to achieve the SDGs.

MSMEs are enfolded globally for the impact they have created on society as a huge job creator. The International Council for Small Business (ICSB) also recognized the MSMEs globally for their contributions

Table 1.2 MSMEs and SDGs

No.	*Contribution toward achieving SDGs*	*Goals in which MSMEs would have a considerable impact*
1	By providing employment	Goal 1: Poverty Alleviation Goal 5: Gender Equality Goal 8: Decent Work and Economic Growth Goal 9: Industry Innovation and Infrastructure Goal 10: Reduced Inequalities
2	By sector-specific operational efficiency	Goal 2: Zero Hunger Goal 7: Affordable and Clean Energy Goal 9: Industry Innovation and Infrastructure Goal 12: Responsible Consumption and Production Goal 17: Partnership for Goals
3	By Corporate Social Responsibility activities	Goal 3: Good Health and Well-being Goal 4: Quality Education Goal 6: Clean Water and Sanitation Goal 11: Sustainable Cities and Communities
4	By ethically and lawfully following laws and bylaws	Goal 13: Climate Action Goal 14: Life Below Water Goal 15: Life on Land Goal 16: Peace Justice and Strong Institutions

Source Blue Orchard Academy, SMEs, and SDGs, 2017 & Role of MSMEs in Achieving Sustainable Development Goals, 2019. https://doi.org/10.18231/2454-9150.2019.0189

to innovation and improving economic conditions. For 2019, ICSB sees a major shift in the landscape for MSMEs (Table 1.4).

The MSMEs are vital in achieving the SDGs. The efforts to enhance access to finance for SMEs across the world are an important element in the execution of SDGs. They are very crucial means in the implementation of SDG no. 8 (decent work and economic growth) and SDG 9 (industry, innovation, and infrastructure.) (www.un.org/en/about-un/).

Table 1.3 Role of MSMEs in fulfilling the SDGs

Goal	*How it is going to achieve SDGs?*
1: No Poverty (End poverty in all its forms everywhere)	MSMEs create employment opportunities that facilitate people to come out of the vicious circle of poverty. Informal enterprises too benefit the poor and downtrodden people. We can end poverty by strengthening MSMEs in rural and semi-urban areas, where the poverty rate is high
2: Zero Hunger (End hunger, achieve food security and improved nutrition, and promote sustainable agriculture)	Small-sized farmers and agro-based MSMEs directly contribute to the goal of meeting the food requirement of the people. There is a vast scope for promoting agro-based and artisan-based enterprises to solve this issue. We can stop hunger by promoting income-based and food-based micro and small units
3: Good Health and Well-Being (Ensure healthy lives and promote well-being for all at all ages)	MSMEs are healthcare service providers and bridge the healthcare gap through better-quality healthcare services to the people. These units will take care of the well-being of people of all ages
4: Quality Education (Ensure inclusive and equitable quality education and promote lifelong earning opportunities for all)	MSMEs will provide quality technical and vocational education, and make the people get access to earnings for their livelihood. Good education facilities transform people to live better with good working skills
5: Gender Equality (Achieve gender equality and empower all women and girls)	A considerable number of MSMEs are owned by women and employ mostly women. Inclusive finance can bridge the gender gap and bring women into the mainstream of economic development. Women entrepreneurship development empowers women and brings gender equality
6: Clean Water and Sanitisation (Ensure availability and sustainable management of water and sanitation for all)	MSMEs provide water and sanitation services. Agricultural and food supply chain MSMEs can conserve and provide clean water, which leads to sustainable water management

(continued)

Table 1.3 (continued)

Goal	*How it is going to achieve SDGs?*
7: Affordable and Clean Energy (Ensure access to affordable, reliable, sustainable, and modern energy for all)	MSMEs have the potential for becoming more energy-efficient. MSMEs can be incentivized by larger enterprises to integrate sustainable practices into their operations. MSME finance can contribute to Goal 7, by providing affordable and clean energy to the people
8: Decent Work and Economic Growth (Promote sustained, inclusive, and sustainable economic growth, full and productive employment, and decent work for all)	MSMEs contribute significantly to GDP and economic growth. MSMEs create jobs and have the potential for promoting decent work and entrepreneurship. MSMEs finance is a target for Goal 8, which stimulates economic activity for the prosperity of the people
9: Industry, Innovation, and Infrastructure (Build resilient infrastructure, promote inclusive and sustainable industrialization, and foster innovation)	Small and Medium enterprises are directly relevant to the goal. MSME finance for small enterprises will contribute to the goal and promote innovation and resilience in fracture. It promotes inclusive and sustainable industrialization which paves the way to foster technological innovation
10: Reduced Inequalities (Reduce inequality within and among countries)	MSMEs promote inclusive economic growth and have the potential to regenerate under-served geographic areas. MSMEs provide higher incomes to low-income and marginalized groups across the countries. The higher productivity of MSMEs can help to reduce wage parity across the world
11: Sustainable Cities and Communities (Make cities and human settlements inclusive, safe, resilient, and sustainable)	MSMEs provide employment and economic growth in cities. It promotes sustainable and inclusive economic growth for well-being and inclusive settlement of the people with safety
12: Responsible Consumption and Production (Ensure sustainable consumption and production patterns)	MSMEs as a group have significant cumulative social and environmental impacts. It promotes sustainable production and consumption which are very essential for the betterment of people and the planet as well

(continued)

Table 1.3 (continued)

Goal	*How it is going to achieve SDGs?*
13: Climate Action (Take urgent action to combat climate change and its impacts)	MSMEs have the potential to combat climate change and its bad impact on the people. MSMEs can contribute to mitigating climate change
14: Life Below Water (Conserve and sustainably use the oceans, seas, and marine resources for sustainable development)	Small-scale fishery and marine-based enterprises are directly relevant to the goal. Promoting MSMEs capabilities leads to developing the sustainability of value chains
15: Life on Land (Protect, restore, and promote sustainable use of terrestrial ecosystems, sustainably manage forests, combat desertification, halt and reverse land degradation, and halt biodiversity loss)	Agribusinesses and smallholder farmers have a critical role to play in conserving land-based ecosystems. It promotes the sustainable use of terrestrial ecosystems, sustainable forest management, reverse land degradation, and biodiversity loss across the world
16: Peace, Justice, and Strong Institutions (Promote peaceful and inclusive societies for sustainable development, provide access to justice for all, and build effective, accountable, and inclusive institutions at all levels)	MSMEs have the potential for promoting good governance principles. MSMEs have a role in conflict/violence resolution and prevention. New business models and opportunities for MSMEs resulting from the 2030 agenda and SDGs for the protection of people and the planet
17: Partnerships for the Goals (Strengthen the means of implementation and revitalize the global partnership for sustainable development)	MSMEs are a source of partnership to implement SDGs. MSMEs are a source of data for SDG monitoring and reporting. It revitalizes the global partnerships for sustainable development the world over

Source UN-DESA (2020)

Growth Prospects and Challenges of MSMEs

Since global financial meltdown 2008, the MSMEs gained lot of momentum and progress world over. The growth in MSMEs in the UK, France, and Sweden, recorded a considerable progress in job creation as reported by the OECD and World Bank. More than 50% of MSMEs surveyed, revealed that there is an increase in employment in short term in all those units with the age of less than three years (Liu, 2020). According to the OECD and the World Bank joint study on inclusive global value chains, the MSMEs development in developing and least developed ccountries (LDCs) dominates in the labor-intensive sectors,

Table 1.4 Top ten trends for MSMEs predicted by International Council for Small Business (ICSB)

No.	*Trend*
10	MSMEs to meet the SDGs
9	Access to finance and new options available
8	Attracting a new workforce for MSMEs
7	Customization is the new hot topic for MSMEs
6	MSMEs redefining the innovation process
5	Digital MSMEs go global
4	MSMEs are about doing business in a humane way
3	MSMEs and navigating trade
2	The impact of 1.8 billion who will transform the future
1	The power of belief

Source ICSB (2019)

and possesses low entry barriers including agriculture and agro-based, manufacturing, and service sectors as well. Other interesting observation in this regard is that most of the MSMEs in developing countries are in informal sector and higher use of knowledge-based capital is noticed in middle-income and emerging countries, which lead to value addition.

Despite merits of MSMEs, this sector is not free from problems across the world. MEMEs are encountered with multifaceted problems in various sectors and segments like limited access to finance, cumbersome procedure and regulatory framework, lack of symbiosis, lack of information flow on marketing, technological changes, demand and supply imbalance, supply chain disruptions, production capacity gaps, skills up gradation among workforce, compliance with environment, lack of basic infrastructure, limited access to ICT and technological changes, limited initiatives in nurturing entrepreneurship and women entrepreneurship promotion, and so on. In LDCs and developing economies, the MSMEs are confronting these problems which are affecting their functioning for contributing considerably for achieving SDGs. Nurturing young women entrepreneurship and fostering business and entrepreneurs capacity leads to alleviate poverty and gender equality. The UN Women report emphasized that if the women are given more importance in entrepreneurial activity and allow them to participate equally on par with men in entrepreneurship landscape, the global economy would be expected to rise to US$ 28 trillion by 2025 (Women Entrepreneurs can drive economic growth, 2017). Most of the micro-enterprises are basically family-owned businesses with a limited skills and expertise in their respective works which results in low productivity and quality of work, hampers the competitiveness in the market. As this trend leads to low wages and salaries and job security as well, it's dire need of the day to formulate structural and regulatory framework of MSMEs, which is conducive for job creation and to achieve SDGs.

COVID-19 Pandemic

As the globe reels under the onslaught of COVID-19, the major contentions floating around are about the global economic slump and uncertainties in the global political and economic orders. The pandemic, by itself, and also through various economic, social, and political avenues, will affect global developmental objectives at a much broader scale. The impacts will be observed more prominently in the UN SDGs.

The year 2020 has been incredibly challenging for the global community. The spread of the novel coronavirus, known as COVID-19, has led to an unparalleled health crisis in countries across the world. The crisis has had unprecedented and serious impacts on all aspects of how people communicate, work, produce, trade, consume, and live. The economic ramifications of the pandemic quickly became apparent, and MSMEs have been on the front lines. With workers and customers staying indoors, and supply chains tested by shutdowns, the MSMEs that provide 70% of jobs in countries around the world and about half of the economic activity have been put under stress.

Small companies tend to be vulnerable during an economic crisis, in part because they have fewer resources with which to adapt to a changing context. The ITC COVID-19 Business Impact Survey gathered evidence on how the pandemic affected 4,467 companies in 132 countries. One-fifth of SMEs are at risk of shutting down permanently within three months. *China, the EU, and the US account for over 60% of supply chain trade. This has had an impact on economies everywhere.* African exporters may lose more than $2.4 billion in global industrial supply chain exports, due to factory shutdowns in China, the EU, and the United States.

These types of enterprises are responsible for significant employment and income generation opportunities across the world and have been identified as a major driver of poverty alleviation and development. MSMEs tend to employ a larger share of the vulnerable sectors of the workforce, such as women, youth, and people from poorer households—populations with high vulnerability in times of COVID-19. MSMEs can sometimes be the only source of employment in rural areas. As such, MSMEs as a group are the main income provider for income distribution at the "base of the pyramid" (International Trade Centre, 2020).

The coronavirus pandemic is causing large-scale loss of life and severe human suffering globally. It is the largest public health crisis in living memory, which has also generated a major economic crisis, with a halt in production in affected countries, a collapse in consumption and confidence, and stock exchanges responding negatively to heightened uncertainties (OECD, 2020). The most recent OECD forecast (27 March) indicates that the initial direct impact of the shutdowns could be a decline in the level of output of between one-fifth to one-quarter in many economies with consumers' expenditure potentially dropping by around one-third.

In recent times, several international organizations have issued studies on aspects of the economic impact of the coronavirus pandemic. For instance, UNCTAD has calculated that the impact of the coronavirus in China has cost global value chains 50 billion USD in exports. Whereas in early March UNCTAD expected FDI could shrink by 5–15% on 26 March the forecast was revised to a reduction by 30–40% in 2020–2021 (UNCTAD, 2020).

The ILO estimates the impact of COVID-19 to result in a rise in global unemployment of between 5.3 million ("low" scenario) and 24.7 million ("high" scenario), signaling that "sustaining business operations will be particularly difficult for MSMEs" (ILO, 2020). Like the OECD, the IMF has published several reflections on the expected effect and policies required. In particular, these highlight that compared to the 2008 global financial crisis, this time the decline in services appears much greater, reflecting the consequences of lockdowns and social distancing, especially in urban settings. The IMF World Economic Outlook released on 14 April forecasts a 3% decline in global GDP in 2020, with considerable further downward risk (IMF, 2020). Several banks and institutes have also made strong negative adjustments to their GDP growth forecasts for 2020 (Deutsche Bank, 2020).

UN MSMEs Day (27 June): The Great Lockdown and Its Impact on Small Business

Small businesses, including those run by women and young entrepreneurs, are being hit hardest by the economic fallout of the pandemic. Unprecedented lockdown measures enacted to contain the spread of the coronavirus have resulted in supply chain disruptions and a massive drop in demand in most sectors. To continue playing their crucial role in creating decent jobs and improving livelihoods, small businesses depend more than ever on an enabling business environment, including support for access to finance, information, and markets.

Let's not forget that these enterprises, which generally employ fewer than 250 persons, are the backbone of most economies worldwide and play a key role in developing countries.

According to the data provided by the ICSB, formal and informal MSMEs make up over 90% of all firms and account, on average, for 70% of total employment and 50% of GDP.

Realizing the crucial role of MSMEs in achieving SDGs, the UN General Assembly has decided to designate the 27th of June of every year as "International Day for MSMEs" to raise public awareness of their contribution to sustainable development and the global economy. The delegation of Argentina, the 193 member body, has invited all stakeholders including the Member States, UN entities, and civil society organizations to observe the day and create public awareness of their contribution to the 2030 agenda for Sustainable Development (https://www.un.org/en/observances/micro-small-medium-businesses-day).

Conclusion

The MSME sector is one of the most crucial sectors with huge potential to affect the progress of most of the 17 SDGs. It gained a lot of importance in achieving SDGs in developed economies in general and in developing economies, in particular, owing to its inherent characteristics of inclusive growth and sustainable development. The most flexible nature and capabilities of MSMEs make them a key player in achieving these goals. The employment generation through MSMEs will often directly benefit the poor and vulnerable, particularly women and youth, thereby directly reducing poverty, increasing economic status, and reducing the severity of hungry. Further, it leads to nurturing entrepreneurship, which is an engine of inclusive growth and stimulus for economic prosperity and empowering people. It makes a substantial contribution to achieving the SDGs, which are directed toward better prospects for people and the planet as well. Keeping in view its vital role in achieving inclusive growth and sustainable development, the UN has assigned top priority and key role to play in this prodigious task.

As the present global COVID-19 pandemic turbulent situation has disrupted badly the functioning of MSMEs, it is the dire need of the day to promote this sector and make it viable in all dimensions. All the developing economies including India have taken several measures to promote MSMEs by starting various flagship programs and initiatives in the context of COVID-19. Governments around the world realize that MSMEs act as a lynchpin connecting the COVID-19 pandemic to the global economic recession. Besides addressing the health crisis, they have scrambled to alleviate the impact of COVID-19 on small businesses, introducing policies to help them cope with the short-term financial risks and long-term business implications. Apart from the various governmental measures initiated globally, the MSME sector is not free from myriad problems and it requires more stimulus to strengthen it further. It requires creating a congenial atmosphere for the MSMEs to prosper and access the new technology and global markets for their survival. It is a dire need to revitalize the SMEs to play an active and vital role in achieving SDGs by 2030 as per UN targets.

The experts believe that the new technology will be the "new normal" post COVID-19 for MSMEs. All the businesses are now looking forward to new kinds of technology and to building a robust digital-driven delivery capability. The drive toward digital transformation and greater use of disruptive technologies may affect physical employment opportunities. The emerging disruptive technologies may replace the human workforce. Further, it needs to focus more on upskilling the workforce to cope with the rapid pace of change and supporting technology in MSMEs. Therefore, a new strategy should be adopted more effectively to create alternative employment opportunities to meet the SDGs successfully. Let us hope for the best.

References

Blue Orchard. (2017). *SMEs and SDGs: Supporting Small and Medium Enterprises to achieve Sustainable Development Goals*. https://www.blueorchard.com/wp-content/uploads/Blue-Orchard-pages-11-12.pdf

Deutsche Bank. (2020). *Deutsche Bank economists forecast a "severe recession" due to Covid-19*. https://www.db.com/newsroom_news/2020/deutsche-bank-economists-forecast-severe-recession-due-to-COVID-19-en-11507.htm

ICSB. (2019). *Annual global Micro-, Small and Medium-sized Enterprises report 2019*. International Council for Small Business.

IFC. (2019). *SME Finance Forum annual report, 2019*. International Finance Corporation, a World Bank Group. https://www.smefinanceforum.org/sites/default/files/SMEFF%20ANNUAL-REPORT-2019_FINAL.pdf

IFC. (2020). *MSME Finance Gap*. SME Finance Forum, International Finance Corporation (IFC), a World Bank Group. https://www.smefinanceforum.org/data-sites/msme-finance-gap

ILO. (2019). *Info stories. The power of small: Unlocking the potential of SMEs*. https://www.ilo.org/infostories/en-GB/Stories/Employment/SMEs#power-of-small

ILO. (2020). *COVID-19 and the world of work: Impact and policy responses*. https://www.ilo.org/wcmsp5/groups/public/---dgreports/---dcomm/documents/briefingnote/wcms_738753.pdf

IMF. (2020, April). *World Economic Outlook, the great lockdown*. https://www.imf.org/en/Publications/WEO/Issues/2020/04/14/weo-april-2020

ITC. (2020) SME COMPETITIVENESS OUTLOOK 2020, Geneva: International Trade Centre, smeco2020forepubv2pdf (intracen.org)

Lessidrenska, T. (2019). *SMEs and SDGs: Challenges and opportunities*. OECD Development Matters, OECD. https://oecd-development-matters.org/2019/04/23/smes-and-sdgs-challenges-and-opportunities/

Liu, C. K. (2020). *The role of Micro-Small and Medium Enterprises in achieving SDGs* (Policy Brief). UN Department of Economic & Social Affair (UN-DESA). https://sdgs.un.org/sites/default/files/2020-07/Policy_Brief_MSMEs_and_SDGs.pdf

OECD. (2017). *Unlock the potential of SMEs for the SDGs*. https://oecd-development-matters.org/2017/04/03/unlocking-the-potential-of-smes-for-the-SDGs/

OECD. (2020, June). *OECD Economic Outlook: Corona Virus (Covid19): The world economy on a tight rope*. Organisation of Economic Cooperation and Development. https://www.oecd.org/economic-outlook/june-2020/

UNCTAD. (2020). *World Investment Report 2020*. United Nations Conference on Trade and Development. https://unctad.org/webflyer/world-investment-report-2020

UN-DESA. (2020). *Micro-, Small and Medium-sized Enterprises (MSMEs) and their role in achieving the Sustainable Development Goals (SDGs)*. Department of Economic and Social Affairs.

World Bank. (2018). *World Bank annual report*. https://documents.worldbank.org/en/publication/documents-reports/documentdetail/630671538158537244/the-world-bank-annual-report-2018

WTO. (2020, June 3). *Helping MSMEs navigate the COVID-19 crisis*. https://www.wto.org/english/tratop_e/covid19_e/msmes_report_e.pdf

PART I

The Contribution of MSMEs to the Achievement of the SDGs

CHAPTER 2

Role of MSMEs in the Achievement of SDGs in India in the Current Scenario

H.M. Thakar

Introduction

MSMEs have played an important role in the Indian economy. It has contributed not only to economic development but also to the country's social development. They are the growth engine of entrepreneurial development in the country. They have provided employment opportunities to the people, particularly rural people. Further, they are the source of innovation. There are about 633.88 lakh unincorporated non-agricultural MSMEs in the country engaged in various economic activities (MSME, 2021). This figure excludes those MSMEs that are registered under various acts. MSMEs contribute 30% to GDP and 40% to exports (MSME, 2021). The sector employs about 11 crore people second only to agriculture.

H.M. Thakar (✉)
Department of Commerce and Management, Shivaji University, Kolhapur, India
e-mail: hmthakar1962@gmail.com

H. Dasaraju and T.T.H. Tambunan (eds.), *Role of Micro, Small and Medium Enterprises in Achieving SDGs*, Sustainable Development Goals Series, https://doi.org/10.1007/978-981-99-4829-1_2

In India, the definition of MSMEs is related to investment in fixed assets and annual turnover. As per the new definition of MSMEs, the MSMEs are classified into the following categories:

a. micro-enterprise (MIEs) where the investment in plant and machinery does not exceed Rs. 1 Crore and annual turnover does not exceed Rs. 5 Crores,
b. small enterprise (SEs) where the investment in plant and machinery does not exceed Rs. 10 Crores and annual turnover does not exceed Rs. 50 Crores, and
c. medium enterprise (MEs) where the investment in plant and machinery does not exceed Rs. 50 Crores and annual turnover do not exceed Rs. 250 Crores.

Meaning of Sustainable Development Goals

The Brundtland Report described sustainable development as "development that meets the needs of the present without compromising the ability of future generations to meet their own needs." These goals cover economic, social, and environmental aspects. UNDP has set 17 SDGs that are to be attained by all the nations whether they are developed, developing, or least developed countries by 2030 and MSMEs play an important role in the attainment of these goals.

MSMEs in India play an important role in the achievement of SDs by 2030. Now we would explain the role of MSMEs in the achievement of SDGs:

1. Sustainable Development Goal 1—Elimination of Poverty

   All the countries in the world whether they are developed or developing, face the problem of poverty. The problem of poor people is serious in developing or least developed countries. There are 730 crore people in the world and nearly 30% of them are living below the poverty line. As far as India is concerned, as per the NITI report, based on a survey of 176,946 households in urban areas and 425,563 households in rural parts, the poverty ratio is 32.75% in rural areas and 8.81% in urban areas. It is expected that by 2030, poverty should be reduced by half of the proportion of

Table 2.1 People living in extreme poverty in India during 2021–2022

Total population	1,392,506,368
People living in extreme poverty	97,697,747
Male	44,806,455
Female	52,891,292

Source https://www.theglobalstatistic.com/poverty-in-india-statistics

men, women, and children of all ages. Table 2.1 shows the poverty data of people during 2021–2022.

Table 2.1 shows that 5.28 crore females population and 4.48 crore males population in India live in extreme poverty. According to Global Multidimensional Poverty Index (MPI), India is 62nd among 107 countries with an MPI score of 0.123, and 27.9% population is identified as multi-dimensionally poor.

The MSMEs play an important role in the elimination of poverty in India, as pointed out earlier, MSMEs have employed about 11 crore people. These MSMEs have contributed around 6.11% of the manufacturing GDP and 24.63% of the GDP from the service sector. Table 2.2 shows the employment generated by MSMEs in rural and urban areas.

Table 2.2 Estimated employment in MSME sector

Broad activity	*Employment (in lacs)*		*Total*
	Rural	*Urban*	
Manufacturing	186.56 (51.76%)	173.86 (48.24%)	360.42 (100%)
Electricity[a]	0.06 (75%)	0.02 (25%)	0.08 (100%)
Trade	160.64 (41.49%)	226.54 (58.51%)	387.18 (100%)
Other services	150.53 (41.56%)	211.69 (58.44%)	362.22 (100%)
Total	497.78 (44.85%)	612.10 (55.15%)	1109.89 (100%)

Note [a]Non-captive electricity generation and transmission
Source GoI (2021)

Table 2.2 indicates that MSMEs provided employment to 497.78 lac rural people which is 44.85% that consists of manufacturing, non-captive electricity generation, trade, and other services. Further MSMEs provided employment to 55.15% of urban people covering 48.24% of people in the manufacturing sector, 25% people in the non-captive electricity generation, 58.51% in the trade sector, and 58.44% in other sectors. The Government of India has launched the MGNREGA scheme in 2005 to provide a minimum of 100 days of wage employment to every household for the alleviation of poverty. The MSMEs in seasonal industries can provide employment to poor people for at least 100 days to eliminate poverty. Since the MSMEs are the second largest employer in the country, they can reduce the poverty level in India up to 2030 by generating more and more employment opportunities.

2. Sustainable Development Goal 2—Erase Hunger

In many developing countries and third-world countries, a sizeable population faces the problem of hunger and malnutrition. According to FAO estimates in "The State of Food Security and Nutrition in the World, 2020 Report" (FAO, 2020), 189.2 million people, that is 14% of the total population are undernourished in India, that covers 51.4% of women the age 14–50 years of age and 34.7% children under 5 years of age. The Global Hunger Index 2020 ranks India at 101 out of 116 countries on the basis of three leading indicators—prevalence of wasting and stunting in children under 5 years, under 5 child mortality rate, and the proportion of undernourished in the population. With a score of 27.5, India has a level of hunger that is serious. To achieve this goal, it becomes necessary for the MSMEs to erase the problem of hunger and malnutrition. The Agricultural MSMEs comprising agriculture and allied sectors can play an important role in reducing the problem of malnutrition, achieving food security, and promoting sustainable agriculture. The agriculture and allied sectors including animal husbandry, fishing, and dairying achieved a growth of 3.9% in 2021–2022 over the preceding year of 2020–2021. This sector is the largest employer of the workforce, that accounted for 54.6% of the total workforce. In India and Rwanda, food waste-reducing technologies minimize post-harvest losses and have resulted in reduced losses by over 60%—and increased smallholder farmers' incomes by more than 30%. For the promotion of sustainable agriculture, the Agriculture MSMEs should double the agricultural productivity by 2030, through technologies and practices that do not put stress on the resource base. In 2020–2021, the agricultural productivity for wheat was 3424 kg per hectare and for rice was 2749 kg per hectare. The MSMEs can go for organic farming, natural farming, agroforestry, precision farming, or conservation agriculture so that agricultural productivity will not affect the environmental resources on one hand and also increase agricultural productivity. This will solve the problem of hunger, malnutrition, and food insecurity by 2030.

3. Sustainable Development Goal 3—Establish Good Health and Well Being

   Goal 3 focuses on the establishment of good health and well-being of human beings. The MSMEs play a vital role in eliminating epidemics, non-communicable diseases, tobacco control, and reducing the occurrence of accidents at the workplace. For the health, safety, and welfare of the employees, the Factory Act 1948 is applicable in India. The MSMEs have to strictly comply with the provisions given in the Factory Act 1948. Recently in the wake of the COVID-19 pandemic, many MSMEs organized a free camp vaccination for their employees. As part of their Corporate Social Responsibility, the MSMEs can organize free medical and vaccination camps, provide free ambulance facilities, maintain blood banks, and provide lodging and boarding facilities to poor patients near famous hospitals. As far as tobacco control in the workplace is concerned, the report of WHO Global Report (2012) shows that 7% of all deaths (for ages 30 and above) in India are attributable to tobacco. Further, tobacco use is a major risk factor for four non-communicable diseases (NCDs)—cardiovascular disease, cancer, chronic lung cancer, and diabetes—putting people with these conditions at higher risk for developing severe illnesses when affected by COVID-19. The MSMEs should strictly adhere to the guidelines issued by the Ministry of Health and Family Welfare regarding the prohibition of tobacco in the workplace by the stakeholders. The Ministry of MSME has mandated sanitizing the workplace so as to prevent the COVID-19 pandemic. In addition to that, MSMEs have to ensure that all their employees are medically insured. They should provide a list of hospitals to the employees near the workplace that are giving treatment for COVID-19.

4. Sustainable Development Goal 4—Provide Quality Education

   This is a fact that every year lot of people are graduating from college, but many of them are not getting jobs, they are aspiring. This is because there is no relationship between the education given to them in colleges and the skills required in the job market. As per reports of the eighth edition of the Indian Skills Report (ISR) released in April 2021, only 46.82% of B.E./B. Tech graduates, 46.59% MBA graduates, 42.72% B.A. graduates, 40.3% B.Com. graduates, 30.34% of B.Sc. graduates, and 22.42% of MCA graduates are employable. It indicates a wide gap between the education that is imparted in different universities/institutions in the country and the practical knowledge and skills required in the job market. Apart from that, if we consider the Gross Enrollment ratio in India, as per the reports of the All India Survey of Higher Education (AISHE), this ratio in 2021–2022 was around 29% for the 18–23 age group. This indicates that a vast majority of students drop out of college. The new Education Policy targets a 100% Gross Enrollment Ratio in the next 10 years (up to 2030). MSMEs can provide professional and technical training to job aspirants so as to cater to the need of the industries. For this purpose, the MSMEs can avail of financial assistance from the Department of MSME for organizing training programs for the job aspirants. MSMEs particularly in rural areas can start primary and secondary schools to increase the Gross Enrollment Ratio so as to meet 100 per percentage by 2030.

5. Sustainable Development Goal 5—Enforce Gender Equality

   Though the Constitution of India has granted equal rights to men and women, still inequality prevails. As per the Global Gender Gap Report 2021 published by the World Economic Forum (WEF, 2021), India is ranked 140 out of 156 countries with a score of 0.625 (out of 1). Actually, gender equality prevents violence again women and empowers women to take part in the decision-making process, and makes them independent. As far as women entrepreneurs and women employees in MSMEs are concerned. The number of women entrepreneurs in India is 13.5 million to 15.7 million. The percentage of women entrepreneurs in 2021 in

India was 20.37%. The breakup of male and women entrepreneurs is given in Table 2.3.

It is clear from Table 2.3 that the maximum number of women entrepreneurs are there in the micro sector is 20.44% followed by 5.26% in the small sector and 2.67% in the medium sector. The question arises: why there is less number of women entrepreneurs in India? It is because of a lack of awareness of accessing finance, lack of social and institutional support, lack of access to professional networks, lack of physical safety at the workplace, lack of mobility, lack of education, and family responsibilities. For getting finance from various government schemes, women entrepreneurs should be aware of the various schemes started by the Government of India for them. These schemes are as follows:

- Mudra Loan for women
- Annapurna Scheme
- Shree Shakti Yojana
- Dena Shakti Scheme
- Bhartiya Mahila Bank Business Loan
- Mahila Udyam Nidhi Yojana
- Orient Mahila Vikas Yojana Scheme
- Cent Kalyani
- Udyogini Scheme

The MSMEs can also arrange awareness programs for women entrepreneurs about the various government schemes.

Table 2.4 indicates that the percentage of women employees in MSMEs in India is only 23.87% which consists of a rural workforce of 27.62% and a 20.82% urban workforce. This is because of the problems faced by women employees in MSMEs. Some of these

Table 2.3 The percentage distribution of enterprises owned by male/female entrepreneurs (category-wise)

Category	*Male*	*Female*	*All*
Micro	79.56	20.44	100
Small	94.74	5.26	100
Medium	97.33	2.67	100
All	79.63	20.37	100

Source GoI (2022)

problems are exploitation, fewer wages, unconducive working environment, no job security, sexual harassment, and lack of basic amenities at the workplace. MSMEs can provide more and more job opportunities for women depending on the skills they are processing. They can improve the terms of employment for women workers, especially those who are considered unskilled. They can provide life skills training on the job or through skills training institutes to women workers and employees to build their self-efficacy, negotiate gender constraints, improve their communication and comprehension to follow complex tasks, and manage their time and problems effectively. They can provide good physical facilities at the workplace and equal remuneration and incentives and rewards for the good performance of women employees.

6. Sustainable Development Goal 6—Provide Clean Water and Sanitation

   This goal is related to providing clean and safe water to all and also providing safe sanitation to all. This is what is referred to as WASH (water, sanitation, and health) facility. Wherever government machinery fails to provide WASH to the people, MSMEs can play an important role in providing such facilities to the people. The World Bank estimates an annual need of $114 billion between now and 2030, three times the current investment levels to achieve the target of WASH of SDG 6 and MSMEs can fill up this gap. The MSMEs can avail the financial assistance from the Ministry of Environment regarding setting up Common Effluent Treatment Plants (CETP), by which they can control the pollution occurred due to the disposal of toxic effluents. The financial assistance consists of 20% entrepreneur's contribution to project cost,

Table 2.4 Distribution of workers by gender in rural and urban areas (no. in lacs and percentage)

Sector	*Female*	*Male*	*Total*
Rural	137.50	360.15	497.65
	(27.62)	(72.38)	(100)
Urban	127.42	484.54	611.96
	(20.82)	(79.82)	(100)
Total	264.92	844.69	1109.61
Share %	(23.87)	(76.13)	(100)

Source GoI (2022)

25% of the Central Government Subsidy of project cost, 25% of the State Government Subsidy of project cost, and the remaining 30% of the loan from financial institutions of the project cost. The MSMEs can provide clean drinking water and efficient sanitation facilities in slum areas in the big and metropolitan cities in India. The MSMEs can help the Government of India in the programs for providing clean water and sustainable sanitation standards like Swachha Bharat, Atal Mission for Rejuvenation and Urban Transformation (AMRUT), and Jal Jeevan. Further, MSMEs can prevent the disposal of dumping of solid and liquid wastes into rivers, lakes, and ponds and thus save the environment from pollution. The MSMEs should educate their employees about the efficient use of water so that there will not be a waste of water.

7. Sustainable Development Goal 7—Grow Affordable and Clean Energy

   The main focus of this goal is to provide affordable and clean energy to all. According to NITI Aayog—energy from renewable resources—wind, water, solar biomass, and geothermal energy is inexhaustible and clean and constitutes 15% of the global energy mix. The MSMEs can play an important role in providing affordable and clean energy to the people by adopting green technology. Green technology is the technology that reduces human impacts on the natural environment. It reduces the carbon footprints of the business firms, reduces waste, conserves water, and consumes reduced energy in comparison to traditional technologies. For that, MSMEs can avail the benefits of various schemes of the Government of India like Technology Upgradation and Quality Certification's ZED certification schemes, A Scheme for Promoting Innovation, Rural Industry and Entrepreneurship (ASPIRE), Credit Linked Capital Subsidy for Technology Upgradation (CLCSS) and Design Clinic for Design Expertise to MSMEs. Further, the key sustainable finance providers to MSMEs are banks, corporations, international financial institutions, institutional investors, and international organizations through financial instruments, climate funds, green bonds, impact finance, social bonds, microfinance, SIDBI Sustainable Financing Scheme for funding, NABARD, and Make in India. MSMEs involved in the production of Solar power plants, renewable energy, green machinery, waste management, Electric vehicles (EVs), clean energy, and energy conservation can avail the sustainable finance. The Union Budget 2022–2023 targets carbon dioxide emissions from coal-fired plants and pollution from burning of crop residue by proposing that biomass pellets of 5–7% of the fuel in the thermal plants will be co-fired. It is estimated that it will reduce about 38 MMT of annual carbon dioxide emissions. On the other hand, farmers and villagers will get income and job opportunities in making biomass pellets from agriculture residue and farm waste. MSMEs can be actively involved in this process by reducing air pollution and providing job opportunities to rural people.

8. Sustainable Development Goal 8—Create Decent Work and Economic Growth

   The main focus of this goal is to create decent work and the economic growth of the country. As mentioned at the beginning of the chapter, MSMEs contribute around 30% toward the GDP, which consists of 6.11% of manufacturing activities and 24.63% of service activities, and 45% toward the export of items like readymade garments, leather goods, processed foods, and engineering goods. The MSMEs have contributed significantly to the economic development of the country in terms of output, employment, export, income, and investment. The MSMEs in India have employed 11 crore people (refer to Table 2.2) which is 45% in the rural sector and 55% in the urban sector. Table 2.5 shows the contribution of MSMEs to India's Gross Value Added (GVA) and Gross Domestic Product (GDP) at current prices from 2014–2015 to 2018–2019.

Table 2.5 Share of Gross Value Added (GVA) of MSME in all India GDP

Year	*Total MSME GVA (Rs. in crores)*	*Growth (%)*	*Total GVA (Rs. in crores)*	*Share of MSME in GVA (%)*	*All India's GDP (Rs. in crores)*	*Share of MSME in all India GDP (%)*
2014–2015	3,658,196	–	11,504,279	31.80	12,467,959	29.34
2015–2016	4,059,660	10.97	12,574,499	31.28	13,771,874	29.48
2016–2017	4,502,129	10.90	13,965,200	32.24	15,391,669	29.25
2017–2018	5,086,493	12.98	15,513,122	32.79	17,098,304	29.75
2018–2019	5,741,765	12.88	17,139,962	33.50	18,971,237	30.27

Source Central Statistics Office (CSO), Ministry of Statistics and Programme

Table 2.5 shows that the share of MSME in GVA has increased continuously from 31.8% in 2014–2015 to 33.5% in 2018–2019, and their share in All India GDP has increased from 29.34% in 2014–2015 to 30.27% in 2018–2019. It indicates that MSMEs have maintained a growth rate of over 10%.

As far as providing decent work conditions is concerned, the MSMEs in India are mostly less skilled and labor-intensive units. Therefore they cannot provide good compensation, facilities, job security, and working conditions to the employees like that of the organized sector. Thus the working conditions are not that much conducive for the employees. The International Labour Organization (ILO) defines decent work as "Work that is productive, delivers a fair income with security and social protection, safeguards basic rights, offers equality of opportunity and treatment, prospects of personal development and the chance for recognition and to have your voice heard" (26). Efforts have to be made to improve the working conditions of employees of MSMEs. This is because many of the MSMEs are there in the informal sector due to cumbersome regulations. Though India has improved its World Bank Ease of Doing Business Rank from 130 in 2016 to 63 in 2020, there is a long way to go for this ease of business to be extended to small and medium-scale businesses as well (27).

9. Sustainable Development Goal 9—Increase Industry, Innovation, and Infrastructure

   The main focus of this goal is to promote sustainable industrialization, innovation, and infrastructure. With around 63.4 million units producing more than 8000 products, MSMEs contribute 6.11% of GDP in the manufacturing sector and 24.63% of GDP in the service sector in India. The total contribution toward GDP is 30%. The total Gross Value Added by MSMEs is 33.5% (see again Table 2.5). The total export of MSMEs is around 45%. The MSMEs have employed more than 11 crore people (see again Table 2.2). Thus, MSMEs have contributed to the industrial development of the country. The output of the MSMEs has continuously increased for the last 5 years i.e. from 2014–2015 to 2018–2019. As per the data from the Ministry of MSME, up to November 26, 2021, the Udyam Registration Portal registered 5,767,734 MSMEs, consisting of 5,441,220 (94.34%)

micro-enterprises, 293,555 (5.09%) small enterprises, and 32,959 (0.57%) medium enterprises.

With respect to innovation based on Global Innovation Index 2021, India stands at 46th among 132 economies. The Global Innovation Index (GII) ranks world economies according to their innovation abilities. India has slightly improved its ranking from 48th in the previous year. India's ranking has been consistently rising in the last few years. From 81 in 2015, it has moved to 46 in 2021. It takes second place in the lower middle-income group. The MSMEs in India have to go for the innovation of their product, process, marketing practices, and technology so as to be competitive in the global environment. The Ministry of MSMEs in India had initiated various schemes and programs like an Incubation scheme, Technology upgradation, Quality certification, Credit Linked Capital Subsidy for Technology Upgradation, Science and Technology Scheme for Coir, etc., to promote and support the untapped creativity of individual and to promote adoption of latest technologies in manufacturing as well as knowledge-based innovative MSMEs. The MSMEs can avail the benefits of these schemes and programs. In 2022, the Government of India has come up with an MSME Innovative Scheme, which is a combination of Incubation, Design intervention, and IPR protection initiatives to enhance MSME competitiveness. The MSMEs can take benefit from these schemes.

As far as infrastructure development is concerned, MSMEs can play an important role in India. For that, the MSMEs can go for the cluster development approach for enhancing productivity and competitiveness as well as capacity building through the Ministry of MSME, Government of India. This will lead to the creation of infrastructural facilities like power distribution network, water, telecommunication, drainage, and pollution control facilities, roads, banks, raw materials, storage, and marketing outlets, common service facilities, and technological backup services for MSEs in the new/existing industrial estates/areas (32). Till 2019–2020, NABARD had sanctioned Rs. 65,838 crore for various infrastructure projects under agriculture and irrigation, roads and bridges, social sectors, etc. Disbursement against ongoing projects was Rs. 56,432 crore. Further, NABARD has also financed projects under Pradhan Mantri Awas Yojana (PMAY), Dairy Processing and

Infrastructure Development Fund (DIDF), Warehouse Infrastructure Fund (WIF) NABARD Infrastructure Development Assistance (NIDA), and Swachha Bharat Mission-Gramin (SBM-G) (33). The Government of India in its Union Budget 2022–2023 has initiated the Prime Minister's Development Initiative for North-East India, PM-DevINE, which will fund infrastructure and social development according to the needs of the region (34). The MSMEs can avail of financial assistance from these schemes and contribute in infrastructure development (GoI, 2020, 2021).

10. Sustainable Development Goal 10—Reduce Inequality

    This goal focuses on reducing inequality. MSMEs can play a vital role in filling up the gap between urban and rural economic conditions. As indicated in Table 2.2, MSMEs have provided employment to 44.85% of people, and thus contributed to the reduction of poverty and enhance the standard of living of rural people. As far as a gender issue in MSMEs is concerned. As shown before in Table 2.3, more and more efforts are to be made to increase the number of women entrepreneurs which is merely 20.37% now. Table 2.6 represents the ownership of MSMEs on the basis of social category.

    It is clear from Table 2.6 that 66.27% of MSMEs are owned by socially backward classes comprised of SC, ST, and OBC categories. Thus, MSMEs play an important role in bridging the gap between the gap in social classes. MSMEs have to increase the share of SC and ST categories which is only 12.45% and 4.1%, so as to attain SDG 10 by 2030.

11. Sustainable Development Goal 11—Mobilize Sustainable Cities and Communities

Table 2.6 Percentage distribution of enterprises by social group of owner in rural and urban areas

Sector	*SC*	*ST*	*OBC*	*Others*	*Not known*	*All*
Rural	15.37	6.70	51.59	25.62	0.72	100
Urban	9.45	1.43	47.80	40.46	0.86	100
All	12.45	4.10	49.72	32.95	0.79	100

Source GoI (2022)

The main focus of this goal is to provide sustainable cities and the communities living in these cities to be safe and healthy. Since the MSMEs in India provide employment to around 11 crore people, it is expected that they should provide affordable housing facilities to their employees. They can also provide public transportation systems in urban areas, road safety, waste management, and durable buildings. The Government of India has decided to build affordable housing under Pradhan Mantri Awas Yojana in rural and urban areas to be completed by 2023. The budgetary allocation made in the Union Budget 2022–2023 is Rs. 48,000 crores for this purpose. The MSMEs can avail the benefits of that and can play an active role in building affordable houses for needy people. As far as waste management is concerned, the Ministry of Environment has initiated a scheme for setting up Common Effluent Treatment Plants (CETP) by the MSMEs. Since many of the MSMEs cannot afford to set up the effluent treatment plant individually, they can form a cluster and set up common effluent treatment plants. The nature of financial assistance is—Entrepreneurs' contribution is 20% of the project cost, State subsidy is 25% of the project cost, Central subsidy is 25% of the project cost, and the remaining 30% of the project cost is a loan from financial institutions. The MSMEs can avail of this facility and prevent the affluence of toxic material in the river, lake, and ponds.

With reference to road safety programs, many activities are covered like publicity campaigns on road safety, providing training to drivers, research and development of traffic control, organizing quizzes, essays, painting and seminars on road safety, etc., for which the Ministry of Roadways provide the financial assistance up to Rs. 5 lacs. The MSMEs can participate in these activities and provide road safety to the people. Further, the Ministry is also providing an ambulance and a crane under the scheme of the National Highway Accident Relief Services Scheme (NHARSS) to the NGO.

12. Sustainable Development Goal 12—Influence Responsible Consumption and Production

The main focus of this goal is to promote green public procurement and adopt sustainable business practices by the MSMEs in India. For adopting sustainable business practices, the Ministry of MSME, Government of India introduced Lean Manufacturing Competitive (LMCS) Scheme in 2014. Bhattacharya and Ramchandran (2021) conducted a study on randomly selected 23 units consisting of brass, pump, machinery, packaging, auto, textile, and garments to see the quality of benefits achieved by MSMEs post-implementation of lean. The result showed a reduction in cost, a reduction in waste, an improvement in quality, and an increase in lead time. Further various non-tangible benefits have been realized like improved workers' performance, workers' health and safety besides increased customer satisfaction. Thus MSMEs in India have to adopt sustainable business practices by going for lean manufacturing techniques.

As far as green public procurement is concerned, MSMEs can play an important role. Green procurement consists of purchasing products and services that cause minimal adverse environmental impacts. Green public procurement is used by many OECD countries. India's own limited experience in the procurement of LED lamps, star rating of electrical appliances, etc., has been encouraging. The Government of India in the Union Budget has allocated Rs. 19,500 crores for the Production Linked Incentive Scheme for the manufacture of high-efficiency solar modules so as to attain the goal of cleaner energy, low carbon emissions, and sustainable development. The MSMEs can avail the benefit of this scheme.

13. Sustainable Development Goal 13—Organize Action to Combat Climate Change and Its Impacts

This goal concentrates on action that is to be taken to combat climate change and its impact on the environment. As pointed out in SDG 7, the MSMEs can adopt Green Technology to conserve the environment. Climate change may result in heavy/deficient rainfall, flood/drought situation, an increase in the sea level, and rising temperature. The ocean warming and increased stratification of the upper ocean caused by global climate change will lead to a decline in dissolved oxygen with amplification of ocean productivity, nutrient cycling, and marine habitat. MSMEs have to take a number of steps and strategies to prevent climate change and its adverse impact on society:

- Identify risks for physical assets and business operations posed by climate change.

- Analysis and evaluation of risks and sensitivities.
- Assessing business opportunities resulting from climate change.
- Development of measures to build resilience and exploit the most promising opportunities.
- Analysis and evaluation of proposed measures, selecting the most adequate.
- Implementing and monitoring these measures and installing a risk management system.

On the basis of the above, MSMEs can formulate the following strategies:

- Protective Strategies—in which MSMEs can take steps to protect against the adverse impact of climate change.
- Accommodating Strategies—in which MSMEs can increase their capacity to cope with the adverse impact of climate change.
- Retreat Strategies—in which MSMEs focus on reducing the exposure to the adverse impact of climate change.

The Government of India wanted to fulfill half of the country's energy requirements by 2030 through the use of renewable energy. Also, the Government of India has set the target of net-zero carbon emissions by 2070. The MSMEs can achieve these objectives through the upgradation of infrastructure and technology. As far as the financial requirements of MSMEs are concerned, the MSMEs can avail the financial assistance from the Ministry of Micro, Small and Medium Enterprises, Government of India.

Regarding the energy audit of MSMEs that indicates energy conservation, the Bureau of Energy Efficiency (BEE) carried out the situational analysis in selected 25 SME clusters in 2016–2017, and the total energy savings, in 988 units in 26 (25 clusters + Firozabad) cluster quantifies to Rs. 15.58 crores per annum with an investment of Rs. 28.06 crores.

14. Sustainable Development Goal 14—Develop Life Below Water

The main focus of this goal is to conserve life below sea water which is a reduction in marine pollution and to establish sustainable development of marine and coastal ecosystems. In other words, to stop overfishing, illegal, unreported, and unregulated fishing, prevent waste management or littering in the seawater to prevent the lives of fish and other creatures below the seawater. The MSMEs can play an important role in the reduction of marine pollution. Because the fishery sector in India supports the livelihood of over 28 million people particularly covering the marginalized and vulnerable communities. This sector is having a share of 7.56% of global production. It contributes 1.24% to the country's GVA and 7.28% to the agricultural GVA. The Government of India has launched Pradhan Mantri Matsya Sampada Yojana (PMMSY) in May 2020 that provides a financial package to the fishermen and also emphasizes the use of optimal water management to achieve "more crop more drop" and improved quality of fish and fish products and address the critical gaps in the value chain through technology infusion. Further, the MSMEs can adopt sustainable practices like eco-labeling and traceability systems for catching fish.

15. Sustainable Development Goal 15—Advance Life on Land

    This goal focuses on sustainable management of freshwater, deforestation, and wetlands, restoring degraded soil and land, preventing the expansion of deserts, and maintaining the ecosystem. India has the tenth largest area in the world. India ranked third globally in increasing its forest area from 2010 to 2020. India's forest cover has increased by more than 3% from 2011 to 2021. The forest cover in 2019 increased to 21.67% of the total geographical area from 21.35% in 2015. NABARD has provided financial assistance to the tune of Rs. 1822 crore on 40 projects. MSMEs can avail the benefits of these schemes so as to converse the natural resources, revive natural water resources to counter rainfall deficit, and increase the green cover through better farming practices integrated with mountain farming and climate-resilient crop cultivation. The Government of India through the Union Budget 2022–2023 has highlighted energy conservation, agroforestry, and circular economy for attaining sustainable development. For that Government will provide financial assistance to farmers and MSMEs who shall go for agroforestry. As far as the circular economy is concerned, it is based upon recycling and reuse of resources rather than taking resources from the environment to make the products and dump the waste. The Union Budget 2022–2023 stress on circular economy for sustainable and greener development in the future. The MSMEs should take the initiatives for the circular economy so as to conserve the environment and for the sustainable and greener development of the country.

16. Sustainable Development Goal 16—Guarantee Peace, Justice, and Strong Institutions

    The main purpose of this goal is to reduce violence, end exploitation, and promote ethical business practices. The MSMEs can play an important role in this. As pointed out in SDG 1, the MSMEs provide employment to poor people and thus deviate them from the path of violence and crime. Further, the MSMEs can adopt ethical business practices and thus reduce the exploitation of the employees working in these organizations. They have to follow the provisions given in various Acts enacted by the Government of India from time to time so as to stop the violence against women, and also prevent child labor in their organization. They can organize short-term training programs for the employees to stop the violence against women, children, and people from depressed classes.

17. Sustainable Development Goal 17—Build Global Partnerships for the Sustainable Development

    This goal focuses on building global partnerships in the area of science, technology, finance, innovation, and capacity building by MSMEs for sustainable development. Since the Indian MSMEs are having the problem of obsolete technology, they can have to tie up with foreign players or institutions by which they can access modern and sophisticated technology. As far as the financial requirement of MSMEs is concerned, World Bank in 2021, approved $500 million for raising and accelerating the MSMEs performance (RAMP) program to address the liquidity and credit needs of millions of MSMEs in India affected by the COVID-19 pandemic. Apart from that, the RAMP program will promote technology-based solutions, green investments, and access to services for women entrepreneurs.

Conclusion

It can be concluded in the end, that MSMEs can play an important role in the achievement of Sustainable Development Goals in India due to their contribution toward GDP and export earning, providing employment to poor people, and thus reducing poverty and low investment cost. Though MSMEs have contributed a lot to the GDP, export earnings, and employment, still they have to do a lot in the achievement of Sustainable Development Goals by 2030. It is a fact that MSMEs in India have a number of weaknesses or drawbacks like limited finance, limited access to national and international markets, obsolete technology, lack of skilled manpower, working in isolation, and lack of coordination between big industrial units and MSMEs. The efforts are to be made to remove these drawbacks by the government as well as MSMEs themselves for the successful achievement of Sustainable Development Goals in India.

References

Bhamra, A., Shanker, H., & Niazi, Z. (2015, August). *Achieving the Sustainable Development Goals in India: A study of financial requirements and gaps.* Report submitted to Ministry of Environment, Forest and Climate Change, Government of India. https://www.devalt.org/images

Bhattacharya, I., & Ramchandran, A. (2021). Lean manufacturing techniques—Implementation in Indian MSMEs and benefits realized thereof. *Indian Journal of Engineering & Materials Sciences, 28*, 89–101. https://www.IJEMS/Indrajit/Bhattacharya1.pdf

FAO. (2020). *The state of food security and nutrition in the world* (2020 Report). Food and Agriculture Organisation.

GoI. (2019). *Annual report of MSME 2019–2020.* Government of India.

GoI. (2020). *NABARD annual report 2019–2020.* Government of India. https://www.annual-report-2020-21-full-report-nabard.pdf

GoI. (2021). *Annual report of MSME 2020–2021.* Government of India.

GoI. (2022). *Annual report of MSME 2020–2021.* Government of India.

Mahesh, K. M., Aithal, P. S., & Sharma, K. R. S. (2022). Impact of sustainable finance on MSMEs and other companies to promote green growth and sustainable development. *International Journal of Applied Engineering and Management, 6*(1), 60–76.

MCA. (2011). *Facing the impacts of climate change, Indian MSMEs and adaption.* Discussion paper by GIZ and SIDBI, Ministry of Corporate Affairs. https://www.giz.de/giz2013-climate-risk-study

MSME. (2021). *MSME schemes, 2015*. Ministry of Micro, Small and Medium Enterprises, Government of India. https://www.MSME_Schemes_English_0.pdf

MSPI. (2021). *Sustainable Development Goals-National Indicator Framework*, Progress Report 2021, Government of India, Ministry of Statistics and Programme Implementation, National Statistical Office.

Priya, D., Priya, N., & Tara, S. (2014). *Gender issues in Micro, Small & Medium Enterprises (MSME) sector in India*. Report submitted to World Bank by International Centre for Research on Women (ICRW).

Rahul, S., Chandra, A., Singh, M., Milton, T., Mandala, G. N., & Goda, N. R. (2021). Role of micro finance in eradication of poverty in India: Perspectives and operation. *Turkish Journal of Computer and Mathematics Education, 12*(9), 1933–1945.

Sobir, R. (2019). *Micro-Small & Medium sized Enterprises (MSMEs) and their role in achieving the Sustainable Development Goals*. Report submitted to DESA, Department of Economic and Social Affairs, United Nations. https://sustainabledevelopment.un.org/documents

WEF. (2021). *The Global Gender Gap Report 2021*. The World Economic Forum.

WHO. (2012). *2012 global progress report on implementation of the WHO Framework Convention on Tobacco Control*. World Health Organization. https://apps.who.int/iris/handle/10665/79170

CHAPTER 3

The Potential Role of MSMEs in Achieving SDGs in Indonesia

Tulus T.H. Tambunan

Introduction

It has been recognized worldwide that MSMEs play a vital role in economic development. In developing and least developed countries, including small island developing states, MSMEs are the majority of firms. Thus, they are very important for job creation, poverty alleviation, improvement of income distribution, development of the manufacturing industry, rural economic development, growth of export especially manufactured goods, and GDP growth. They also provide business opportunities to women, the unemployed, and less educated youths. MSMEs are important because they are labor-intensive (Tambunan, 2021). By extrapolating data from the World Bank's Enterprise Surveys, a report from the International Finance Corporation (IFC, 2017) shows that there are close

T.T.H. Tambunan (✉)
Center for Industry, SME and Business Competition Studies, Universitas Triskti, Jakarta Barat, Indonesia
e-mail: ttambunan56@yahoo.com

H. Dasaraju and T.T.H. Tambunan (eds.), *Role of Micro, Small and Medium Enterprises in Achieving SDGs*, Sustainable Development Goals Series, https://doi.org/10.1007/978-981-99-4829-1_3

to 162 million formal MSMEs in developing and least developed countries, of which 41 million are microenterprises (MIEs), and 21 million are small and medium enterprises (SMEs). Countries like Brazil, China, and Nigeria, contribute 67% to the total number of MSMEs, which is equivalent to 109 million enterprises. There are close to 12 million MSMEs in China alone, which represent 56% of all MSMEs in developing countries. China also has 44 million MIEs, which represent 31% of all MIEs in developing countries. There is a large concentration of MSMEs in the East Asia region (64 million), followed by Sub-Saharan Africa, which has 44 million enterprises, the majority of which (97%) are MIEs.

In developing economies in the Asia and Pacific (AP) region, for instance, more than 98% of companies were considered MSMEs, with more than half of the economies, including Indonesia, holding a share of more than 99% (Table 3.1). This share has remained constant over the past decade for all economies. Based on how each economy defined its MSMEs and the availability of the most recent data, nearly 150 million businesses in the region was considered MSMEs, representing around 99.8% of all businesses in the region. It is important to note that what is considered MSMEs in one economy may not be considered MSMEs in other economies given the fact that economies in the region define MSMEs differently.

Table 3.1 Number of MSMEs in developing economies in the AP region[a]

Economy	*Total number (million)*	*% of total enterprises*	*Year*
Brunei Darussalam	5.9	97.2	2017
Chile	944.9	98.6	2017
China	21,921.1	99.6	2017
Indonesia	64,194.1	99.99	2018
Malaysia	907.1	98.5	2015
Mexico	4,169.7	99.7	2018
Papua New Guinea	49.5	13.0	2016
Peru	1,899.6	99.5	2017
Philippines	920.7	99.6	2017
Singapore	262.6	99.5	2018
China Taipei	1,466.2	97.6	2018
Thailand	3,077.8	99.8	2018
Viet Nam	507.9	98.1	2017

Note [a]The number of MSMEs was rounded up
Source APEC (2020)

Given their vital role discussed above, especially for poverty eradication, mostly among women, the UN has assigned a great role to MSMEs for taking a lead in achieving most of the economic-related SDGs, including promoting inclusive and sustainable economic growth, increasing productive employment opportunities and decent work, especially for the poor and vulnerable, particularly women and youth, advancing sustainable industrialization and innovation, and creating a positive push for a higher quality of life, better education and good health for all (OECD, 2017, cited from Dasaraju et al., 2020). At least theoretically, as they are the greatest generator of employment and business opportunities in developing and least developed countries, MSMEs are the backbone for these countries to achieve the SDGs. But, for MSMEs to be able to play this role they must be highly competitive and able to grow rapidly in a sustainable manner.

With the above background, by analyzing secondary data, this chapter discusses the potential role of MSMEs in achieving SDGs in Indonesia by 2030.

Achievement of SDGs in Indonesia

At the time this chapter was written, it has been almost 7 years since the world adopted the SDGs into their respective country's development plans. However, based on several reports issued by several international institutions such as The Sustainable Development Goals Report 2021, the progress made by countries in the world has not been as expected. Although the results of achieving the SDGs have not been as targeted or expected, it can be seen that there is a gradual increase in each goal and target indicator, especially regarding poverty reduction, improving maternal health, and gender equality, but progress is not fast or broad enough. Since the coronavirus hit, much of that progress has declined or been reversed (https://www.goodnewsfromindonesia.id/2021/07/21/perjalanan-program-sdgs-di-indonesia-apa-saja-achievement).

Indonesia's own ranking in 2017 had dropped to 100 but experienced an increase in points of 62.9. In that year, Indonesia still had a red list of most targets in each of the SDGs indicators or goals. Meanwhile, the indicator targets that are clean from the red list are the first goal (SDG 1) and the thirteenth goal (SDG 13). There is one SDGs indicator in which Indonesia has a green color, namely SDG 13 or the goal of Handling Climate Change, while there are 7 other indicators in red which indicate the main challenges in these indicators, a total of 8 indicators in orange which means there are significant challenges in and 1 yellow indicator which means there are challenges with low intensity in the implementation of these indicators.

In 2018, Indonesia's ranking increased to 99 with 62.8 points with only two indicators whose target details were clear of red marks, namely SDG 1 and SDG 13. However, in that year there was a decline in the level of achievement which was marked by none of the SDGs indicators in green, indicating that it has been achieved. The indicator with red color has actually increased to 8 indicators (previously 7), there are 7 indicators with orange color and 2 indicators with yellow color. In 2019, Indonesia was ranked 102nd with 64.2 points. The achievement of Indonesia's SDGs has increased, namely, the indicator with yellow color increased to 3 indicators, 6 indicators in orange, and 8 indicators in red. In the 2019 report, SDG 4 and SDG 12 have increased by changing the color of the dashboard from orange to yellow, while SDG 1 has decreased. In 2020, Indonesia is ranked 97th with 66.3 points. According to the report, the achievement of Indonesia's SDGs faces major challenges on 9 indicators labeled in red (originally 8 indicators) with the addition of SDG 14, 5 indicators with significant challenges and 3 indicators with persistent challenges. In 2021, Indonesia's ranking and points in The Sustainable Development Report are the same as the previous report, which is ranked 97th with 66.3 points. The implementation of the SDGs in the report shows no sign of a decrease in points from each indicator. The drop in points is indicated by a down arrow in red. The visualization of the 2021 SDGs dashboard is exactly the same as the 2020 SDGs dashboard, but there are differences in the details of each indicator target (https://www.gatra.com/news-485592-ekonomi-peringkat-index-sdgs-2020-di-indonesia-increase.html).

Based on the 2021 Southeast Asia SDGs Index, the country with the nickname White Elephant is in the top position in achieving the 2021 SDGs with 74.19 points. Vietnam trailed in second place with 72.85 points, Malaysia in third with 70.88 points. Meanwhile, Indonesia ranks 6th in the achievement of the 2021 SDGs in the Southeast Asia region (https://sdgs.bappenas.go.id/wp-content/uploads/2021/02/Roadmap_Bahasa-Indonesia_File-Upload.pdf). Table 3.2 outlines the achievement targets that have been set for each of the 17 goals.

Table 3.2 Targets for achieving a number of SDGs by 2030

Goal	*Achievement 2030*	
	Without SDGs	*With SDGs*
No poverty	Percentage of people living below the national poverty line	
	↓5.73%	↓4.33%
No hunger	Prevalence of population experiencing moderate or severe food insecurity (Food Insecurity Experience Scale (FIES))	
	4.70 FIES	3.30 FIES
	Prevalence of stunting in children under five	
	22.37%	10.0%
A healthy and prosperous life	Maternal deaths per 100,000 live births	
	210	131
	Infant mortality per 100,000 live births	
	15.6	12
	Incidence of Tuberculosis (TB) per 100,000 population	
	261	65
	Percentage of smoking in adolescents (age 10–18 years)	
	15.95%	7.5%
	Prevalence of Obesity in Adult Population (Age 18 years)	
	47.5%	21.8%
	Total Fertility Rate (TFR)	
	2.42	2.1
	National Health Insurance Coverage	
	97.5%	100%
Quality education	Proportion of 4th graders who achieve minimum proficiency in reading	
	67.2%	
	Proportion of 4th graders achieving minimum proficiency in mathematics (trend projection)	
	35.5%	
	Proportion of 9th graders who achieve min. reading skills	
	36.7%	50.0%
	Proportion of 9th graders who achieve math proficiency	
	25.0%	38.0%
	Gross enrollment in primary school education	

(continued)

Table 3.2 (continued)

Goal	*Achievement 2030*	
	Without SDGs	*With SDGs*
	65.51%	
	Gross enrollment in higher secondary education	
	86.88%	90.55%
	Gross enrollment in tertiary education	
	43.85%	60.84%
	Net ratio of girls to boys in higher secondary education	
	104.26	102.31
	Lower net ratio of girls to boys in secondary education	
	113.01	98.00
Gender equality	Proportion of women aged 20–24 who married or started a family before the age of 18	
	10.03%	6.94%
Clean water and proper sanitation	Percentage of households with access to improved drinking water	
	75.8%	100%
	Percentage of households with access to better sanitation	
	100%	
Clean and affordable energy	Electric power consumption per capita	
	2,035 kWh	3,201 kWh
	Renewable energy mix	
	12.1%	26.1%
Decent work and economic growth	Real GDP growth rate per capita	
	4.4%	5.4%
	Unemployment rate	
	4.7%	3.8%
Industry, innovation and infrastructure	Proportion of population served by mobile broadband services	
	92.5%	100%
Decrease inequality	Gini coefficient	
	0.379	0.363
Sustainable cities and communities	Proportion of households with access to adequate and affordable housing	

(continued)

Table 3.2 (continued)

Goal	*Achievement 2030*	
	Without SDGs	*With SDGs*
	52.18%	68.06%
Handling climate change	Greenhouse gas emission intensity	
	367.78 Ton CO2e/bill. Rupiah	261.06 Ton CO2e/bill. Rupiah
Land ecosystem	Proportion of forest cover to total land area	
	40.0%	45.5%
Partnership to achieve goals	Ratio of tax revenue to GDP	
	13.5%	15.1%
	Proportion of individuals using the internet	
	81%	89%

Source Bappenas (sdgs.bappenas.go.id)

The Importance of MSMEs

It has been recognized worldwide that MSMEs play a vital role in the economic development of all countries in the world. Especially in developing and least developed countries, as they are the majority of enterprises, MSMEs have a crucial role to play because of their potential contributions to poverty reduction, improvement of income distribution, employment creation, industrial development, rural development, GDP growth, and export diversification and growth. They also provide business opportunities to women, unemployed, and less educated youths. The growth of MSMEs is also strategic in achieving resilience and stronger recovery in these countries after the COVID-19 crisis.

By extrapolating data from the World Bank's Enterprise Surveys, a report from the International Finance Corporation (IFC, 2017) shows that there are close to 162 million formal MSMEs in developing and least developed countries, of which 141 million are microenterprises (MIEs), and 21 million are small and medium enterprises (SMEs), Three countries, i.e., Brazil, China, and Nigeria, contribute 67% to the total number of MSMEs, which is equivalent to 109 million enterprises. There are close to 12 million MSMEs in China alone, which represents 56% of all MSMEs in developing countries. China also has 44 million MIEs, which represents 31% of all MIEs in developing countries. There is a large concentration of enterprises in the East Asia region (64 million), followed by Sub-Saharan

Africa, which has 44 million MSMEs, the majority of which (97%) are microenterprises.

Data from the Indonesian State Ministry of Cooperatives and SMEs (Menegkop & UKM) as well as the Central Statistics Agency (BPS) showed that there were approximately 39.765 million MSMEs which represents 99.8% of the total business establishments in Indonesia in 1997. The number was observed to be growing every year except in 1998 when the Asian financial crisis of the 1997–1998 period hit Indonesia which caused the Indonesian rupiah (IDR) exchange rate to depreciate against the United States dollar (USD) by more than 200%. Table 3.4 shows that almost 100% of total companies in Indonesia are from this category of enterprises and they generated approximately 97% of total employment in the country.

Potential Role of MSMEs

Indonesia is committed to successfully implementing the SDGs by achieving the 2030 development agenda. In this regard, Indonesia's Presidential Regulation no. 59/2017 concerning the implementation of SDGs in Indonesia mandated the Ministry of National Development Planning of the Republic of Indonesia to provide the Roadmap of SDGs Indonesia. The Roadmap of SDGs Indonesia was developed through a long process and discussion by involving multi-stakeholder participation, ensuring that the contents of this roadmap reflect all stakeholders' aspirations with rigorous exercises. The roadmap defines issues and projections of main SDGs indicators in each goal, including its forward-looking policies to achieve such targets. There are around 60 selected indicators to include in the full version of the roadmap. From the projection exercises and intervention scenarios of the indicators, it is obvious that the achievement of such targets needs a strong collaboration among stakeholders and commitments in both activities and financing, as the gaps remain for achieving the ambitious 2030 agenda (Bappenas, 2019).

Among the stakeholders are MSMEs, as his group of enterprises is the key contributor to economic development in Indonesia. MSMEs have been assigned an important role due to their huge potential for solving socio-economic challenges facing the country. Their most assigned role is to create a business as well as employment opportunities for the especially low-educated population, unemployed youths, poor households, women,

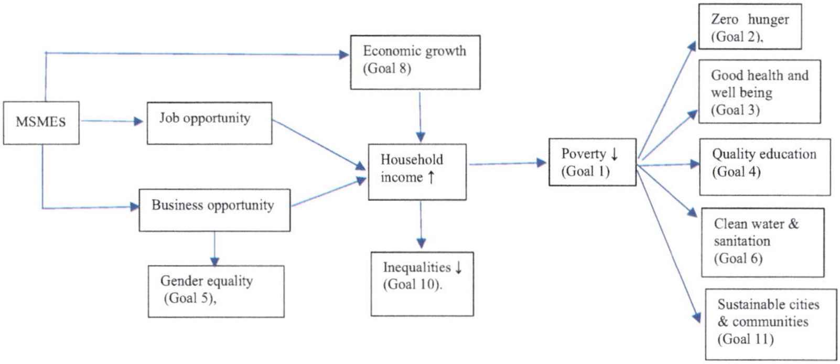

Fig. 3.1 The gradual process of the role of MSMEs in achieving various SDGs

and other socially underprivileged groups. By fulfilling this role, the enterprises will make a great contribution to poverty alleviation in the country. Therefore, their vital contribution to the broader socio-economic objectives, including employment generation and poverty alleviation makes them a key priority area for achieving the SDGs in Indonesia.

There are 17 SDGs ratified by the United Nations General Assembly in September 2015, which were presented as "a universal call to action to end poverty, protect the planet and ensure that all people enjoy peace and prosperity", which are targeted to be achieved by 2030. As shown in Table 1.2, MSMEs involved in specific sectors could help achieve many objectives under different goals (Blue Orchard, 2017, cited from Dasaraju et al., 2020).

From an economic perspective, the direct contribution of MSMEs to the achievement of SDGs is to create job and business opportunities, including for married women from poor/low-income households in rural areas, and economic growth. In turn, economic growth, in theory at least, increases the income of the lowly educated workforce, along with increased employment and business opportunities reducing poverty. By reducing the number of poor people, several other goals can also be achieved (Fig. 3.1).

Goal 1: Poverty Alleviation

Goal 1 is to end poverty in all its forms everywhere. The contribution of MSMEs to achieve this goal is by generating employment and business opportunities that facilitate people to come out of the vicious circle of poverty. As shown in Table 3.3, almost 100% of total companies in Indonesia are from this category of enterprises and they generated approximately 97% of total employment in the country. The employment effect of MSMEs can be direct as well as indirect. The direct effect is the number of people working in MSMEs, whereas the indirect effect is the creation of jobs in other linked sectors. Just a simple example: a small firm produces fertilizer for agriculture. In this case, it can be said that the growth or sustainability of production in the agricultural sector is highly dependent on that firm, assuming there are no other firms that also make fertilizers and there is no import of fertilizer. Suppose that the firm has 20 workers which can be considered the direct employment effect, while in the agricultural sector there are 100 laborers, which is the indirect employment effect. Then in total, the firm has generated as many as 120 people. It means that the total employment effect of MSME is the sum of their direct effects plus indirect effect.

Based on official statistics, Table 3.4 shows poverty changes in Indonesia from 1976 to 2021. As can be seen, the percentage of poverty at the national level decreased significantly from about 40% to almost 17.5% during the period 1976–1996, and the largest decrease occurred

Table 3.3 Number of MSMEs and their workers by sub-category in Indonesia, 2018–2019

Description	*Unit of measure*	*2018*		*2019*	
		Total	*Share (%)*	*Total*	*Share (%)*
MSMEs	Unit	64,194,057	99.99	65,465,497	99.99
–MIEs		63,350,222	98.68	64,601,352	98.67
–SEs		783,132	1.22	798,679	1.22
–MEs		60,702	0.09	65,465	0.10
LEs		5,550	0.01	5,637	0.01
MSMEs + LEs		64,199,607	100.00	65,471,134	10.00
MSMEs	People	116,978,631	97.00	119,562,843	96.92
–MIEs		107,376,540	89.04	109,842,384	89.04
–SEs		5,831,256	4.84	5,930,317	4.81
–MEs		3,770,835	3.13	3,790,142	3.07
LEs		3,619,507	3.00	3,805,829	3.08
MSMEs + LEs		120,598,138	100.00	123,368,672	100.00

Source Menegkop & UKM (http://www.depkop.go.id/)

Table 3.4 Poverty in Indonesia, 1976–2021

Year	*Total number of poor people (million)*			*Percentage of poverty (%)*		
	Urban	*Rural*	*National*	*Urban*	*Rural*	*National*
1976	10.00	44.20	54.20	38.79	40.37	40.08
1980	9.50	32.80	42.30	29.04	28.42	28.56
1984	9.30	25.70	35.00	23.14	21.18	21.64
1987	9.70	20.30	30.00	20.14	16.14	17.42
1990	9.40	17.80	27.20	16.75	14.33	15.08
1996	9.42	24.59	34.01	13.39	19.78	17.47
1998	17.60	31.90	49.50	21.92	25.72	24.23
1999	15.64	32.33	47.97	19.41	26.03	23.43
2000	12.30	26.44	38.74	14.60	22.38	19.14
2001	8.60	29.27	37.87	9.76	24.84	18.41
2002	13.30	25.09	38.39	14.46	21.10	18.20
2003	12.20	25.14	37.34	13.57	20.23	17.42
2004	11.30	24.85	36.15	12.13	20.11	16.66
2005[a]	12.40	22.70	35.10	11.68	19.98	15.97
2006[b]	14.49	24.81	39.30	13.47	21.81	17.75
2007[b]	13.56	23.61	37.17	12.52	20.37	16.58
2008[b]	12.77	22.19	34.96	11.65	18.93	15.42
2009[b]	11.91	20.62	32.53	10.72	17.35	14.15
2010[b]	11.10	19.93	31.02	9.87	16.56	13.33
2011[c]	10.95	19.06	30.01	9.09	15.59	12.36
2012[c]	10.51	18.20	28.71	8.60	14.70	11.66
2013[c]	10.39	18.21	28.60	8.42	14.28	11.46
2014[c]	10.51	17.22	27.73	8.34	14.17	10.96
2015[c]	10.65	17.86	28.51	8.29	14.21	11.13
2016[c]	10.49	17.27	27.76	7.73	13.96	10.70
2017[b]	10.67	17.10	27.77	7.72	13.96	10.64
2020[c]	12.04	15.51	27.55	7.88	13.20	10.19
2021[b]	12.18	15.37	27.54	7.89	13.10	10.14

Note [a]February; [b]March; [c]September
Source BPS (http://www.bps.go.id) (data processed from the National Social Economic Survey (SUSENAS)); BPS (2016a, 2016b, 2021a)

during the 1970s to early 1980s with 13 percentage points, whereas during the 1981–1993 period the rate of decline was only about 16 percentage points. However, the rate of decline in the number of poor people tended to slow down starting in 2004.

Table 3.4 gives a clear impression that the role of MSMEs in achieving Goal 1 is especially important in rural areas because, as can be seen, both the number and percentage of poor people in rural areas are greater than those in urban areas. And indeed, the number of MSMEs, especially micro and small enterprises (MSEs), is higher in rural than in urban areas. So, if rural MSEs can develop well with high productivity and competitiveness, this will greatly help reduce poverty in rural areas and thus also at the national level.

Next, Table 3.5 shows the number and percentage of poor people in Indonesia vary by province. From the end of the 1998–1999 Asian financial crisis until the present Jokowi era, there have been several provinces whose poverty levels have decreased, while in other provinces the poverty conditions have remained worse or even gotten worse. This variation in changes in poverty between provinces is due to differences between provinces in many aspects, such as the rate of economic growth (low or high) and its nature (whether it is labor-intensive which means that economic growth creates many new job opportunities or capital intensive which means that the growth of employment opportunities is lower than the rate of economic growth), economic structure (whether the economy is still agrarian or dominated by non-primary sectors, especially the manufacturing industry which is the largest sector among all economic sectors in contributing value added to the economy), infrastructure conditions (quantitative and qualitative), the regional impact of an economic crisis (which is largely determined by, among other things, the economic openness of the province to the wider regional or international economy, and the province's readiness to an economic shock). Apart from these factors, the implementation at the provincial level of anti-poverty programs, especially during times of crisis, by central and local governments also greatly determines the success of a province in reducing poverty.

For instance, in 2013 (March), the average poverty rate in western Indonesia was around 10.01%, with the highest rate in Aceh at 17.6% and the lowest in DKI Jakarta at 3.55%. Meanwhile, in the eastern region of the country, an average of about 17.25% was recorded, with the highest in Papua at 31.13% and the lowest in North Maluku at 7.50%. This structure has remained relatively unchanged until recently (2021) based on the data available. By island, Maluku and Papua had the highest poverty rates, namely 20.66%; followed by Bali and Nusa Tenggara with 13.84%, Sulawesi at 10.29%, Sumatra at 10.15%, Java at 9.67%, and the lowest was Kalimantan 6.09%. This structure indicates the existence of economic inequality in Indonesia between the relatively more developed western regions (especially Java) and the relatively backward regions of the east.

Table 3.5 Poverty in Indonesia by province, 1990–2021 (%)

Province	*1990*	*1999*	*2002*	*2005*	*2006*	*2008*	*2012[a]*	*2013[b]*	*2017[b]*	*2021[b]*
Aceh	15.9	14.8	29.8	28.7	28.3	23.6	18.6	17.6	16.9	15.33
North Sumatera	13.5	–	15.8	14.7	15.0	12.5	10.4	10.1	10.2	9.01
West Sumatera	15.0	16.7	11.6	10.9	12.5	10.6	8.0	8.1	6.9	6.63
Riau	13.7	13.2	13.6	12.5	11.9	10.8	8.1	7.7	7.8	7.12
Jambi	–	14.0	13.2	11.9	11.4	9.3	8.3	8.1	8.2	8.09
South Sumatera	16.8	26.6	22.3	21.0	20.99	17.7	13.5	14.2	13.2	12.84
Bengkulu	–	23.5	22.7	22.2	23.0	19.1	17.5	18.3	16.5	15.22
Lampung	13.1	19.8	24.1	21.4	22.8	20.9	15.7	14.9	13.7	12.62
Bangka Belitung	–	29.1	11.6	9.7	10.9	7.9	5.4	5.2	5.2	4.90
Kep. Riau	–	–	–	10.97	12.2	8.7	6.8	6.5	6.1	6.12
DKI Jakarta	7.8	–	3.4	3.6	4.6	3.9	3.7	3.6	3.8	4.72
West Java	13.9	4.0	13.4	13.1	14.5	12.7	9.9	9.5	8.7	8.40
Central Java	17.5	19.8	23.1	20.5	22.2	18.99	14.98	14.6	13.0	11.79
DI Yogyakarta	15.5	28.5	20.1	18.95	19.2	18.0	15.9	15.4	13.0	12.80
East Java	14.8	26.1	21.9	19.95	21.1	18.2	13.1	12.6	11.8	11.40
Banten	–	29.5	9.2	8.9	9.8	8.2	5.7	5.7	5.5	6.66
Bali	11.2	8.5	6.9	6.7	7.1	5.9	3.95	3.95	4.3	4.53
West Nusa Tenggara	23.2	33.0	27.8	25.9	27.2	23.4	18.0	17.97	16.1	14.14
East Nusa Tenggara	24.1	46.7	30.7	28.2	29.3	25.7	20.4	20.0	21.9	20.99
West Kalimantan	27.6	26.2	15.5	14.2	15.2	10.9	7.96	8.2	7.9	7.15
Central Kalimantan	–	15.1	11.9	10.7	11.0	8.4	6.2	5.9	5.4	5.16
South Kalimantan	21.2	14.4	8.5	7.2	8.3	6.2	5.0	4.8	4.7	4.83
East Kalimantan	–	20.2	12.2	10.6	11/4	8.5	6.4	6.1	6.2	6.54
North Kalimantan	–	–	–	–	–	–	–	–	7.22	7.36

(continued)

Table 3.5 (continued)

Province	*1990*	*1999*	*2002*	*2005*	*2006*	*2008*	*2012*[a]	*2013*[b]	*2017*[b]	*2021*[b]
North Sulawesi	14.9	18.2	11.2	9.3	11.5	9.8	7.6	7.9	8.4	7.77
Central Sulawesi	–	28.7	24.9	21.8	23.6	20.6	14.9	14.7	14.1	13.00
South Sulawesi	10.8	18.3	15.9	14.98	14.6	13.4	9.8	9.5	9.4	8.78
Southeast Sulawesi	–	29.5	24.2	21.5	23.4	19.4	13.1	12.8	12.8	11.66
Gorontalo	–	–	32.1	29.1	29.1	20.5	17.2	17.5	17.7	15.61
West Sulawesi	–	–	–	–	20.7	15.3	13.0	12.3	11.3	11.29
Maluku	–	46.1	34.8	32.3	33.0	29.2	20.8	19.5	18.5	17.87
North Maluku	–	–	14.0	13.2	12.7	11.5	8.1	7.5	6.4	6.89
West Papua	–	–	–	–	41.3	33.5	27.0	26.7	25.1	21.84
Papua	–	54.8	41.8	40.8	41.5	35.5	30.7	31.1	27.62	26.86

Note [a]September, [b]March
Source BPS (http://www.bds.go.id) and BPS (2016a, 2016b, 2021a, b)

This imbalance could also be seen in the role of the provinces in the formation of the country's GDP, where the largest contribution originated from Java (above 50%), and in the second rank was Sumatra (above 20%).

From the data shown in Table 3.5, it is obvious that poverty in Indonesia was greater in the east than in the west. Following Government Regulation No. 131 of 2015 concerning the determination of underdeveloped areas for the 2015–2019 period, the number of underdeveloped areas in Indonesia reached 122 and most of them were in the eastern region. There were five provinces in the eastern region with the most disadvantaged areas, namely Papua, West Papua, Central Sulawesi, West Kalimantan, and West Nusa Tenggara (NTB) (Tono, 2017).

Next, Table 3.6 shows the percentage and number of poor people by island, and the highest, 20.65%, was recorded in the Maluku and Papua region while the lowest, 6.16%, was in Kalimantan. It is also important to note that the majority of poor people in the country are on the island of Java because more than 50% of the country's total population lives on this island. This is consistent with the data from the 2016 National Economic Census which showed that the majority of MSMEs in Indonesia are in Java which is the most densely populated island and the center of national economic activities with a focus on the manufacturing industry, trade, finance, construction, agriculture, and services in the country.

Table 3.6 Percentage and number of poor people on the island in Indonesia, September 2020

	Percentage of poor people			*Number of poor people (000 men)*		
	Urban	*Rural*	*Total*	*Urban*	*Rural*	*Total*
Sumatera	8.80	11.34	10.22	2,306.81	3,759.37	6,066.18
Java	8.03	13.03	9.71	8,105.76	6,646.27	14,752.03
Bali and Nusa Tenggara	8.99	18.18	13.92	633.96	1,482.53	2,116.49
Kalimantan	4.72	7.51	6.16	375.55	640.56	1,016.11
Sulawesi	5.95	13.45	10.41	477.07	1,584.44	2,061.51
Maluku and Papua	5.49	28.51	20.65	139.34	1,398.02	1,537.36
Indonesia	7.88	13.20	10.19	12,038.50	15,511.19	27,549.69

Source BPS (http://www.bps.go.id)

This means most of the companies in the non-agricultural sector reaching 16.2 million units are on this island and dominated by MSEs with 15.9 million units which are approximately 61% of all non-agricultural MSEs as indicated in Table 3.7. Moreover, most of the MLEs in non-agricultural sectors which are up to 291.7 thousand units or 65.2% of the total in the country were also found in Java while Papua and Maluku Islands considered being the least developed regions have a very low number of non-agricultural businesses as indicated by only 451.9 thousand MSEs and 7.5 thousand MLEs which are estimated at 1.8% and 1.7% of the total non-agricultural business in the country. It was also observed from field observations that most MSEs are in rural areas while most MLE are in urban or semi-urban areas although the exact figure is not known due to the unavailability of data.

One of the challenges for Indonesia concerning Goal 1 of the SDGs is how to reduce poverty in the poorest part of Indonesia. The eastern part of Indonesia has always been poorer than the western part. Papua, for instance, with an average of 25.4% poverty rate would need a special intervention so that the policy would be effective. Moreover, Indonesia is prone to natural disasters and when that happens, the poverty rate could hike sharply. Furthermore, aligned with reducing the number of the poor, it must also be sure that people already above the poverty line would not fall again into poverty. Poverty eradication policy, thus, should go about macroeconomic stability (growth and inflation management, particularly on food pricing), and at the micro-level, it should go about enhancing the social protection program and productive economy (Bappenas, 2019).

Table 3.7 Distribution of MSEs and MLEs by island, 2016 (%)

Island	*MSEs*	*MLEs*
Java	60.7	65.2
Sumatera	18.6	16.6
Sulawesi	8.1	5.6
Kalimantan	5.1	6.0
Bali and Nusa Tenggara	5.7	4.9
Papua and Maluku	1.8	1.7
Total	100.0	100.0

Source http://www.depkop.go.id/berita-informasi/data-informasi/data-umkm/; http://www.lisubisnis.com/2016/12/perkembangan-jumlah-umkm-di-indonesia.html

Goal 2: Zero Hunger

Goal 2 is zero hunger through the achievement of food security and improved nutrition and promotion of sustainable agriculture. In other words, the role of MSMEs in preventing hunger in an area is through two sides at once, namely from the demand side: people's incomes increase so that the ability to buy food increases, and from the supply side: the production of agricultural commodities by MSMEs increases so that food stocks are available in the market (Fig. 3.2).

Households that experienced food insecurity varied among provinces in Indonesia. In 2017, households in East Nusa Tenggara were at the most risk of experiencing food insecurity (31.8%) while Bangka Belitung possessed the lowest risk (3.8%). This figure shows the persisting issue of access disparity among Indonesian households across regions in accessing affordable food and nutrition. Map of Food Security and Vulnerability from the Ministry of Agriculture shows that "a high number of households without access to electricity" and "a high number of villages with no proper road/water access" were among the main characteristics that were associated with high levels of vulnerability to food insecurity. Hence, improving the economic access to food, including continued investments in infrastructure, especially to improve food access among poor households, is essential to the progress of food security in Indonesia (Bappenas, 2019).

In 2013, 48% of children under five from families in the poorest fifth of the population were stunted compared to 29% in the richest fifth group. This gap might be explained by the unequal access to improved sanitation and drinking water, health care, and high nutrient foods as well as inadequate access to care and feeding practices among households of different socio-economic statuses and geographic conditions on (Bappenas, 2019).

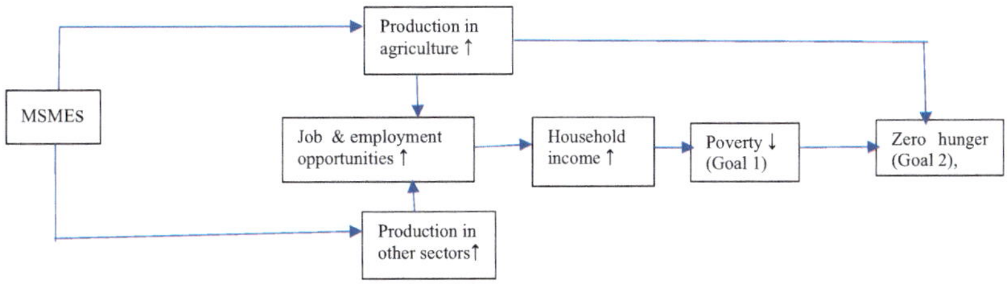

Fig. 3.2 The role of MSMEs in achieving Goal 2

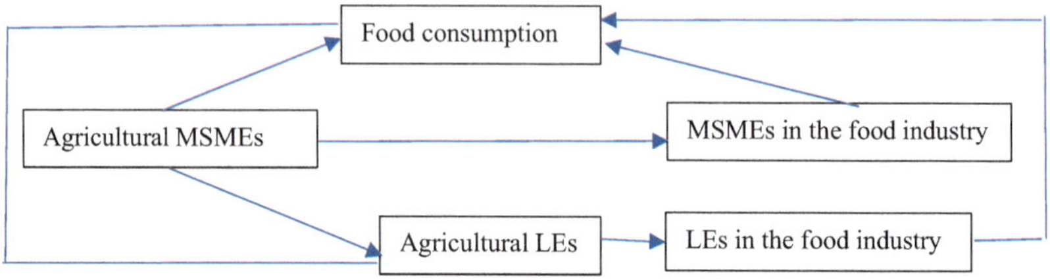

Fig. 3.3 The role of agricultural MSMEs in achieving food security

The MSMEs have a crucial role here namely to meet the food requirement of the people. In this case, two groups of MSMEs are very important, i.e., MSMEs in the agricultural sector (i.e., farmers or farming businesses) and those in the manufacturing industry. To achieve this goal, MSMEs in the first group should, on one hand, increase their productivity and quality of their output, and on the other hand, arrange cooperation or production linkages among themselves, e.g., micro, small, and medium-sized farming, and with large counterparts (e.g., small-sized livestock owners supply animal fertilizers to medium or large-sized farmers), as well as with MSMEs (and large enterprises/LEs) in the agro-based or food processing industry (Fig. 3.3). Data from 2017 on MSEs in the manufacturing industry shows that of the number of MSEs in the manufacturing industry of 4,464,688 companies, 1,538,117 companies or about 96% were in the food industry group. So, the food industry group is indeed the main industry for these enterprises in the manufacturing sector (Fig. 3.4).

Goal 3: Good Health and Well-Being

Goal 3 is to ensure healthy lives and promote well-being for all ages. The most relevant aspect of healthy lives or well-being for the role of MSMEs is maternal mortality per 100,000 live births, and this in Indonesia is still a serious problem. As explained in a 2015 report from The Indonesia Secretariat for Sustainable Development Goals, Ministry of National Development Planning/National Development Planning Agency, maternal mortality in Indonesia was the highest among ASEAN countries. According to the report, the number was 4.6 times higher than Malaysia, 1.8 times higher than Thailand, and even 1.3 times higher than the Philippines. The main causes of maternal mortality varied from health issues to socio-economic issues. Women who were married at a younger

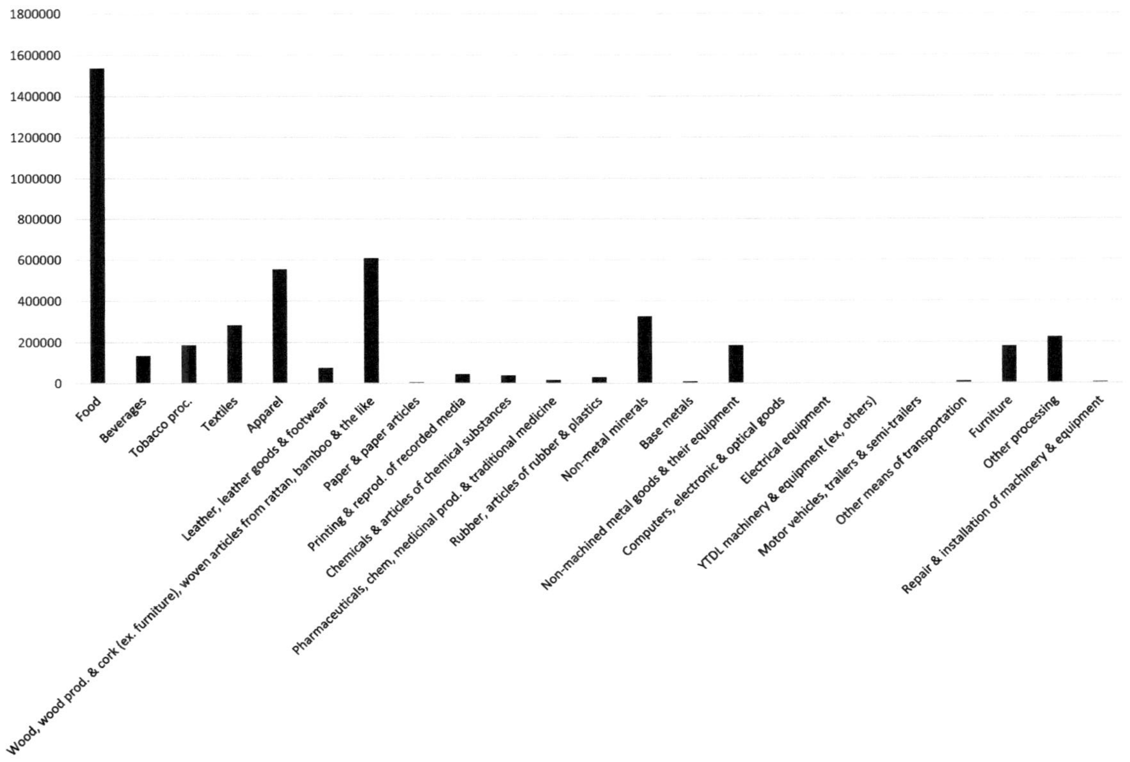

Fig. 3.4 Number of MSEs in the manufacturing industry by industry group (ISIC) Indonesia, 2017 (%) (*Source* BPS 2018)

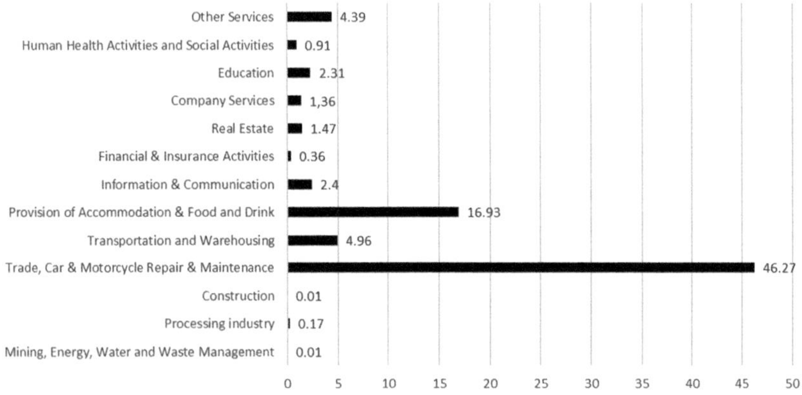

Fig. 3.5 Percentage of MSEs by sector, Indonesia, 2016 (%) (*Source* BPS 2017)

age (who mostly came from poor families in rural areas) have a higher risk of mortality during childbirth, and there was still a gap in the coverage of skilled birth attendance across the region. The report shows that the coverage in the Java-Bali region was 52% while in other regions was only 42%. As given in the report, lack of access to quality healthcare, especially in rural areas was also a serious problem. Access to quality healthcare was not only hampered by the absence of healthcare providers but also related to the geographical barrier, particularly for people living in remote areas (Bappenas, 2019).

Based on data from the 2016 National Social Economic Survey (SUSENAS), there were 239,236 MSEs in human health and social activities or around 4.4% of total MSEs in all sectors (Fig. 3.5). Although they were not so many compared to MSEs in some other sectors, especially in the trade sector which accounted for 46%, they still can play an important role to achieve Goal 3 by bridging the healthcare gap through better-quality healthcare services to the people.

Goal 4: Quality Education

Goal 4 is to ensure inclusive and equitable quality education for all. According to Indonesian statistics, the gross enrollment rate at the higher secondary level in 2019 reached 83.98%, or about 4.1% higher compared to the rate of 0.68% in 2018. Whereas at the lower secondary level it was 90.57, slightly decreased from 91.52. In primary school, the rate in 2019 reached almost 107.50, which also slightly dropped from 108.61 in 2018. The gross enrollment rate that does not reach 100% indicates that there are out-of-school children who either did not enroll in or dropped out of higher secondary education. In line with the condition at the lower secondary level, a significant gap was still found between regions in which Papua accounted for the lowest gross enrollment rate at 67.94% in the same year.

Indonesia's net ratio of girls to boys in almost all education levels, except primary education, shows higher participation of girls than boys. This figure is supported by Ministry of Education and Culture data that shows a higher number of drop-outs in boys where the number of male students who dropped out from lower and higher secondary education was nearly thrice and twice of female drop-outs. At the higher secondary level, the following provinces exhibited a high gap in girls to boys net ratio: West Sumatera 124.15, West Nusa Tenggara 86.18, and Gorontalo 143.63. A more glaring disparity was found in tertiary education where nearly one-third of provinces in Indonesia record a net ratio of girls to boys at the range of 120–160 (Bappenas, 2019).

To achieve this goal, MSMEs have also an important role to play by providing quality technical and vocational education and making the people get access to earnings for their livelihood. As shown in Fig. 3.4, around 2.3% of total MSMEs in 2016 were in the education sector. They are more important in rural areas where boys and children are more likely to drop out of school than girls and their peers who live in urban areas. Also, the education gap is still found between rural and urban areas as well as across income groups (Bappenas, 2019).

Goal 5: Gender Equality

Goal 5 is to achieve gender equality and empower all women and girls. Indonesia is the fourth most populous country in the world and has one of the highest literacy rates in Asia (females 93.59% and males 97.17%). Although the country has put in place several laws, regulations, and, programs that provide support to girls and women, including the *Jaminan Kesehatan Nasional* (JKN), one of the world's largest national health insurance program currently provides free healthcare to almost 3/4 of the population and has included significant investments to reduce maternal mortality, legal barriers in some key areas coupled with a patriarchal culture and religious conservatism, continue to prevent girls and women from fulfilling their rights. Tax and inheritance laws, for example, discriminate against women and legislation to protect women against sexual harassment and domestic violence is weak or goes unenforced (https://data.em2030.org/countries/Indonesia/).

According to the 2016 Women's Health and Life Experiences Survey, one in three women aged 15–64 years in Indonesia reported that she had experienced physical and/or sexual violence in her lifetime. Women also faced legal barriers and discrimination in the economy: at 51% in 2017, Indonesia's female labor-force participation rate was well below that for males (around 80%) and lower than average for countries at a comparable stage of development (https://indonesia.unfpa.org/sites/default/files/pub-pdf/2016_SPHPN_%28VAW_Survey%29_Key_Findings1_0.pdf).

According to a study, the main drivers of low female labor-force participation in Indonesia were marriage, having children under the age of two in the household, low educational attainment (below upper-secondary and tertiary levels), and a changing economic structure that has seen a decline in the sector of agriculture as a result of migration from rural to urban areas, in particular (AIPEG, 2017).

Based on the 2016 Economic Census, in general, the percentage of MSE entrepreneurs in all sectors owned by females was lower than that owned by men, around 42.84% versus 57.16%. However, there were variations between provinces, and the highest was in the province of South Sumatra which reached almost 48% (Fig. 3.6). The result of the Census also shows that the sector or business area in which the percentage of women entrepreneurs was higher than that of men was the human health and social activities sector, namely at 63.68%. On the contrary, the lowest level of the women entrepreneurs was in the category of transportation and warehousing, namely only 2.22%. This showed that there were different trends in choosing beta women woman and men's jobs. Women tend to avoid too much physical work.

Yet from the Census, in terms of the generated turnover, MSEs owned by men show a different composition from those owned or managed by women. Most MSEs owned by men (48.28%) were in the business groups with a turnover of between 300 million and 2.5 billion Indonesian rupiahs (INR). Meanwhile, the majority of MSEs owned by women (59.98%) were in business groups with a turnover of less than 300 million INR. Apart from the assumption that business leaders or company managers

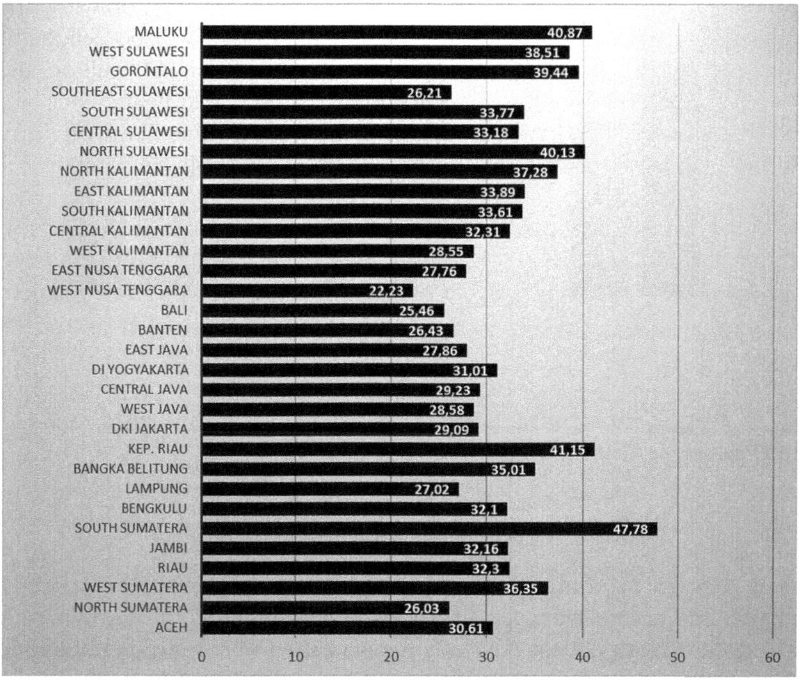

Fig. 3.6 Percentage of MSEs in all sectors owned by female by province, Indonesia, 2016 (%) (*Source* BPS)

should be men, several other things led to the notion that women's ability to lead a business is still below men's. One thing men said is that women are considered less able to take risks and less aggressive in their movements. However, this is still a matter of debate among researchers.

With this, MSEs could bridge the gender gap, especially in employment and business opportunities, and bring women into the mainstream of the country's economic development. MSEs could function as a place for women's entrepreneurship development which will further empower women and brings gender equality.

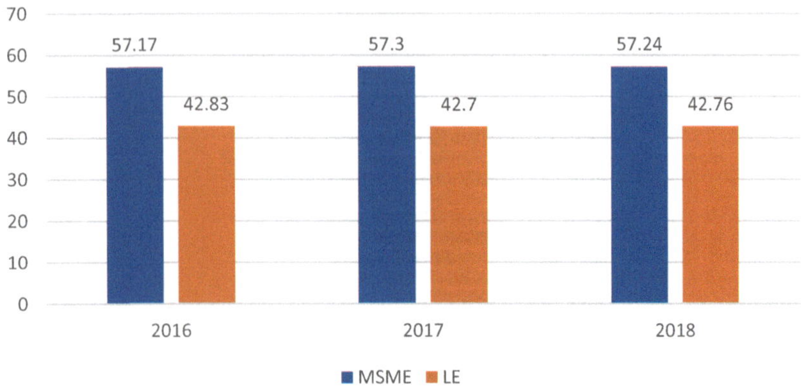

Fig. 3.7 Share of GDP by business size in Indonesia, 2016–2018 (constant 2000 prices; %) (*Source* Menegkop & UKM and BPS)

Goal 8: Decent Work and Economic Growth

Goal 8 is to promote sustained, inclusive, and sustainable economic growth and decent work for all. In many developing countries, not only economic growth rate and real income per capita were low due to, among other factors, a slow process of economic structural transformation from agriculture to the manufacturing industry that has higher productivity, but these countries also experienced high open or underemployment and low productivity. In Indonesia, even after successfully decreasing the unemployment rate to 4.99% in the first quarter of 2019 from 5.28% in the third quarter of 2019 or much higher before 2019 (https://tradingeconomics.com/indonesia/unemployment-rate), underemployment and low productivity of jobs persist.

Concerning economic growth, MSMEs in developing countries have contributed significantly to the formation of GDP and hence economic growth. In Indonesia, these enterprises contributed more than 50% of the country's GDP (Fig. 3.7).

Goal 9: Industry, Innovation, and Infrastructure

Goal 9 is to promote industrialization, innovation, and infrastructure. Concerning industrialization, Indonesia began to experience very rapid industrialization from the late 1960s, at the beginning of the 'New Order' era (1966–1998) with the manufacturing industry growing at more than 10% per annum for most of the subsequent three decades. Initially, this growth was import-substituting in nature, but from the mid-1980s after the end of the 'oil booming' era, a successful transition to export-orientated industrialization was engineered. The most important exported goods were textile and garments, footwear, furniture from wood, rattan, and bamboo, and some leather products with MSMEs as the leading producers. In the period 1997–1998 that rapid growth was halted by the Asian financial crisis, which resulted in a peak-to-trough growth collapse of more than 20 percentage points together with a significant drop in the country's GDP growth by minus 13%. Subsequently, growth recovered, but the effects on the nature and drivers of industrial growth appear to be profound (Aswicahyono et al., 2011; Ishida, 2003).

Table 3.8 presents data on the structure of output by economic sectors, which shows the share of the manufacturing industry in the country's GDP was around 33–38% during the period 2000–2021.

In this respect, MSMEs have two tasks to carry out in contributing to industrialization. First, as producers of final consumption goods including import substitution goods. Second, as producers or suppliers, or vendors of semi-finished or intermediate goods, production tools, machinery, and processed raw materials for final goods or assembling industries through

Table 3.8 Structure of output by economic sectors in Indonesia, 2000, 2005, 2010, and 2021(%)

Sector	*Distribution*			
	2000	*2005*	*2010*	*2021*[a]
Agriculture, livestock, forestry, and fishery	11.38	8.63	9.06	14.30
Mining and quarrying	7.29	6.81	7.18	9.55
Manufacturing industry	38.96	37.41	33.34	19.15
Electricity, gas, and water supply	1.13	1.56	2.54	1.17
Construction	8.43	10.17	13.15	10.39
Trade, hotel, and restaurant	14.67	12.85	13.81	15.29
Transport and communication	5.60	7.00	7.32	8.27
Financial, real estate, and business services	5.97	6.19	4.77	8.70
Services	6.56	9.37	8.82	13.18
Total	**100.00**	**100.00**	**100.00**	**100.00**

Note [a]Quarter III
Source BPS (2020, 2021b)

subcontracting arrangements. With these two roles, indirectly, MSMEs could contribute to the reduction of the country's heavily dependent on imports. Thus, combining these two tasks would strengthen the MSMEs' role in promoting industrialization. Figure 3.8 shows the growth of the MSE output index by a group of industries in 2009 which consisted of MSIs in both task groups.

Regarding innovation, based on the Global Innovation Index (GII) 2020, Indonesia is ranked 85th out of 131 world economies. This ranking has not changed since 2018. But, despite that, Indonesia has a better performance in innovation outputs than innovation inputs in 2020. This year Indonesia ranks 91st in innovation input, lower than last year's 87 and 2018s 90. For innovation output, Indonesia ranks 76th. This position is higher than last year at 78 and lower than in 2018 which was 73. Indonesia ranks 9th out of 29 countries with lower-middle-income group economies and ranks 14th out of 17 countries in Southeast Asia, East Asia, and Oceania. That way, Indonesia is not included in the list of the 10 best countries in Asia. Countries that are in the top 10 include Singapore (ranked 8th), South Korea (10), Hong Kong (11), China (14), Japan (16), Australia (23), New Zealand (26), Malaysia (33), Vietnam (42), and Thailand (44). The GII ranks the world economy according to the country's innovation capability. The assessment is based on 80 indicators, which are grouped into innovation inputs and innovation outputs. GII aims to see the multidimensional aspects of innovation (Cornell University et al., 2020).

As MSMEs have an important role to play in promoting economic growth and creating new employment opportunities, consequently these enterprises are responsible for much of the innovation which leads to new higher-value products and services. As evidently in many countries, MSMEs are the essential elements of the economy which is responsible for driving innovation and competition. These enterprises function as an important source of national innovation activities (Kamariah et al., 2014; Lin et al., 2007; van Auken et al., 2008).

Concerning infrastructure, MSMEs also have a role to play, namely as the suppliers of materials required for infrastructure development projects, especially for medium-sized projects in rural areas such as roads, bridges, street lighting facilities, warehouses, and market centers.

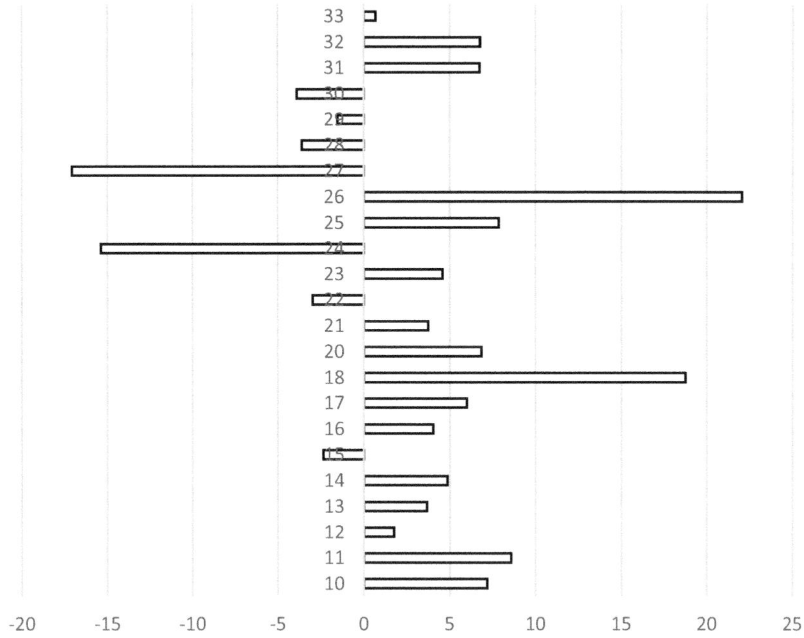

Fig. 3.8 Growth of the MSE Output Index by industry group in Indonesia, 2019 (2010 = 100) (*Note* 10: food, 11: beverages, 12: tobacco processing, 13: textiles, 14: apparel, 15: leather, leather goods, and footwear, 16: wood, wood products, and cork (excluding furniture), woven articles from rattan, bamboo and the like, 17: paper and paper articles, 18: printing and reproduction of recorded media, 20: chemicals and articles of chemical substances, 21: pharmaceuticals, chemical medicinal products, and traditional medicine, 22: rubber, articles of rubber and plastics, 23: non-metal minerals, 24: base metals, 25: non-machined metal goods and their equipment, 26: computers, electronic and optical goods, 27: electrical equipment, 28: YTDL machinery and equipment (excluding others), 29: motor vehicles, trailers, and semi-trailers, 30: other means of transportation, 31: furniture; 32: other processing; 33: repair and installation of machinery and equipment. *Source* BPS [database])

Goal 10: Reduced Inequalities

Goal 10 is to reduce inequality. In Indonesia, the level of inequality as measured by the Gini ratio had risen sharply since 2004 and reached a peak at the end of 2013 at 0.413. However, starting in 2015, the rate gradually declined. In September 2018, the ratio reached 0.384, decreasing by 2.9 points in 5 years. In March 2020, the ratio was 0.381. This figure was an increase of 0.001 points when compared to the Gini ratio in September 2019 which amounted to 0.380 and decreased by 0.001 points compared to that in March 2019 which was 0.382 (Fig. 3.9).

A Gini ratio of 0.384 is classified as a moderate gap. This Gini ratio is relatively lower compared to many other developing countries, showing a more equal income distribution in Indonesia. As a ratio, Gini has two magnitudes that must be controlled such as middle-and high-income groups. The Gini ratio is influenced by the economic dynamics and it can be out of line with the poverty level. The ratio can be very low, but it does not provide an overview of society's welfare in general. Nevertheless, over the past 4 years, the poverty rate and the Gini ratio in Indonesia have declined (Bappenas, 2019).

The inequality emerges more apparent in the urban area where the urban Gini ratio reached 0.401 while the rural Gini ratio was 0.324 in 2018. The ratio in urban areas in March 2020 was recorded at 0.393, an increase compared to September 2019 at 0.391 in March 2019 at 0.392.

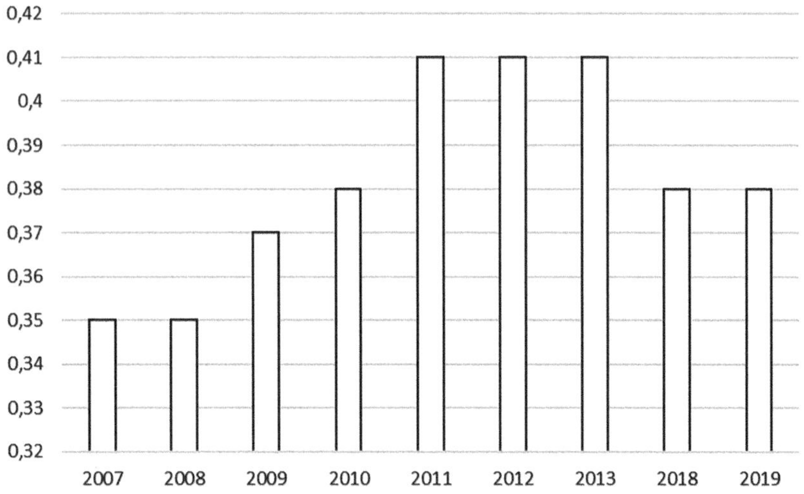

Fig. 3.9 Inequality in Indonesia (Gini ratio), 2007–2019 (*Note* All data were recorded in March in the particular years. *Source* BPS [database])

The Gini ratio in rural areas in March 2020 was recorded at 0.317, an increase compared to 0.315 in September 2019 and unchanged compared to March 2019 which amounted to 0.317.

Regionally, provinces with high economic activities also show higher income inequality. Income inequality was the highest in Yogyakarta, Southeast Sulawesi, and West Java where the Gini ratio reached 0.441, 0.409, and 0.407, respectively. Meanwhile, Bangka Belitung with 0.281, North Sumatera with 0.318, and North Kalimantan with 0.303 had the lowest Gini ratio among other provinces. In March 2020, the province with the highest Gini ratio was again Yogyakarta with 0.434. Meanwhile, the lowest Gini ratio was recorded in Bangka Province Belitung with 0.262 (Table 3.9). Compared with the national Gini ratio of 0.381, there were eight provinces with a higher Gini ratio, namely Yogyakarta (0.434), Gorontalo (0.408), West Java (0.403), DKI Jakarta (0.399), Papua (0.392), Sulawesi Southeast (0.389), South Sulawesi (0.389), and West Papua (0.382).

For MSMEs to achieve this particular goal, they could contribute in two ways, namely by creating employment directly as well as indirectly via production and marketing linkages with other firms (i.e. producers and traders/distributors) and by increasing their productivity. Through these two ways, the gap between low-income group and middle-to-high-income groups of the population could be narrowed.

Goal 11: Sustainable Cities and Communities

This goal is to make cities and human settlements inclusive, safe, resilient, and sustainable. The key issue here is the proportion of households in urban or cities with access to decent and affordable housing. This in turn is closely related to income. At least theoretically, higher income depends on better job opportunities. So, its related hypothesis is that in cities where there are productive employment and business opportunities the proportion of households having decent and affordable housing is higher than that in cities with stagnant economies and large parts of the population unemployed.

In this case, the role of MSMEs is very clear namely to provide productive employment and high-value-added business opportunities that together generate economic growth in cities. In cities in developing countries such as Indonesia and India, compared to LEs, MSMEs are the

Table 3.9 Inequality in Indonesia by province, March 2019 and March 2020 (Gini ratio)

Province	*2019*	*2020*
Aceh	0.319	0.323
North Sumatera	0.317	0.316
West Sumatera	0.306	0.305
Riau	0.334	0.329
Jambi	0.321	0.320
South Sumatera	0.331	0.339
Bengkulu	0.340	0.334
Lampung	0.329	0.327
Bangka Belitung	0.269	0.262
Kep. Riau	0.341	0.339
DKI Jakarta	0.394	0.399
West Java	0.402	0.403
Central Java	0.361	0.362
DI Yogyakarta	0.423	0.434
East Java	0.370	0.366
Banten	0.365	0.363
Bali	0.366	0.369
West Nusa Tenggara	0.379	0.376
East Nusa Tenggara	0.356	0.354
West Kalimantan	0.327	0.317
Central Kalimantan	0.336	0.329
South Kalimantan	0.334	0.332
East Kalimantan	0.330	0.328
North Kalimantan	0.295	0.292
North Sulawesi	0.375	0.370
Central Sulawesi	0.327	0.326
South Sulawesi	0.389	0.389
Southeast Sulawesi	0.399	0.389
Gorontalo	0.399	0.408
West Sulawesi	0.365	0.364
Maluku	0.324	0.318
North Maluku	0.312	0.308
West Papua	0.386	0.382
Papua	0.394	0.392

Source BPS (http://www.bds.go.id)

leading sector in promoting sustainable and inclusive economic growth and thus inclusive settlement of the people with safety in cities.

Main Constraints

The question now is: are MSMEs, especially MSEs able to play their role optimally? Of course, the answer is: it totally depends on the key resources they have, such as capital, technology, and skilled human resources. Unfortunately, here lies the problem. Evidence from many developing countries, including Indonesia, shows that the development or growth of MSMEs has been hindered by many obstacles which differ in intensity based on regions, rural and urban areas, sectors, and even between companies in the same sector. However, there are some problems considered to be common to all MSMEs in any country, especially developing countries and these include limited working capital and investment; difficulties in marketing, distribution, and procurement of raw materials and other inputs; limited access to information about market opportunities and others; limited skilled personnel or low quality of human resources; low technological capabilities; high transportation and energy costs; limited communication; high costs due to complicated administrative and bureaucratic procedures, especially in business licensing; and uncertainty due to unclear or uncertain economic regulations and policies.

The data from the 2010 National Survey on MSIs showed that approximately 78% of all MSIs which represents 2,732,724 units experienced difficulties in running their businesses and the most prevalent ones were associated with funding, marketing, and raw materials with 806,758 units, 495,123 units, and 483,468 units respectively. Moreover, the MSIs in the food industry had the greatest difficulties with 745,824 units (34.96%) which include those related to the capital with 255,793 units, raw materials with 206,309 units, and marketing with 146,185 units. The same trend was also observed in the data from the 2017 National Survey on MSIs as indicated in Fig. 3.10.

Concerning funding, there were several special credit schemes for MSIs during the period with some still existing but most of the respondents, especially those in rural areas, said they have never received credit from banks or other financial institutions. This means they depended entirely on their own money or savings, financial assistance from relatives, and loans from informal money lenders to finance their business activities. Some of the reasons provided for not dealing with banks include not having information on the existence of such special credit schemes, some tried to apply but were rejected because their business was deemed unfit for funding, some resigned due to complicated administrative procedures

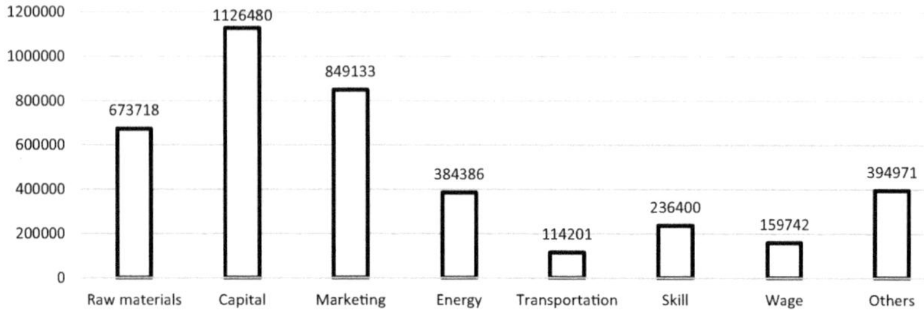

Fig. 3.10 Types of MSIs constraints, 2017 (*Source* BPS 2018)

or inability to fulfill requirements including the provision of guarantees or collateral such as house or land certificates. Others did not wish to borrow from formal financial institutions because they felt uncomfortable or afraid of defaulting (BPS, 2018).

Concerning marketing, the MSIs generally did not have the resources to seek, develop, or expand their markets. Instead, they relied heavily on their trading partners such as mobile traders, collectors, or trading houses to market their products, also on consumers visiting their production sites, and through production linkages or subcontracting arrangements with larger enterprises which contribute a small percentage to the marketing efforts.

Conclusion

From the discussion above, there is no doubt that MSMEs have a great potential to contribute to the success of achieving the SDGs in Indonesia. The question now is, can this business group realize that role? This question is very important because, like in many other developing countries, MSMEs, especially MSEs, in Indonesia face many obstacles, especially limited access to funding, skill, and advanced technology. Without full access to these three key production resources, they can't contribute significantly to Indonesia's efforts to achieve the SDGs by 2030.

For their contribution to be meaningful for the achievement of the SDGs, MSMEs must not only be sustainable but also be able to increase their production and expand their market. Increased production and market expansion, especially in the global market, will have a positive impact on increasing employment opportunities combined with an increase in workers' real income (which is referred to as an increase in productive employment opportunity), and in turn, it will reduce the poverty rate significantly, compared to the number of MSMEs that continues to grow but with low productivity and competitiveness.

References

AIPEG. (2017, June). *Women's economic participation in Indonesia*. The Australia Indonesia Partnership for Economic Governance, Department of Foreign Affairs and Trade (DFAT) and Monash University's Centre for Development Economics and Sustainability (CDES), Canberra. https://www.monash.edu/business/cdes/our-research/publications/publications2/Womens-economic-participation-in-Indonesia-June-2017.pdf

APEC. (2020, April). *Overview of the SME sector in the APEC region: Key issues on market access and internationalization*. APEC Policy Support Unit, Singapore: Asia-Pacific Economic Cooperation Secretariat.

Aswicahyono, H., Hill, H., & Narjoko, D. (2011, September). *Indonesian industrialization. A latecomer adjusting to crises* (Working Paper No. 2011/53). The World Institute for Development Economics Research (WIDER), the United Nations University (UNU).

Bappenas. (2019). *Roadmap of SDGs Indonesia: A highlight*. Ministry of National Development Planning/National Development Planning Agency.

BPS. (2016a, December). *Penghitungan dan Analisis Kemiskinan Makro Indonesia*. Badan Pusat Statistik.

BPS. (2016b, July 18). *Profil Kemiskinan di Indonesia Maret 2016* (Berita Resmi Statistik No.66/07/Th.XIX). Badan Pusat Statistik.

BPS. (2017, June). *Profil Industri Mikro dan Kecil 2016*. Badan Pusat Statistik.

BPS. (2018, June). *Profil Industri Mikro dan Kecil 2017*. Badan Pusat Statistik.

BPS. (2020). *Statistik Indonesia 2020*. Badan Pusat Statistik.

BPS. (2021a, July 18). *Profil Kemiskinan di Indonesia Maret 2021* (Berita Resmi Statistik No.66/07/Th.XIX). Jakarta: Badan Pusat Statistik.

BPS. (2021b, November 5). *Pertumbuhan Ekonomi Indonesia Triwulan III-2021* (Berita Resmi Statistik No. 83/11/Th. XIV). Badan Pusat Statistik.

Cornell University, INSEAD, & WIPO (2020). *The Global Innovation Index 2020: Who will finance innovation?*

Dasaraju, H., Somalaraju, K., & Kota, S. M. (2020). MSMEs in developing economies and their role in achieving Sustainable Development Goals in the context of Covid19: A theoretical exposition. *International Journal of Small and Medium Enterprises and Business Sustainability*, *5*(2), 93–120.

IFC. (2017). *MSME Finance Gap*. International Finance Corporation.

Ishida, M. (2003). *Industrialization in Indonesia since the 1970s* (IDE Research Papers 5). Institute of Developing Economies, Japan External Trade Organization (JETRO).

Kamariah, I., Omar. W. Z. W., Soehod, K., Senin, A. A., & Akhtar, Ch. S. (2014). *Role of innovation in SMEs performance: A case of Malaysian SMEs*. proceedings of the 2014 International Conference on Economics and Business Administration II (EBA '14), Prague, Czech Republic.

Lin, C. Y. Y., & Chen, M. Y. C. (2007). Does innovation lead to performance? An empirical study of SMEs in Taiwan. *Management Research News, 30*(2), 115–132.

Tambunan, T. T. H. (2021). *The role of MSMEs in achieving SDGs in developing economies*. Eliva Press.

Tono, S. (2017, September 22). Pembangunan Bias Daratan. *Kompas Newspaper*, Friday, p. 6.

van Auken, H., Madrid-Guijarro, A., & Garcia-Perezde-Lema, D. (2008). Innovation and performance in Spanish manufacturing SMEs. *International Journal of Entrepreneurship and Innovation Management, 8*(1), 36–56.

CHAPTER 4

MSMEs: The Growth Engine of India to Achieve Sustainable Development Goals and Self-Reliance

Meenu Singla

Introduction

The MSME sector is the most dynamic sector which contributes highly to country's manufacturing output, employment and exports. It comprises of more than 80% of the total enterprises of the economy with a labour intensity higher than the large enterprise. MSMEs are significant pillars of Indian economy contributing considerably to the growth and development of Indian economy (Joshi et al., 2020). These produce more than 6000 varied products providing employment opportunities to around 70 million people and contribute about 45% of manufacturing production and 40% to exports (www.ibef.org). At present, the MSME sector is largely led by microenterprises. This signifies the creation of new opportunities for earning and gaining self-reliance. Therefore, this paper is focused

M. Singla (✉)
Mehr Chand Mahajan DAV College for Women, Chandigarh, India
e-mail: drmeenucommerce@gmail.com

H. Dasaraju and T.T.H. Tambunan (eds.), *Role of Micro, Small and Medium Enterprises in Achieving SDGs*, Sustainable Development Goals Series, https://doi.org/10.1007/978-981-99-4829-1_4

on the role of MSMEs in the economic growth of the country and the attainment of Sustainable Development Goals (SDGs) laid by the United Nations (UN) to transform the world along with the various challenges faced by the sector and their possible remedies are suggested.

Micro, Small, Medium Enterprises (MSMEs) are entities that are involved in production, manufacturing and processing of goods and commodities (OI, 2020). The concept of MSME was first announced by the Government of India through the Micro, Small and Medium Enterprises Development (MSMED) Act, 2006 which is classified into two divisions.

1. **Manufacturing enterprises**—engaged in the manufacturing or production of goods in any industry
2. **Service enterprises**—engaged in providing or rendering services

The Ministry of Micro, Small and Medium Enterprises is the apex executive body for the formulation and administration of rules, regulations and laws for the promotion and development of MSMEs in India. It supports the MSME sector in the endeavours of formalisation, access to knowledge services, improving competitiveness, reducing imports, availability of skilled manpower, access to latest technology and marketing (Joshi et al., 2020). The objective of the framework is to unlock the potential of India's MSME sector in the endeavour for becoming the growth engine for the economy with a vision to reach $5 trillion generating mass employment and becoming dominant player of *Aatmanirbhar Bharat* (OI, 2020; www.ibef.org). The Ministry is committed towards an agenda of inclusive development and has taken various initiatives and measures to ensure that weaker sections benefit from its work both demographically and geographically.

MSMEs are classified as per their turnover and investment. The new classifications as per the *Aatmanirbhar Bharat Abhiyan Scheme* in 2020 are given below:

Size of the enterprise	*Investment and annual turnover*
Micro	Investment < Rs. 1 crore; Turnover < Rs. 5 crores
Small	Investment < Rs. 10 crores; Turnover up to Rs. 50 crores
Medium	Investment < Rs. 20 crores; Turnover up to Rs. 100 crores

Source MSME Annual Report 2021–22

The Ministry of MSME runs numerous schemes targeted at

a. providing credit and financial assistance,
b. skill development training focussing on employment opportunities,
c. infrastructure development,
d. marketing assistance in domestic as well as export markets,
e. upgradation of technology and quality,
f. Other Services for the MSMEs which include the growth in demand in the domestic market, expertise assistance in project profile at various stages of project development, product development, design innovation and modernisation of the sector.

Role of MSMEs in the Indian Economy and the Attainment of SDGs

The Micro, Small and Medium Enterprises (MSMEs) are contributing significantly to the expansion of entrepreneurial endeavours through business innovations. It is the backbone of the Indian economy that promotes substantially the economic development of the nation. India has approximately 6.3 crore MSMEs (Table 4.1) that are a viable source of income in the manufacturing industry. The MSMEs are termed as the 'engine of growth' widening their domain across sectors of the economy, producing diverse range of products and services to meet domestic as well as global market demands (Soni et al., 2022). The MSMEs in India are playing a prominent role in the development of the country by contributing to almost 30% of India's GDP (FICCI-EY Re-imagining India's M&E sector, Ministry of Micro Small and Medium Enterprises, GeM Portal) providing large employment opportunities to more than 11.10 crore people (Table 4.4), scaling manufacturing capability at comparatively lower capital cost than large industries, industrialisation of rural and backward areas (Table 4.1 and Fig. 4.1) thus, reducing regional imbalances (Tables 4.2 and 4.3), assuring more equitable distribution of national income and wealth (Fig. 4.3). It has an enormous share of 40% in the export market and 45% in the manufacturing sector (www.ibef.org). Hence, they are of paramount importance for overall economic development of India.

As per the National Sample Survey (NSS) 73rd round, conducted by National Sample Survey Office, Ministry of Statistics and Programme

Table 4.1 Estimated number of enterprises (in lakhs)

Activity category	*Rural*	*Urban*	*Total*	*Share (%)*
Manufacturing	114.14	82.50	196.65	31
Trade	108.71	121.64	230.35	36
Other services	102.03	104.86	206.88	33
All	324.88	309	633.88	100

Source MSME Annual Report 2021–22 (Data-National Sample Survey Office, Ministry of Statistics & Programme Implementation)

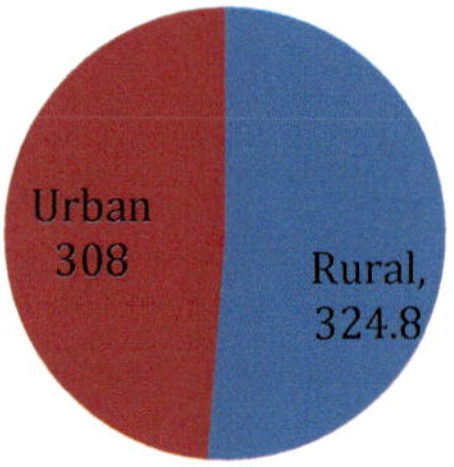

Fig. 4.1 Rural and urban MSMEs (*Source* MSME Annual Report 2021–22)

Table 4.2 Percentage of enterprises owned by male/female in rural and urban areas

Sector	*Male*	*Female*	*All*
Rural	77.76	22.24	100
Urban	81.59	18.42	100
All	**79.63**	**20.37**	**100**

Source MSME Annual Report 2021–22

Table 4.3 Percentage of enterprises owned by male/female entrepreneurs (category-wise)

Activity category	*Male*	*Female*	*All*
Micro	79.56	20.44	100
Small	94.74	5.26	100
Medium	97.33	2.67	100
All	**79.63**	**20.37**	**100**

Source MSME Annual Report 2021–22

Implementation during the period 2020–2021, there were 633.88 lakh unincorporated non-agriculture MSMEs in the country engaged in different economic activities (Table 4.1 and Fig. 4.2) out of which 51.25%

Table 4.4 Estimated employment in the MSME sector (activity-wise)

Activity category	*Rural*	*Urban*	*Total*	*Share (%)*
Manufacturing	186.56	173.86	**360.41**	**32**
Trade	160.64	226.54	**387.18**	**35**
Other services	150.59	211.71	**362.29**	**33**
All	**497.78**	**612.10**	**1109.89**	**100**

Source MSME Annual Report 2021–22

are in rural areas and 48.75% in urban areas (Fig. 4.1). The dominance of ownership is by males who owned 79.63% of enterprises as compared to female-owned enterprises 20.37% (Tables 4.2 and 4.3). There is no significant difference in pattern of ownership in urban and rural areas (Table 4.2). The enterprises owned by different social groups reveal micro sector comprises 66.42% owned by socially backward groups. MSME sector has created 11.10 crore jobs (Tables 4.4 and 4.5) out of which 497.78 crore are in rural areas and 612.10 in the urban areas providing employment to 76% males and 24% females (Table 4.6). Table 4.7 reveals the state-wise number of enterprises and the employment opportunities. The above data clearly depicts the role of MSMEs in the reduction of disparities based on region and gender in varied sectors of the country where Uttar Pradesh and West Bengal tend to be the exemplary states benefitting from the initiative (*MSME Annual Report 2021–22*).

The 30% contribution of MSMEs sector to the country's GDP is a significant help for the Indian economy (Economic Survey 2020–21). The MSMEs are acknowledged to be the backbone of the nation acting as an 'engine of growth' (Fig. 4.4). The significant increase in the number of MSMEs has solved the financial issues of the nation and is further

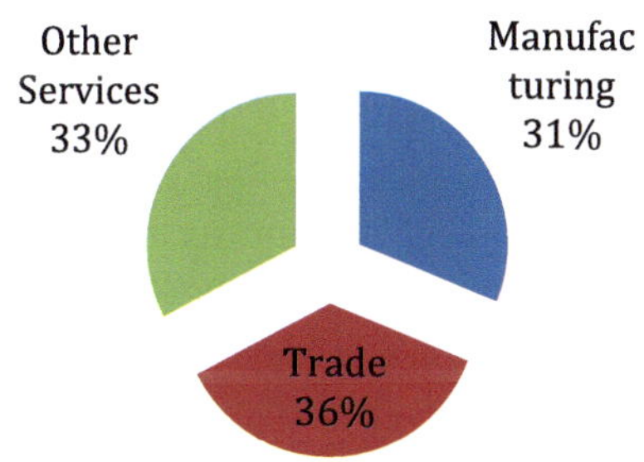

Fig. 4.2 MSMEs by nature of activity (*Source* MSME Annual Report 2021–22)

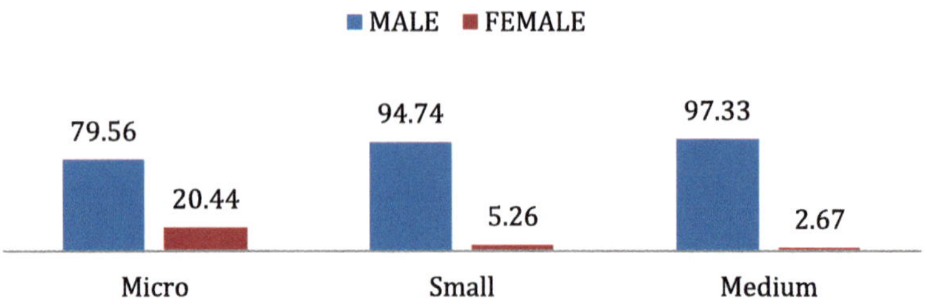

Fig. 4.3 Percentage distribution of enterprises category wise (*Source* MSME Annual Report 2021–22)

Table 4.5 Employment distribution and type of enterprises in rural/urban areas (in lakhs)

Sector	*Micro*	*Small*	*Medium*	*Total*	*Share (%)*
Rural	489.30	7.88	0.60	497.78	45
URBAN	586.88	24.06	1.16	612.10	55
All	1076.19	31.95	1.75	1109.89	100

Source MSME Annual Report 2021–22

Table 4.6 Distribution of workers by gender in rural and urban areas

Sector	*Male*	*Female*	*All*	*Share (%)*
Rural	360.15	137.50	**497.78**	45
Urban	484.54	127.42	**612.10**	55
All	844.68	264.92	**1109.89**	100
Share (%)	76	24	100	

Source MSME Annual Report 2021–22

supporting to transform the world by achieving the UN-laid SDGs as quoted in Fig. 4.5. Favourable industrial policies creating an ease of doing business are the key drivers of the MSMEs (Soni et al., 2022). Liberalised FDIs and technology boosted their competitiveness through enhanced operations by connecting the global value chains. The requirement of less amount of investment, flexibility in operations and enhanced government assistance has transformed the innovative ideas of budding entrepreneurs into actual products with a multiplier impact of handling

Table 4.7 State-wise distribution of estimated number of MSMEs

State/UT	*Number*				*Employment*		
	Micro	*Small*	*Medium*	*Total*	*Male*	*Female*	*Total*
Andhra Pradesh	33.74	0.13	0	**33.87**	21.01	34.98	**55.99**
Arunachal Pradesh	0.22	0	0	**0.23**	0.11	0.29	**0.41**
Assam	12.10	0.04	0	**12.14**	1.78	16.37	**18.15**
Bihar	34.41	0.04	0	**34.46**	4.79	48.26	**53.07**
Chhattisgarh	8.45	0.03	0	**8.48**	4.07	12.79	**16.86**
Delhi	9.25	0.11	0	**9.36**	2.41	20.59	**23**
Goa	0.70	0	0	**0.70**	0.41	1.20	**1.60**
Gujarat	32.67	0.5	0	**33.16**	13.71	47.44	**61.16**
Haryana	9.53	0.17	0	**9.70**	2.78	16.27	**19.06**
Himachal Pradesh	3.86	0.06	0	**3.92**	1.13	5.29	**6.43**
Jammu & Kashmir	7.06	0.03	0	**7.09**	1.50	9.37	**10.88**
Jharkhand	15.78	0.10	0	**15.88**	5.57	19.34	**24.91**
Karnataka	38.25	0.09	0	**38.34**	19.73	51.11	**70.84**
Kerala	23.58	0.21	0	**23.79**	13.77	30.86	**44.64**
Madhya Pradesh	26.42	0.31	0.01	**26.74**	10.13	38.61	**48.80**
Maharashtra	47.60	0.17	0	**47.78**	17.97	72.77	**90.77**
Manipur	1.80	0	0	**1.80**	1.40	1.52	**2.92**
Meghalaya	1.12	0	0	**1.12**	0.72	1.19	**1.91**
Mizoram	0.35	0	0	**0.35**	0.28	0.34	**0.62**
Nagaland	0.91	0	0	**0.91**	0.59	1.18	**1.77**
Odisha	19.80	0.04	0	**19.84**	8.37	24.87	**33.26**
Punjab	14.56	0.09	0	**14.65**	4.24	20.55	**24.80**
Rajasthan	26.66	0.20	0.01	**26.87**	8.01	38.31	**46.33**
Sikkim	0.26	0	0	**0.26**	0.14	0.31	**0.45**
Tamil Nadu	49.27	0.21	0	**49.48**	32.27	64.45	**96.73**
Telangana	25.94	0.10	0.01	**26.05**	15.24	24.91	**40.16**
Tripura	2.10	0.01	0	**2.11**	0.44	2.51	**2.95**
Uttar Pradesh	89.64	0.36	0	**89.99**	27.27	137.92	**165.26**
Uttarakhand	4.14	0.02	0	**4.17**	0.69	5.91	**6.60**
West Bengal	88.41	0.26	0.01	**88.67**	43.51	91.95	**135.52**
A & N Islands	0.19	0	0	**0.19**	0.10	0.29	**0.39**

(continued)

Table 4.7 (continued)

State/UT	*Number*				*Employment*		
	Micro	*Small*	*Medium*	*Total*	*Male*	*Female*	*Total*
Chandigarh	0.56	0	0	**0.56**	0.12	1.17	**1.29**
Dadra & Nagar Haveli	0.15	0.01	0	**0.16**	0.07	0.29	**0.36**
Daman & Diu	0.08	0	0	**0.08**	0.02	0.12	**0.14**
Lakshadweep	0.02	0	0	**0.02**	0.01	0.02	**0.03**
Puducherry	0.96	0	0	**0.96**	0.57	1.27	**1.84**
All	**630.52**	**3.31**	**0.05**	**633.88**	**264.92**	**844.68**	**1109.89**

Source MSME Annual Report 2021–22

the unemployment and underemployment problem of the country (Joshi et al., 2020). Various Government schemes like Prime Minister Employment Generation Program (PMEGP), Credit Guarantee Trust Fund for Micro and Small Enterprises (CGTMSE), Interest Subvention Scheme of Reserve Bank of India, Credit Linked Capital Subsidy Scheme to upgrade technology, Gram Udyog Vikas Yojna to promote village industries are focussed to attain self-reliance and sustainable development through the MSMEs (Economic Survey 2020–21).

The MSME sector has proved to be highly dynamic in the Indian economy promoting the growth and development of various product segments and industries. It establishes direct and indirect association between MSMEs and the SDGs. This is in consensus with the direct or indirect attainment of so many SDGs laid by UN like Poverty eradication (Goal 1), Zero hunger (Goal 2), Good health and well-being (Goal 3), Quality Education (Goal 4), Gender Equality (Goal 5), Clean Water and Sanitation (Goal 6), Affordable and Clean Energy (Goal 7), Decent Work and Economic Growth (Goal 8), Industry Innovation and Infrastructure (Goal 9), Reduced Inequalities (Goal 10), Building Sustainable Cities and Communities (Goal 11), Responsible Consumption and Production (Goal 12) and Partnership for Goals (Goal 17). The impact of MSMEs in achieving sustainable development is highlighted through the theme MSMEs—Small Business, Big Impact and has set June 27 as International MSME Day (Verma & Nema, 2019).

Goal 1 (Fig. 4.5) focuses to eradicate and reduce poverty by mobilising resources, strengthening policies ensuring social protection and

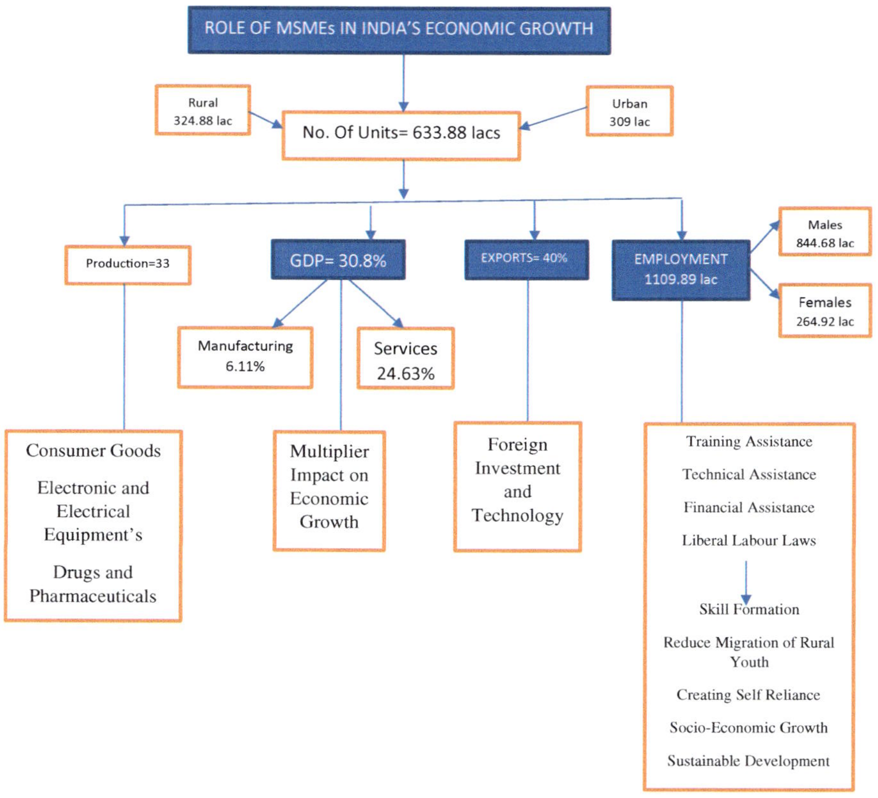

Fig. 4.4 Role of MSMEs in India's economic growth (*Note* Author's Construct. *Source* MSME Annual Report 2021–22)

Fig. 4.5 Sustainable Development Goals (*Source* www.un.org/sustainabledevelopment)

access to economic resources. The significant role of MSMEs in creating employment opportunities as a source of livelihood including women and youth to reduce the regional and gender disparities has proved to be a main driver to fight against poverty and integrate them in MSME value chain (Joshi et al., 2020). MSMEs Sector has a potential to strengthen the socio-economic growth of the country by resolving the unemployment problem locally thus, reducing the inequalities (Goal 10) in terms of economic imbalance geographically (Soni et al., 2022). The provision of employment opportunities will further reduce the problems related to overpopulation in the urban areas through the inclusive sustainable growth of MSME sector. This ambitious push to eliminate poverty through SDGs offers opportunities with a large cumulative impact (UNDESA).

Goal 2 focuses to end hunger, eliminating all forms of malnutrition, and increasing farmers' productivity is attained through agriculture focussed MSMEs that tend to be labour intensive. Improved nutrition status and awareness programmes can create a healthy workforce contributing to improved productivity and further, promote food consumption.

Goal 3 ensures healthy lives and promote well-being and is attained by providing better quality healthcare services by MSMEs at low cost by managing operational inefficiencies and re-energising them at work (Karlstrom, 2018).

Goal 4 focuses on provision of equitable childhood, vocational and tertiary education. The diverse skill requirements of MSMEs are also a challenge which is handled by the Ministry through its various training institutions to enhance the skills of the local people as per the requirements while producing diverse range of products and services demanded at home and globally at a comparative cost advantage. MSMEs through work-based learning programmes provide continuous learning opportunities to enhance technical and competency skills of youth and women for obtaining decent jobs, i.e. Goal 8.

Goal 5 focuses on achieving gender equality by ending discrimination and empowering women and girls by promoting their participation in decision-making through the access to economic resources. Women-owned MSMEs act as a source of economic empowerment to female entrepreneurs and have trained them in various skills to be at par with men (MSME Annual Report, 2021).

MSMEs to a great extent help in resolving water and sanitation issues i.e. Goal 6. They help to provide clean water to the residents and support them in improving water efficiency and sanitary management.

Similarly, MSMEs with their innovative methods are providing affordable and clean energy i.e. Goal 7 and are getting certification from the Ministry for providing energy-efficient products e.g. UJALA scheme.

Goal 9 is to build a strong infrastructure, nurture innovation and encourage sustainable industrialisation. MSMEs played a significant role to boost industrialisation and the entrepreneurial activities lead to low-cost innovations. They also created good infrastructure in their operational areas and got engaged in building schools, hospitals, roads, etc. as a part of their social responsibility.

MSMEs with the diverse range of operations help to build sustainable cities and communities, i.e. Goal 11 connecting the areas with good transportation, water and electricity supply. Also, Goal 12 focussing on responsible consumption and production is attained with the development of efficient technology and reduction in wastage through the MSMEs operations. Goal 17 leads to attain sustainable development for the nation through the Partnership of all formal and informal enterprises within the country by creating a better value chain (MSME Annual Report, 2021).

MSMEs involved in specific sectors could help achieve many objectives under different goals namely, by providing employment, by operational efficiency and by performing Corporate Social Responsibility (CSR) activities. By employing people in their establishments MSMEs help directly in achieving goals like Poverty Alleviation (Goal 1), Gender Equality (Goal 5), Decent Work and Economic Growth (Goal 8), Industry Innovation and Infrastructure (Goal 9) and Reduced Inequalities (Goal 10). MSMEs contribute to the attainment of certain goals by their sector-specific operational efficiency which includes Zero hunger (Goal 2), Affordable and Clean Energy (Goal 7), Responsible Consumption and Production (Goal 12) and Partnership for Goals (Goal 17). MSMEs also contribute to SDGs by performing Corporate Social Responsibility (CSR) activities in various forms to the goals of 'Good health and well-being' (Goal 3), Quality Education (Goal 4), Clean water and sanitation (Goal 6) and building Sustainable cities and communities (Goal 11). The MSMEs have crucial roles in attaining the targets of some goals by ethically and lawfully following laws and by-laws and using the resources of locality in a judicial manner. Under this the major contributions are towards Climate Action (Goal 13), Life Below Water (Goal 14), Life on Land (Goal 15) and Peace Justice and Strong Institutions (Goal 16).

Challenges and Way Forward for India

Although the MSMEs are flourishing at a faster pace the proper implementation of plans and measures taken to improve the MSMEs is expected to bring a paradigm shift in recent years. Challenges of MSMEs along with their possible remedies are suggested below.

- Along with the financial access, timely and adequate loans at subsidised rates are required to be made available for initial investments and future technology upgradations.
- Government has initiated to upgrade the youth through various vocational skills and collaborating with technical and research institutes like BARC, CSIR and NRDC to enhance the employee accessibility to technology through E-literacy mission by upskilling and reskilling manpower.
- Rural enterprises are needed to be developed with mobile training facilities and ICT infrastructure to match the global standards and be cost-competitive.

- Large-scale concerted efforts are required to attain self-reliance through the effective implementation of digital MSME schemes and review the legal compliance to promote the 'Ease of doing business'.
- Rural-agro-based industries could be supported by minimising the logistic costs by providing incentives to set the warehouses, grading and packaging, solar plants and herbal extractions in rural areas or nearby farms.
- Non-availability of specific corridors for transportation at affordable cost and speedy delivery could be addressed by Dedicated Freight Corridors (DFC) of Roadways, Railways, and Inland waterways to optimise transport strategy and generate cost savings that will help the eco-system.
- Non-availability of labourers because of reverse migration could be handled through schemes like MNREGA and "One nation, One Ration Card" system to facilitate inter-state ration card portability.
- To promote Micro units in the rural areas besides industrial infrastructure and Rural Industrial Zones, social infrastructure to provide mass quality public education, stronger health system, entertainment facilities and shopping facilities need to be planned.
- Research and Development need to be scaled up to promote energy-efficient technologies and solar energy applications for sustainable growth.
- Digitisation of MSMEs along with the access to new technology and its adoption could be created through online platforms. Virtual trade fairs, exhibitions, online training programmes can be organised to provide the right platforms for digital marketing.
- Subsidised ERP and other designing software or apps be made available to MSME to be compatible with the vendors and partners through technology system.
- The Common Service Centre needs to be popularised to support the Rural and Agro-based MSME in forward and backward market linkages. The procurement of raw materials at competitive cost and logistics challenges due to supply chain issue should be addressed.
- Non-availability of important raw materials to meet the needs of MSMEs can be assisted by NSIC, KVIC and Coir Board for various clusters like SFURTI and artisans. Special focus may be given for raw material assistance for products related to import substitution.
- Laptops, internet dongle and smart phones may be provided to all Gazetted Officers to update with trends and technology.

- Adoption of modern manufacturing tools and techniques like Block chain, Artificial Intelligence and Data analytics should be promoted for progressive results.
- Export eco-system for MSMEs can be promoted through unification of Export compliances, affordable transportation and logistics facilities, affordable testing and certifications, reduction of turnaround time through digitisation of export clearances and boosting exports through shopping festivals and trade fairs.
- Weaker sections like SC, ST and women entrepreneurs should be adequately motivated and incentivised to adopt latest technologies and quality tools to avail the benefits earmarked under Public Procurement Policy (PPP).
- Nano household enterprises with lesser restrictions should be categorised separately to promote inclusiveness and be exempted from statutory clearances.

Conclusion

The MSME sector is the backbone of the Indian economic structure and has a significant role in safeguarding it from the global economic shocks and adversities. MSMEs have a vibrant demography which needs to be analysed to understand the prevailing conditions, strength and gaps to facilitate suitable policy interventions to uplift the sector and attain the SDGs which act as a blueprint to attain a sustainable future for all in the world by addressing the global challenges. The economic activities which have a competitive advantage should be encouraged and developed with necessary financial support. Schemes should be devised to support those sectors which are yet to be nurtured. Financial support from state and central governments and favourable policies can push India to become a high-growth manufacturing-based economy. Efforts are required to create comprehensive policy to improve the country's business climate which is conducive to the development of MSMEs. India can attain the goal of being self-reliant and achieve sustainable fiscal and socio-economic growth through a clear vision and concrete plans involving MSMEs. India could emerge as a major contributor to attain the varied Sustainable Development goals laid by United Nations.

References

Economic Survey 2020–21.

FICCI-EY Re-imagining India's M&E sector, Ministry of Micro Small and Medium Enterprises, GeM Portal.

Joshi, V. D., Panigrahi, A., & Pitke, M. (2020, December). Vision of "Atma Nirbhar Bharat" role and significance of MSME. *The Management Accountant, 55*, 44–47

Karlstrom, C. (2018). *What impact can SMEs have on the Sustainable Development Goals?* Royal Society for the Encouragement of Arts, Manufactures and Commerce (RSA).

MSME annual report 2021–22, 2020–21

OI. (2020, July 9). Aatmanirbhar Bharat not self-containment: PM assures global investors. *Outlook India*. Retrieved 27 March 2022

Soni, G., Kumar, S., Mahto, R. V., Mangla, S. K., Mittal, M. L., & Lim, W. M. (2022). A decision-making framework for Industry 4.0 technology implementation: The case of FinTech and sustainable supply chain finance for SMEs. *Technological Forecasting and Social Change, 180*, 121686.

UNDESA—Report on MSMEs and the Sustainable Development Goals. www.sustainabledevelopment.un.org; www.un.org/sustainabledevelopment/sustainable-development-goals/

Verma, T., & Nema, D. (2019, March). Role of Micro, Small and Medium Enterprises (MSMEs) in achieving Sustainable Development Goals. *International Journal for Research in Engineering Application & Management (IJREAM), 4*(12). ISSN: 2454-9150.

www.ibef.org, November 2021.

https://sdgs.un.org/topics/capacity-development/msmes

CHAPTER 5

MSMEs and SDGs: Evidence from Bangladesh

Dhanonjoy Kumar and Susela Devi K. Suppiah

Introduction

MSME definitions vary according to the criteria used in different countries, such as the number of employees, asset value, sales and output volume, and so on (Cunningham & Rowley, 2008). MSMEs are making significant contributions to Bangladesh's economic growth. They work as an impetus to boost national income as well as to generate employment opportunities since this sector is labor-intensive and less time-consuming for production with less capital expenditure or lower establishment cost. Like other developing countries, Bangladesh has a great potential for development of the MSME sector. In fact, the MSME sector works as

D. Kumar (✉)
Department of Management, Islamic University, Kushtia, Bangladesh
e-mail: djoysk@mgt.iu.ac.bd

S.D.K. Suppiah
Department of Accounting, Sunway University, Bandra Sunway, Selangor, Malaysia
e-mail: suselas@sunway.edu.my

H. Dasaraju and T.T.H. Tambunan (eds.), *Role of Micro, Small and Medium Enterprises in Achieving SDGs*, Sustainable Development Goals Series, https://doi.org/10.1007/978-981-99-4829-1_5

a catalyst and plays a pivotal role in transforming Bangladesh into an industrially developed country. At present, MSME comprise about 7.8 million and they contribute about 25% to our GDP. The SME Policy 2019 will play a unique role toward fulfilling the election commitments of the present government, such as, providing city-dweller benefits to each village, transforming the youth into skilled manpower and ensuring their employment simultaneously, contributing positively to the further advancement of Bangladesh (Haque & Ahlan, 2018). In order to achieve balanced development through economic, social, and environmental protection of the country, the government announced the SME sector as the main pillar of industrial development in National Industrial Policy 2016. The development of the SME sector will play a significant role in achieving the targets enunciated in the policy and planning documents both national and international like National Industrial Policy 2016, Seventh Five-Year Plan, Vision 2021 and LDC graduation by 2024, SDG 2030 and Vision 2041. As Bangladesh has skilled human resources and intellectual capacities, the SME sector has a good opportunity to grow. However, there is a need for relevant policies to ensure a supportive environment to make this sector vibrant.

Overview of MSMEs in Bangladesh

Bangladesh has its strength in its abundance of labor. Businesses that guarantee the highest level of national welfare, including the creation of jobs alongside sustainable businesses, may capitalize on that strength. Bangladesh's MSME sector dominates the national economy. About 87% of the civilian population is employed by MSMEs, which provide 25% of Bangladesh's GDP. The Bangladeshi government has given MSMEs priority in the National Industrial Policy 2016.

There are 7.81 million economic entities in Bangladesh overall, according to the Bangladesh Bureau of Statistics' 2013 National Economic Census. These economic entities make up 11% of SMEs and about 88% of cottage businesses. Nonetheless, SMEs make up nearly 99% of Bangladesh's formal commercial enterprises. They account for around 75% of non-agricultural employment and 25% of the country's GDP. SMEs are regarded as the main drivers of economic growth in Bangladesh, making up more than 90% of all businesses, compared to 97.60% in India, 99.0% in China, 99.70% in Japan, and 60% in Pakistan, according to the ICAB World Webinar (2021). SME GDP contribution is quite low compared to other emerging economies at only 20.25%. There are 17,384 microenterprises in Bangladesh, of which 15,666 are tiny, 6103 are medium-sized, and 3639 are large-scale (Abdin, 2019).

SMEs, which help to tackle the unemployment problem by creating new innovative job opportunities, are one of the most promising and dominant sectors in private enterprise. Given this, the SME sector makes a significant contribution to economic development by expanding business activity and increasing foreign currency revenues. Bangladesh Bank is continuing its efforts to channel loans to small and medium-sized businesses. The SME sector eventually helps to enhance the lifestyle of the low-income group, empower women, and minimize gender disparities. Bangladesh Bank's refinancing facilities are being maintained in order to help the sector flourish further.

While, Bangladesh's SME sector has the potential to be a major driver of economic growth and job creation, there are a number of market and firm-level concerns preventing this sector from reaching its full potential. The Bangladeshi government and other sector advocates have been working on various programs to help small firms thrive for decades, but large-scale and transformational growth has remained elusive. Nevertheless, in recent decades, the Bangladesh economy has seen significant fundamental changes. The industry sector's overall contribution to GDP has increased gradually. According to Bangladesh Bureau of Statistics (BBS), the broad industry sector's contribution to GDP is expected to be 37.07% in FY 2021–22, up from 36.01% in FY 2020–21. The government is consistently taking synchronized and inclusive initiatives to improve and flourish all of the country's industrial sectors, including manufacturing and fuel for energy security, agriculture and forestry, mineral extraction and processing, tourism and service industry, construction industry, and ICT-based industry.

Furthermore, the government issued the "National Industrial Policy 2016" in order to speed up the country's industrialization. Sustainable and inclusive industrial growth through the generation of productive employment in order to broaden the perspectives of new entrepreneurs; mainstreaming women in the industrialization process; and worldwide market connectivity are among the basic objectives of the Industrial Policy 2016. The government is aggressively pursuing this goal by offering loans and other forms of supplementary assistance through banks and other financial organizations (MoF, 2022).

Table 5.1 MSMEs definition as the National Industrial Policy 2016

Sl. No.	*Industry type*		*The amount of investment (replacement cost and value of fixed assets, excluding land and factory buildings)*	*The number of employed workers*
1	Cottage Industry		Less than ten lakh	The number of employees cannot exceed 15
2	Micro Industry		10 lakh to 75 lakh	16 to 30
3	Small Industry	Manufacturing	75 lakh to 15 crore	31 to 120
		Service	10 lakh to 2 crore	16 to 50
4	Medium Industry	Manufacturing	15 crore to 50 crore	121 to 300
		Service	2 crore to 30 crore	51 to 120
5	Large Industry	Manufacturing	More than 50 crore	More than 300
		Service	More than 30 crore	More than 120

Source National Industrial Policy 2016

In the manufacturing sector, small industries will be defined as businesses with fixed assets excluding land and buildings valued between TK 7.5 million and TK 150 million, or with 31 to 120 employees. In the case of services, "small industry" shall refer to businesses with fixed assets valued between Tk. one million and Tk. twenty million, or with 16–50 employees. In manufacturing, "medium industry" will be defined as businesses having a fixed asset worth of between Taka 150 million and Taka 500 million, or with 121–300 employees. The "middle industry" for services will be defined as businesses with fixed assets valued between Taka 20 million and Taka 300 million, or with 51–120 employees (Tables 5.1 and 5.2).

As seen thus far, the MSME sector is the engine of growth. For this chapter, the focus of the research is on SMEs. The rationale is that lessons learnt from SMEs can be further extended in future studies to the micro businesses.

Table 5.2 Volume and growth rate of the manufacturing sector (at constant prices of 2015–16) (in Crore Tk.)

Type of industry	*2015–16*	*2016–17*	*2017–18*	*2018–19*	*2019–20*	*2020–21*	*2021–22**
Cottage industry	72,127(–)	78,829 (9.29)	84,700 (7.45)	96,704 (14.17)	100,257 (3.67)	110,557 (10.27)	123,543 (11.75)
Small, Medium and micro industry	129,108 (–)	142,102 (10.06)	157,882 (11.10)	174,632 (10.61)	179,325 (2.69)	204,241 (13.89)	228,164 (11.71)
Large industry	221,152 (–)	231,388 (11.08)	257,016 (12.79)	289,885 (0.41)	291,072 (10.61)	321,967 (10.61)	363,416 (12.87)
Total	422,387 (–)	452,319 (7.09)	499,598 (10.45)	561,220 (12.33)	570,654 (1.68)	636,765 (11.59)	715,123 (12.31)

Note Figures in parentheses indicate the rate of growth. *Provisional
Source Bangladesh Bureau of Statistics

Literature Review

SMEs in Developing Countries

SMEs play a vital role in the national economy. They can be found in every sector of the economy, but particularly in the service, catering, wholesale, retail, consumer goods, and food industries (Velinov & Ponomarev, 2016). Indeed, the value of SMEs has gradually increased, and they are now widely regarded as the most important component of the world economy and a major source of economic growth (Androniceanu, 2019; Bădulescu, 2010; Karpak & Topcu, 2010; Meyer & de Jongh, 2018; Siekelova et al., 2019). This is because micro businesses are known for their dynamism, inventiveness, and efficiency, as well as their ability to make quick decisions due to their small size. It's no wonder that many of these companies believe that internationalization and export potential are the most effective methods to stay competitive (Massaro et al., 2017). A dynamic economy with a significant number of small businesses would encourage entrepreneurship (Read & Staines, 2004).

In both industrialized and developing countries, an industrial organization system with a significant number of micro, small, and medium companies (MSMEs) can generate solid economic growth. Small and medium businesses can grow and foster forward linkages between medium and large businesses. It can also help with equitable regional growth, income redistribution, and the creation of distributive jobs, poverty reduction, and innovation, especially in small businesses (Calvino & Virgillito, 2018; Helmsing, 2003; Reeg, 2013).

In most developing nations, micro-and small-firms account for over 95% of all businesses. MSMEs account for more than 90% of all corporations and business operations globally, employing more than 60% of total employment. MSMEs contribute less to GDP in developing nations than they do in affluent countries (Shoma, 2019).

Prior Studies on SMEs in Bangladesh

Many past studies have evidenced a strong correlation between SME development and GDP, which ensures the economic ultimate goal, the relationship between growth and the overall business environment for SMEs overshadows the former relationship, despite the fact that SMEs require low capital per unit of output (Beck et al., 2005; Pervin et al., 2020). Further, Chowdhury et al. (2013) found that SMEs make significant contributions to poverty reduction programs and have the potential to contribute to overall industrial and economic growth, and suggested that Venture Capital, Bank Syndication Schemes, Human Resource Development in financial institutions and government offices, Training and Development, and other tools could be used to alleviate SME financing constraints.

A stable macroeconomic environment, an open trade and investment regime, and a competitive financial sector, according to Oldsman and Hallberg (2002), are the most important ingredients for a thriving private sector. However, with a law and order situation below optimal levels, corruption considerably over acceptable levels, and an uncertain political scenario, Bangladesh's domestic environment does not aid, but rather hampers the growth of SMEs in the country.

According to Bangladesh Economic Review 2022, the industrial sector of Bangladesh's economy has been gradually and consistently expanding over the last few years. Bangladesh Bureau of Statistics (BBS) estimates the contribution of the broad industry sector to GDP stood at 36.01% in FY 2020–21 which increased to 37.07% in FY 2021–22. Among the fifteen sectors of GDP, the broad industry sector includes five sectors namely mining and quarrying, manufacturing, electricity gas, steam and air condition, water supply and construction. The contribution of the manufacturing sector is the highest in GDP. In FY 2021–22 the contribution of manufacturing sector to GDP was 23.36% which increased to 24.45% in FY 2021–22. During FY 2021–22, up to February 2022, the number of medium, small, and cottage industrial units established with direct and indirect assistance of BSCIC are 36, 1269, and 2766 respectively. Total investment in these industrial units is Tk. 1588.93 crore (MoF, 2022).

GDP growth in Bangladesh is accelerate, reaching more than 8% in 2019. At the same time, the proportion of people living in extreme poverty has decreased from 43.5% in 1991 to just 14.3% in 2016. This is an astronomically fast rate of societal growth. Since 1990, child mortality has decreased by nearly 90%, life expectancy has climbed from 58 to 72 years, and female secondary school attendance has gone from 14% in 1990 to nearly 80% now (CPSD, 2021).

Alauddin and Chowdhury (2015) argued that in today's competitive market, Bangladesh's SME sector holds a solid position. SME entrepreneurs are always improving themselves in order to remain competitive and the best. As the experiences of SME finance in Bangladesh demonstrate, a credit delivery system that analyzes borrowers' credit worthiness on a criterion other than fixed asset ownership is crucial. Examining the borrowers' transaction records, determining the worth of movable goods, and so on may be required as part of the appraisal.

Acma (2015) argues that the performance of SMEs did not meet Bangladesh's expectations. It could be due to a lack of institutional finance, a lack of a consumer market for SMEs' products, or the lack of formal entrepreneur development efforts for SMEs. As a result, proper regulations and government initiatives, as well as infrastructure improvements, should be implemented to meet the problems that SMEs encounter in order to promote the sector's development.

Bangladesh Small and Cottage Industries Corporation (BSCIC) plays a vital role in Bangladesh's industrial expansion and poverty reduction, as well as household consumption and poverty reduction. A 1% increase in BSCIC income increases annual household consumption by 0.73%, whereas a 1% increase in non-BSCIC income increases consumption by 0.36% (Haider et al., 2015).

By assisting existing SMEs, creating non-traditional opportunities, generating employment, making the labor force more skilled and effective through training, increasing labor productivity, and improving product and service quality, the SME sector has played a significant role in contributing to economic growth. Small and medium-sized enterprises (SMEs) account for 25% of Bangladesh's GDP and 80% of total industrial jobs, contributing considerably to the country's economic growth rate of 6% to 8% over the last decade (Uddin, 2014).

Islam et al. (2013) found that the government has provided proportionally less support to SMEs in terms of policy or fiscal incentives. Small and medium-sized businesses (SMEs) have made a significant contribution to the development of Bangladesh's economy, both in terms of GDP and job creation. However, the activities of SMEs in Bangladesh have been found to be far below the worldwide norm, particularly in terms of quality assurance, funding allocation, and marketing efforts.

According to Ahmed et al.(2004), SMEs in Bangladesh have failed to ensure the quality of their products and services in both domestic and international markets due to a lack of a national quality policy and an adequate support system, as well as the lack of credibility of the quality certification authority. He also claims that access to capital is one of the most pressing issues for SMEs in Bangladesh. A lack of investment or running cash is one of the most common complaints of practically all Bangladeshi SMEs.

According to Hasan and Islam (2008), banks rarely demonstrate an interest in SME funding. Higher operational expenses, lower returns, and the significant risk associated with SME finance are the reasons for this conservatism. The operational costs are higher due to the tiny loan size, and they demand intense monitoring and oversight. The main reason for the increased risk is that small and medium-sized businesses are much less likely to meet the collateral requirements because they often do not own any real estate. Banks and non-bank financial organizations are sometimes hesitant to loan SMEs on the basis of collateral.

Ahmed and Chowdhury (2009) found SME also encourages the development of entrepreneurial and innovative talents. In addition to alleviating poverty, SMEs can prevent urban migration and enhance income flow in rural areas. As a result, the standard of living in rural areas will improve. SME performance in Bangladesh has been shown to be far below international standards. Although the Bangladeshi government has made certain attempts to promote the growth of SMEs, these efforts are insufficient.

Andalib and Halim (2019) found difficulties and obstacles are appropriately addressed, SMEs in Bangladesh have the ability to grow and contribute more to the economy. The government, apex bodies, agents, financial institutions, and SME entrepreneurs must all work together to help SMEs grow and prosper.

Alauddin et al. (2015) found that SMEs, especially in rural regions, continue to be the engine of economic growth by providing large-scale employment and income-earning opportunities at comparatively modest costs. It also enhances economic efforts to achieve high and long-term growth, which is required to alleviate widespread poverty and socioeconomic inequity. SMEs are important since they are labor-intensive industries that can generate additional job possibilities in industrialized countries. Given the considerable contribution of SMEs to overall growth and sustainable economic development in Bangladesh, it is critical to take the required steps to facilitate their expansion.

Further, Table 5.3 gives the summary of prior studies on SMEs in Bangladesh.

SMEs and SDGs in Developing Countries

SMEs are the backbone of a country's economies and multinational corporations' global supply lines. SMEs have comparatively minor environmental and social impacts on their own. However, the effects are far greater when viewed as a whole; SMEs account for over 90% of all firms. SMEs, according to the World Bank, contribute significantly to GDP and play an important role in job creation. Together with the Paris Agreement on Climate Change, the 2030 Agenda for Sustainable Development and the Sustainable Development Goals (SDGs) have reshaped the global development agenda, presenting unprecedented opportunities for transforming development policy and practice and achieving sustainable development for all. In 2015, the millennium development goals were superseded by the Sustainable Development Goals (SDGs) (United Nations, 2015). The SDGs outline global development priorities up to 2030 and are critical for resolving the global economic, social, and environmental concerns that communities face. They advocate for a wide range of public and private sector measures that can spur economic growth in novel and creative ways. These objectives are intertwined and may be indivisible (Le Blanc, 2015; Nilsson et al., 2016). Indeed, most entrepreneurial activities address multiple SDGs. This emphasizes the significance of institutional elements that influence entrepreneurial talent and, as a result, socioeconomic prosperity (Urbano et al., 2019).

Table 5.3 Summary of prior studies on SMEs

Author	*Findings*
Islam et al. (2013)	The studies mentioned that SMEs have got policy or fiscal incentives based support slightly and disproportionately from the government and it is weak in filling up their needs and demands
Hasan and Jamil (2014)	Discussing with the financiers and other stakeholders, the study takes into consideration. Venture capital may be the best possible long-term funding for SMEs and different international development based funds may decrease the related financial risk
Uddin (2014)	The contribution of the SME sector in this regard is incredible but the performance of SMEs are being below the standard level globally, because of limited financing, high interest rates, poor infrastructure, credit information gap, etc
Alauddin and Chowdhury (2015)	The study stated that SME here demonstrates a credit delivery system whereas fixed asset ownership is crucial. The borrowers' transaction records and the worth of movable goods might be appraised
Alauddin et al., (2015)	The results of this study revealed that SMEs alleviate poverty side by side create employment generation and women's empowerment in the small and cottage industry sectors. Besides, it took part a great proportion in GDP. On the other hand, the performance of SMEs is not satisfactory enough here due to lack of sufficient financing, skilled workers, and political instability
Roy (2016)	The study reported that in Bangladesh, SMEs demands 87% employment of total that filled the 33% of industrial goals. At present, women entrepreneurs are less than 10% of total business entrepreneurs of Bangladesh
Bosri (2016)	This study ascertained that banks and financial institutions do not provide enough SME loans but they give industrial loans. What they provide for SMEs, their recovery rate is very high

(continued)

Table 5.3 (continued)

Author	*Findings*
Jahur (2018)	The study reinforced that SMEs in Bangladesh face the following problems and challenges. These are capital and regulatory factors, technological and bureaucratic factors, entrepreneurial and managerial skills, and SME policies and support services of management
Ali and Islam (2018)	The study portrayed that the growth of SMEs and the employment in this sector are higher than other sectors of Bangladesh as well as credit growth in this sector is higher than others. Though the disbursement of credit has been gradually increasing, the share of industrial credit in this sector is gradually decreasing over the years
Masuduzzaman et al. (2018)	The study showed that there are no significant differences in the efficiency between conventional and Islamic banks based on SMEs related activities. But a slight difference may be noticed by considering the size, sector, and region
Pervin et al., (2020)	SME entrepreneurs in Bangladesh have commercially viable technology but most SME companies suffer from a lack of complementary assets such as skills, market leadership, networking, access to new technologies, strengthening relationships with suppliers, manufacturing plants, marketing capabilities, and distribution channels for commercialization
Rahman and Khondkar (2020)	The study explored that the SME sectors have caused the economic growth by creating non-traditional opportunities, by generating employment, by making the labor force more skilled and effective through training, by enhancing labor productivity and improving product and service quality
Rahman and Kabir (2020)	The study found that Bangladeshi SMEs can concentrate on their core activities while outsourcing the entire technology-related supporting processes to a third-party vendor without incurring much expense

(continued)

Table 5.3 (continued)

Author	*Findings*
Harel (2021)	The findings disclosed a relationship between revenue change in business activity during the pandemic and revenue from subcontracting work. The businesses did not make any changes to their activities during this period and did not suffer a decline in revenue because the rate of revenue from subcontracting work was high. The large businesses that were essential for continuing economic conduct were allowed to continue their activities under some restrictions. Small businesses were also allowed to continue their work without physical contact with their customer during the lockdown periods
Reim et al. (2022)	This study evaluates a lot of challenges which are related to three components of a business model. The three components are- value creation, value delivery and value capture. The study identifies and matches digitalization activities with business model challenges that SMEs face when it is in international markets

SMEs can play a critical part in fostering a sustainable future through ethical business practices, and there is a compelling commercial rationale for them to do so. Companies, and SMEs in particular, must be unconcerned about the community and environment in which they operate in order to be successful. Peaceful surroundings, a healthy workplace, legal clarity, and positive employee interactions are all important components of corporate success. SMEs can begin by ensuring that its vision, mission, strategy, and business model embrace responsible business practices and regard them as vital to the SME's long-term success (Avasilicai, 2009).

According to the United Nations Industrial Development Organization (UNIDO), MSEs are estimated to account for over 90% of private firms and more than 50% of employment and GDP in most African nations. Micro and small businesses (MSEs) have demonstrated job-creating benefits, and the future of jobs will be largely determined by how MSEs respond to the changing economy. The SDGs of the United Nations Global Development Agenda emphasize the necessity of integrated development solutions. This means that for MSEs to make a significant contribution to the SDGs, they must be resource efficient, ensure effective natural resource management, be carbon neutral, and appropriately manage their waste (Grochowski et al., 2016).

SMEs provide an existing base of economic activity that can be expanded and developed; they provide employment opportunities to the poorest and most disadvantaged; they are a potential source of wealth creation, innovation, and tax revenue; they often require fewer formal qualifications and skills; they can improve countries' resilience to external shocks and global price fluctuations through their diversification; and they can improve countries' resilience to external shocks and global price fluctuations through their diversification. For all of these reasons, MSMEs have been highlighted as a critical component of poverty reduction and sustainable development and therefore of meeting the SDGs (CAFOD, 2013).

SME innovation development, social and environmental contribution, strong management and leadership practices, contribution to local network incentives, and NGOs are all examples of SMEs' role in sustainability (Dobreva & Ilieva-Koleva, 2015).

According to Zindiye et al. (2012), sustainable development variables connected with coercive isomorphic pressures have a considerable impact on all three sustainable development dimensions of an economic, environmental, and social character.

SMEs are regarded as good economic growth drivers because of their intended positive impacts in a variety of areas, including: reducing unemployment by creating jobs; protecting the environment by investing in sustainable projects; incorporating social responsibility into their strategic principles; utilizing natural energies, particularly renewable energies, which contribute to lower raw material and manufacturing costs; and contributing to social prosperity by implementing social policies. SMEs face several hurdles in achieving the SDGs, regardless of whether they operate in an internal or external setting (Cherroun & Cherroun, 2022).

According to Noranarttakun and Pharino (2021) the majority of SMEs have a positive impression of the green industry and plan to implement it. Despite SMEs' internal resource capabilities, particularly budgeting, green technology knowledge remains a serious issue, as does the lack of adequate legislative incentives. The global SDGs and the 20-year national strategy focusing on green economy and sustainable development hold enormous promise.

Ethiopia has the highest percentage of microenterprises per capita in developing countries. Investing in SMEs can assist in meeting 60% of the

SDG targets. In the manufacturing sector, micro-enterprises employee 97% of Ethiopians. Informal businesses account for 80% of job growth (Endris & Kassegn, 2022). In Ethiopia, employment in micro-enterprises was assessed as important factors such as poverty reduction, social and economic empowerment of women, and contributing to the local economy and communities through income tax collection.

In recent decades, the knowledge-based economy has expanded the contribution of SMEs to innovate dynamics in the economic sector. The role of SMEs in the transition to highly sustainable production and consumption patterns is critical for the greening of economic development. Innovative SMEs show serious roles in the eco-industry and clean-tech marketplaces in several OECD nations. Trade and investment openness, intellectual property protection, infrastructure, and institutional quality in SMEs' commitment to global markets (Mukole, 2010).

A climate-resilient SME is one that has decreased its environmental footprint and made its activities more environmentally friendly. Delivering competitively priced goods and services that meet human needs and improve quality of life while decreasing environmental consequences and resource intensity (Schmidheiny & Stigson, 2000).

SMEs are important stakeholders in the economy and the broader business ecosystem. To enhance economic growth and to deliver a more inclusive globalization require businesses to be able to adapt and survive in a more open environment. SMEs performed to achieve the SDGs by promoting inclusive and sustainable economic growth, providing employment and decent work for all, promoting sustainable industrialization and fostering innovation, and reducing income inequalities across countries at all stages of development (Angulo-Guerrero et al., 2017).

Recently, SMEs are innovative, new and small businesses that are often the driving force behind the kinds of radical innovations that are critical for economic growth because they can work outside of dominant paradigms, exploit technological or commercial opportunities that have been overlooked by larger firms, or enable the commercialization of knowledge that would otherwise remain un-commercialized in universities and research institutions (Baumol, 2014).

SME participation in the shift to more sustainable production and consumption patterns is critical for economic development to become more environmentally friendly. Small firms' individual environmental footprints may be small, but their aggregate impacts in some industries can exceed those of major corporations. A significant aspect of success in the green transformation is reducing the environmental effects of SMEs by achieving and going beyond environmental compliance with existing rules and regulations in both manufacturing and services. This is especially important for SMEs in the manufacturing sector, which consumes a substantial portion of the world's resources, pollutes the air and water, and generates waste (Vasilescu et al., 2022).

Prior Studies from Bangladesh on SDGs and SMEs

SDG is a multidimensional topic and there have been diverse poverty-related studies in rural and urban areas, and poverty impacts on the economy and society with social changes. SDG, as a relatively new dimension in the world economy for human prosperity and shared development, has multiple areas for impact study, effectiveness, and achievement. The study found that the private sector in Bangladesh, as a vital actor in this massive agenda, has growing importance in various dimensions to bring in positive results in the journey of SDG within the given timeline, creating employment opportunities and reducing poverty and hunger (Patwary, 2020).

Since its inception in 2016, the Citizen's Platform for SDGs, Bangladesh takes part in highlighting the contribution of non-state actors (NSAs) in implementation of the SDGs in Bangladesh. The Platform is helping raise the voice and mobilize the NSAs in this connection toward more effective SDG implementation in the country. More than hundred civil society organizations (CSOs), non-government organizations (NGOs), and private sector organizations are members of the Platform, and are engaged in the SDG process in various capacities (Khatun et al., 2020).

Bangladesh has made tremendous progress toward meeting the Millennium Development Goals (MDGs), earning international praise in the process. It is projected that the government will be able to maintain its progress and achieve SDGs. Because the SDGs encompass a broader range of interconnected goals and indicators, they may pose new challenges for the government. The achievement of the SDGs will be contingent on regional and global cooperation. Bangladesh's standpoint, it is critical to comprehend how South Asian countries can achieve sustainable development through peaceful collaboration and smart diplomacy (Khatun & Saadat, 2021).

SDGs requires relevant development priorities and programming where local governments are best suited to implement policies and programs to provide improved services that can alleviate poverty, reduce inequality, promote climate vulnerability, and promote gender equality. Participatory grassroots local government is essential for achieving SDGs, especially in poor and marginal areas in Bangladesh (Imran et al, 2022). The multi-stakeholder approach needs to be followed for rural sustainable development through SME expansion. The approach refers to collaborative efforts for the development and implementation of the proposed action plan. Like the SDGs, the SME localization plan will be a quick and effective remedy to address the effects of Covid-19 in both the medium and long term (Hossain et al., 2022).

Further, Table 5.4 gives the summary of studies done before on SMEs related to SDGs in Bangladesh.

Research Gap

SMEs also rely on their internal institutional environment, such as employee working performance, organizational culture and structure, innovation, knowledge sharing, and internal networks among employees. Most of the SMEs suffer from lot of problems, such as financial problems, scarcity of raw materials, labor turnover, lack of skills manpower, marketing problems of SMEs products, networking, access to new technologies, strengthening relationships with suppliers, manufacturing plants, marketing capabilities, and distribution channels for commercialization and so on.

The SDGs outline global development priorities up to 2030 and are critical for resolving the global economic, social, and environmental concerns that communities face. They advocate for a wide range of public and private sector measures that can spur economic growth in novel and creative ways. SMEs, however, face several hurdles in achieving the SDGs, regardless of whether they operate in an internal or external setting. Due to SMEs' overall lack of capacity to reach the SDGs, a cluster network strategy to incorporate an SDG-oriented framework into SMEs will enable them to move beyond their individual strategies and overcome the key hurdles when individually dealing with the SDGs.

The majority of SMEs have a positive impression of green industries and plan to implement them. Despite SMEs' internal resource capabilities, particularly budgeting, green technology knowledge remains a serious

Table 5.4 Summary of prior studies on SMEs and SDGs

Author	*Findings*
Tonis (2015)	A descriptive study and secondary sources of data were used. The study revealed that Sustainable development might be achieved by entrepreneurial enterprises with social responsibility; by innovative SMEs led by entrepreneurs that not only respect the 21 Agenda advice but innovate in technologies that reduce the impact of their ecological footprint, or that might regenerate natural resources
Fayomi et al. (2018)	The study used qualitative approach which focused on the opinion and attitude assessment. This study gave clear indications that research as a full concept is not a cliché and it is revolving year-in year-out in our contemporary world of interest as the need for human survival on the planet earth gets more prioritized on the agenda of the various sustainability organizations like UNESCO, United Nations, UNEP, and IISD
Verma (2019)	The study found that the MSME sector is one of the sectors that carries a huge potential to affect the progress of most of these goals. The diversity and flexibility in the operational capabilities of MSMEs make them an important player in this endeavor and thereby making a substantial contribution towards achieving the SDGs
Šebestová and Sroka (2020)	The study confirmed that there are differences in both countries in relation to sustainable development goals, with SME decisions in the Czech SME sector being at level 1.0 of business sustainability, and in Poland being below level 1. Additionally, Czech entrepreneurs have more knowledge and practice in "green" methods than their Polish counterparts

(continued)

Table 5.4 (continued)

Author	*Findings*
Hamburg, (2020)	The qualitative research approach and secondary data sources are used here. The study explored that particular needs are in reskilling workers who have been made redundant through digitalization or other changes in work, and upskilling existing workers who use technology-rich environments. This issue requires long-term investmentand a change in the existing model to make it more dynamic and support SDG4
Patwary (2020)	The study showed that SMEs have grown scattered across the country without clear objectives as they were scarcely directed to grow in a clustered manner. These study findings are expected to frame long-term policies of the government, including an appropriate fiscal budget in the national budget
Haryati et al. (2021)	This study has used a quantitative analysis method using the Structural Equation Model (SEM) through AMOS software version 22. The study found that the development of sustainable SME performance in West Sumatera, it is argued that businesses can make a suitable variety of innovations starting from products, processes, and marketing under the needs of sustainable environmental sustainability
Biryukov et al. (2021)	This study examined trends and opportunities in implementing the concept of sustainable development in the world and in the economy of the Russian Federation at the level of small and medium-sized enterprises. These studies of global trends show a decline in business and economic activity towards the achievement of the SDGs, especially in the SME sector
Noranarttakun and Pharino (2021)	This study was conducted to investigate the strengths, weaknesses, opportunities, and threats (SWOT) of the implementation of GI in Thai SMEs in the sector of electronic products and electrical equipment. The result revealed that the majority of SMEs have a positive perception with green industry and express intention to adopt green industry. The key opportunity is arising from the global SDGs and the 20-year national strategy focusing on green economy and sustainable development

(continued)

Table 5.4 (continued)

Author	*Findings*
Manzoor et al. (2021)	A multi-stage stratified random sampling technique was used to select sites for the study. For data analysis, SPSS (v.25) and as a statistical tool, AMOS (v.23) was used as a statistical tool, and to test the hypotheses. This study showed that SME evolution has a positive and optimistic influence on rural development. The findings of this paper hold significant implications for both the research society and loan-issuing institutions and departments
Yousaf et al. (2021)	A cross-sectional plan was used for the investigation of associations among the constructs of the hypothesized model. The results revealed that the digital orientation, IoT, and digital platforms are major antecedents of sustainable digital innovation. The results also showed that the digital platforms mediate between both digital orientation-sustainable digital innovation link and the IoT-sustainable digital innovation link
Jiménez et al. (2021)	The case study analysis, through observation, surveys, and interviews was used. The study made small steps taken by the CMO in fostering the SDG framework dissemination among the SMEs and in working closely with them to identify cooperation opportunities in order to correctly deliver the SDGs, which will result in a big leap forward in corporate sustainability enhancement for the furniture industry
Cherroun and Cherroun (2022)	Qualitative study and used secondary sources of data. The study argued that due to SMEs' overall lack of capabilities to reach SDGs individually, a cluster network approach to incorporating an SDG-oriented framework into SMEs would allow them to think beyond their individual strategies and overcome the key barriers when individually engaging with the SDGs

(continued)

Table 5.4 (continued)

Author	*Findings*
Endris and Kassegn (2022)	The study found that MSMEs significantly contributed to the sustainable development goals of Ethiopia through creating employment, alleviating poverty, and improving their living standards

issue, as does the lack of adequate legislative incentives. The SDGs and the 20-year national strategy focusing on the green economy and sustainable development hold enormous promise. To close the gap between the implementation of green industry practices and sustainable growth for SMEs, policymakers should promote the implementation of economic-based incentives such as sustainable procurement and voluntary green product certification schemes, as well as pragmatic regulation mechanisms such as green product law and product tax. SMEs are important stakeholders in the economy and the broader business ecosystem. Boosting economic growth and delivering a more inclusive globalization require businesses to be able to adapt and survive in a more open environment as well as participate more actively in the digital transition.

Most of the research used a qualitative approach and secondary sources of data. On the other hand, much of the research using quantitative statistical analysis is used to find out the contribution and forms of participation in solving environmental problems and the formation of sustainable development trends in terms of achieving the SDGs of SMEs. The data collection has been done with the use of structured questionnaires. The questions in the questionnaires include open-ended questions, close-ended questions, multiple-choice questions, and questions based on the five-point Likert scale. This method can generate less missed information compared with online surveys, telephone surveys, and mail surveys. Descriptive statistics, factor analysis, and regression analysis were conducted to prove the hypothesis of the study. The SPSS and Structural Equation Model (SEM) through AMOS software were used for the analysis of the data.

Very few studies have been conducted on the basis of primary data. There are significant limitations without collecting primary data from SME owners about SME contributions to SDGs. It is possible to know the root-level SME problems and achieve SDGs. Most SME owners are unaware of the SDGs' goals and activities properly in their enterprise. Primary data collection and analysis help identify the real barriers to achieving the SDG goals.

Achieving SDGs requires relevant development priorities and programming where local governments are best suited to implement policies and programs to provide improved services that can alleviate poverty, reduce inequality, promote climate vulnerability, and promote gender equality. Participatory grassroots local government is essential for achieving SDGs, especially in poor and marginal areas in Bangladesh. The multi-stakeholder approach needs to be followed for rural sustainable development through SME expansion. The SME owners and integrated authorities are not so aware about the development relationships with SMEs and SDGs. There is limited research (cite studies that examined this phenomenon of SDGS and SMEs) conducted on SMEs and SDGs. So, updated and integrated research works and activities are needed in the present context.

Objectives and the Research Questions of the Study.

Objective

The objective of this study is to:

(i) to examine in what extent SDGs are adopted by SMEs;
(ii) to find out the important problems of SMEs for achieving SDGs; and
(iii) to assess the SMEs' contribution for achieving SDGs.

Research Questions

The study has three research questions, which are the followings:

(i) do SMEs adopt SDGs and how?
(ii) what are the important problems faced by SMEs for achieving SDGs in Bangladesh?
(iii) how do SMEs help to achieve SDGs in Bangladesh?

Methodology

To conduct the research, both qualitative and quantitative research approaches were used. For the purpose of the study, both primary and secondary data were used. Among the various methods of obtaining the primary data, the researcher selected the following: the interview method; the observation method; and the library work method. Secondary data is cheaper and more quickly obtainable than primary data and may also be available when primary data cannot be obtained at all. Researchers utilized it to construct the conceptual framework of this study. Secondary data was gathered from newspaper reports, related books, journals, articles, seminar papers, publications from national and international research institutions, reports from various medical institutions, public records and statistics, various research reports, and so on. All types of data were processed through the SPSS-26. Before feeding the data into the computer, the data was converted into numerical codes, and the details of these codes were recorded in a code book. Some simple statistical tools were used to achieve the objectives of the study. A survey was undertaken and achieved 243 respondents using random sampling, respondents are the SMEs' owners. A five-point Likert-type scale was used to assess SME activities and their contribution to achieving SDGs in Bangladesh.

Data Analysis and Interpretation of Data

Table 5.5 represents the personal profile of the respondents in terms of age, gender, education, marital status, and industry type. The majority of 95 (39.10%) SMEs owners were classified as being between the ages of 30 and 39 years. 88 (36.20%) of SME owners' age group is 40–49 years, 39 (16%) of SME owners' age group is above 50 years and only 21 (8.60%) SME owners' age group is 20–29 years. Among 243 respondents, 222 (91.40%) were male and 21 (8.60%) were female. The education level was the highest at 67 (27.60%). SME owners were graduates and lowest was only 2 (0.80%) of the SME owners were with vocational school degrees. Out of 243 SME owners the highest 219 (90.10%) were married and only 2 (0.80%) were separated/divorced. In terms of number of years of work experience, most of the respondents had 5 years or less (55.9%), followed by 6–10 years (26.1%), and 26 years or above (3.2%). As for the monthly income, 30.3% of the respondents earned between TK. 10,001 and TK. 20,000, 30.3% of the respondents earned between TK. 20,001 and TK.

30,000, 8.5% earned less than TK. 10,000, and 10.6% earned TK. 50,000 and above. In this survey, 123 (50.6%) industries were manufacturing category and 120 (49.40%) were service industry category.

Table 5.6 presents the frequencies and percentages of the size of enterprise in terms of number of employees. It indicates that highest numbers of employees 163 (67.10%) were the employees level (1–15) and lowest numbers of employees 2 (0.80%) were employees level 121–300.

Table 5.7 presents the frequencies and percentages of the size of enterprise in terms of number of asset. It indicates that highest amount of asset 127 (52.30%) were less than 10 lakh taka and lowest amount of asset 24 (9.90%) were amount of asset figure 75 lakh to 15 crore taka.

Table 5.8 presents the frequencies and percentages of the size of enterprise in terms of the figures of employees involved in different working sectors. It indicates that highest number of SME owners doing similar work was 58 (23.90%), another person's private business 94 (38.70%), public listed corporations 6 (2.50%), government employees 3 (1.20%),

Table 5.5 Demographic profile of the respondents

Subject	*Category*	*Frequency*	*Percent*
Age	20–29 years	21	8.60
	30–39 years	95	39.10
	40–49 years	88	36.20
	Above 50 years	39	16.00
Gender	Female	21	8.60
	Male	222	91.40
Education	No formal schooling	3	1.20
	Primary school	14	5.80
	High school level	22	9.10
	High school Pass	28	11.50
	Vocational school	2	0.80
	College level	19	7.80
	College Pass	46	18.90
	Graduate	67	27.60
	Post Graduate	42	17.30
Marital status	Single	22	9.10
	Married	219	90.10
	Separated/divorced	2	0.80
Industry type	Manufacturing	123	50.6
	Service	120	49.4

Table 5.6 What is the size of your enterprise in terms of number of employees?

Number of employees	*Frequency*	*Percent*	*Cumulative percent*
1–15	163	67.10	67.10
16–30	65	26.70	93.80
31–120	13	5.30	99.20
121–300	2	0.80	100.0

Table 5.7 What is the size of your enterprise in terms of asset?

Amount of asset	*Frequency*	*Percent*	*Cumulative percent*
Less than 10 lakh taka	127	52.30	52.30
10 lakh to 75 lakh taka	92	37.90	90.10
75 lakh to 15 crore taka	24	9.90	100.00

overseas workers 5 (2.10%), unemployed 48 (19.80%), full-time students 19 (7.80%) and others were (10 4.10%).

Table 5.9 presents the frequencies and percentages of the size of enterprise in terms of number of SME owners for choosing this business. It indicates that highest numbers of employees want to be own boss 93 (38.30%), can work when and how I want 43 (17.70%), can't find any work 59 (24.30%), can't find any work that I really like 22 (9.10%),

Table 5.8 What were you doing before you began this business?

Subject	*Frequency*	*Percent*	*Cumulative percent*
I run another business doing similar work	58	23.90	23.90
I was an employee in another person's private business	94	38.70	62.60
I was an employee in a publicly listed corporation	6	2.50	65.00
I was an employee in government	3	1.20	66.30
I was an overseas worker	5	2.10	68.30
I was unemployed	48	19.80	88.10
I was a full time student	19	7.80	95.90
Others	10	4.10	100.00

don't have skills for other kinds of work 15 (6.20%), previous employment ended 7 (2.90%), and others were 4 (1.60%).

Table 5.10 presents the frequencies and percentages of the size of enterprise in terms of annual turnover. The highest turnover figure indicates 155 (63.80%) were between Tk. 100,000–1,000,000; 42 (17.30%) between Tk. 1,000,001–2,000,000; 21 (8.60%) between Tk. 2,000,001–3,000,000; 18 (7.40%) between Tk. 3,000,001–4,000,000; 3 (1.20%) between Tk. 4,000,001–5,000,000; and 4 (1.60%) between Tk. 5,000,001 to above.

Table 5.11 presents the frequencies and percentages of the size of enterprise in terms of number of employees facing different types of problems. It indicates that highest numbers of SME owners facing financial problem 55 (22.60%); high-interest rates 21 (8.60%); poor infrastructure 16 (6.6%); credit information gap 17 (7.00%); marketing problem 40

Table 5.9 Which of the following best describes your reason(s) for being in business?

Subject	*Frequency*	*Percent*	*Cumulative percent*
I want to be my own boss/I don't like to work for others	93	38.30	38.30
I can work when and how I want	43	17.70	56.00
I can't find any work	59	24.30	80.20
I can't find any work that I really like	22	9.10	89.30
I don't have skills for other kinds of work	15	6.20	95.50
My previous employment ended	7	2.90	98.40
Others	4	1.60	100.00

Table 5.10 What was the annual turnover of your company in 2021?

Subject	*Frequency*	*Percent*	*Cumulative percent*
Tk. 100,000–1,000,000	155	63.80	63.80
Tk. 1,000,001–2,000,000	42	17.30	81.10
Tk. 2,000,001–3,000,000	21	8.60	89.70
Tk. 3,000,001–4,000,000	18	7.40	97.10
Tk. 4,000,001–5,000,000	3	1.20	98.40
Tk. 5,000,001 to above	4	1.60	100.00

Table 5.11 What is the important problem that you are facing?

Subject	*Frequency*	*Percent*	*Cumulative percent*
Financial problem	55	22.60	22.60
High-interest rates	21	8.60	31.30
Poor infrastructure	16	6.60	37.90
Credit information gap	17	7.00	44.90
Marketing problem	40	16.50	61.30
High employee turnover rate	14	5.80	67.10
Quality assurance	42	17.30	84.40
Technological problem	7	2.90	87.20
Lack of quality manpower	31	12.80	100.00

(16.50%); high employee turnover rate 14 (5.80%); quality assurance 4 (17.30%); technological problem 7 (2.90%); lack of quality manpower 31 (12.80%).

Table 5.12 presents the frequencies and percentages of the size of enterprise in terms of number of competitors. Direct competitors indicate that highest numbers of 130 (53.50%); and indirect competitors 24 (9.90%); partner competitors 11 (4.50%); perceived competitors 78 (32.10%).

Table 5.13 presents the frequencies and percentages of the size of enterprise in terms of the number of most important stakeholders. It indicates that highest numbers of investors were 33 (13.60%); employees were 64 (26.30%); customers were 115 (47.30%); suppliers were 10 (4.10%); communities were 15 (6.20%); trade associations were 6 (2.50%).

Table 5.14 presents the frequencies and percentages of the size of enterprise in terms of number of major challenges for the relationship with customers. It indicates that inexperience in business was 41 (16.90%); lack of technical knowledge 37 (15.20%); poor managerial skills

Table 5.12 What are the enterprise's competitors?

Subject	*Frequency*	*Percent*	*Cumulative percent*
Direct competitors	130	53.50	53.50
Indirect competitors	24	9.90	63.40
Partner competitors	11	4.50	67.90
Perceived competitors	78	32.10	100.00

Table 5.13 Who are the most important stakeholders for your enterprise?

Subject	*Frequency*	*Percent*	*Cumulative percent*
Investors	33	13.60	13.60
Employees	64	26.30	39.90
Customers	115	47.30	87.20
Suppliers	10	4.10	91.40
Communities	15	6.20	97.50
Trade associations	6	2.50	100.00

Table 5.14 What are the major challenges do you face when maintaining the relationship with customers?

Subject	*Frequency*	*Percent*	*Cumulative percent*
Inexperience in business	41	16.90	16.90
Lack of technical knowledge	37	15.20	32.10
Poor managerial skills	34	14.00	46.10
Lack of planning skills	36	14.80	60.90
Market research skills	48	19.80	80.70
Technological change	20	8.20	88.90
All	27	11.10	100.00

Table 5.15 Have you any knowledge about sustainable development goals (SDGs)?

Subject	*Frequency*	*Percent*	*Cumulative percent*
Yes	56	23.00	23.00
No	187	77.00	100.00

34 (14.00%); lack of planning skills 36 (14.80%) market research skills 48 (19.80%); technological change 20 (8.20%); and all were 27 (11.10%).

Table 5.15 presents the frequencies and percentages of the respondents having any knowledge about Sustainable Development Goals (SDGs). It indicates that highest numbers of employees answered have no knowledge about SDGs 187 (77.00%) and lowest numbers of employees answered have knowledge about SDGs 56 (23.00%).

Next, Tables 5.16 and 5.17, respectively, shows frequencies of agree and disagree ratings of 17 SDGs and descriptive statistics of 17 SDGs according to mean hierarchy from 56 respondents.

Table 5.16 Frequencies of agree/disagree ratings of 17 sustainable development goals (SDGs)? ($N = 56$)

Sl. No.	*Subject*	*Strongly agree*	*Agree*	*Neither agree nor disagree*	*Disagree*	*Strongly disagree*	*Total*
1	SDG No. 1: End poverty in all its forms	34	21	0	1	0	56
2	SDG No. 2: End hunger, achieve food security	22	23	9	2	0	56
3	SDG No. 3: Ensure healthy lives	14	27	14	1	0	56
4	SDG No. 4: Ensure inclusive education	9	22	19	6	0	56
5	SDG No. 5: Achieve gender equality	10	29	15	2	0	56
6	SDG No. 6: Ensure availability of water, sanitation	4	15	32	5	0	56
7	SDG No. 7: Ensure access to sustainable energy	5	10	30	11	0	56
8	SDG No. 8: Promote inclusive economic growth	25	29	1	1	0	56
9	SDG No. 9: Build resilient infrastructure	9	21	14	12	0	56
10	SDG No. 10: Reduce inequality among countries	7	33	11	5	0	56
11	SDG No. 11: Make cities safe and sustainable	5	24	20	7	0	56
12	SDG No. 12: Ensure responsible consumption	2	32	19	3	0	56
13	SDG No. 13: Urgent action to combat climate change	2	10	15	29	0	56
14	SDG No. 14: Conserve the ocean	2	9	12	29	4	56

(continued)

Table 5.16 (continued)

Sl. No.	*Subject*	*Strongly agree*	*Agree*	*Neither agree nor disagree*	*Disagree*	*Strongly disagree*	*Total*
15	SDG No. 15: Protect terrestrial ecosystems	1	11	21	20	3	56
16	SDG No. 16: Promote peaceful and inclusive societies	7	30	14	5	0	56
17	SDG No. 17: Build strength partnership	7	39	7	3	0	56

Table 5.18 represents Micro, Small, and Medium Enterprise Cross tabulation and the important problem that is facing. Out of 243 respondents they indicate highest problem is financial problem, 55 (Micro Enterprise 28, Small Enterprise 22 Medium Enterprise 5); Quality assurance problem is 42 (Micro Enterprise 25, Small Enterprise13, Medium Enterprise 4); Marketing problem is 40 (Micro Enterprise 22, Small Enterprise 18, Medium Enterprise 0); Lack of quality manpower is 31 (Micro Enterprise 23, Small Enterprise 7, Medium Enterprise 1); High-interest rates 21 (Micro Enterprise 4, Small Enterprise10, Medium Enterprise 7); Credit information gap is 17 (Micro Enterprise 9, Small Enterprise 7, Medium Enterprise1); Poor infrastructure is 16 (Micro Enterprise 7, Small Enterprise 8, Medium Enterprise1); High employee turnover rate is 14 (Micro Enterprise 5, Small Enterprise 4, Medium Enterprise 5); and the lowest is technological problem; and total is 7 (Micro Enterprise 4, Small Enterprise 3, Medium Enterprise 0).

Table 5.17 Descriptive statistics of 17 SDGs according to mean hierarchy (*N* = 56)

Subject	*N*	*Minimum*	*Maximum*	*Mean*	*Std. deviation*	*Rank according to mean value*
SDG No. 1: End poverty in all its forms	56	2.00	5.00	4.5714	0.59870	1
SDG No. 8: Promote inclusive economic growth	56	2.00	5.00	4.3929	0.62315	2
SDG No. 2: End hunger, achieve food security	56	2.00	5.00	4.1607	0.82631	3
SDG No. 3: Ensure healthy lives	56	2.00	5.00	3.9643	0.76192	4
SDG No. 17: Build strength partnership	56	2.00	5.00	3.8929	0.67900	5
SDG No. 5: Achieve gender equality	56	2.00	5.00	3.8393	0.75743	6
SDG No. 10: Reduce inequality among countries	56	2.00	5.00	3.7500	0.79201	7
SDG No. 16: Promote peaceful and inclusive societies	56	2.00	5.00	3.6964	0.80723	8
SDG No. 4: Ensure inclusive education	56	2.00	5.00	3.6071	0.88787	9
SDG No. 12: Ensure responsible consumption	56	2.00	5.00	3.5893	0.65441	10

(continued)

Table 5.17 (continued)

Subject	*N*	*Minimum*	*Maximum*	*Mean*	*Std. deviation*	*Rank according to mean value*
SDG No. 9: Build resilient infrastructure	56	2.00	5.00	3.4821	1.00889	11
SDG No. 11: Make cities safe and sustainable	56	2.00	5.00	3.4821	0.83101	12
SDG No. 6: Ensure availability of water, sanitation	56	2.00	5.00	3.3214	0.74118	13
SDG No. 7: Ensure access to sustainable energy	56	2.00	5.00	3.1607	0.84803	14
SDG No. 15: Protect terrestrial ecosystems	56	1.00	5.00	2.7679	0.89425	15
SDG No. 13: Urgent action to combat climate change	56	2.00	5.00	2.7321	0.88402	16
SDG No. 14: Conserve the ocean	56	1.00	5.00	2.5714	0.96967	17

Finally, Table 5.19 represents that Micro, Small, and Medium Enterprise Cross tabulation about Sustainable Development Goals (SDGs). Out of 243 respondents, they have knowledge total of 56 (Micro Enterprise 25, Small Enterprise 23, and Medium Enterprise 8) don't have any knowledge about SDGs, 187 (Micro Enterprise102, Small Enterprise 69, Medium Enterprise16).

Table 5.18 What is the important problem that you are facing? *Micro, small, and medium enterprise cross tabulation

Construct		*Micro, small, and medium enterprise*			*Total*
		Micro enterprise	*Small enterprise*	*Medium enterprise*	
What is the important problem that you are facing?	Financial problem	28	22	5	55
	High interest rates	4	10	7	21
	Poor infrastructure	7	8	1	16
	Credit information gap	9	7	1	17
	Marketing problem	22	18	0	40
	High employee turnover rate	5	4	5	14
	Quality assurance	25	13	4	42
	Technological problem	4	3	0	7
	Lack of quality manpower	23	7	1	31
Total		127	92	24	243

Table 5.19 Have you any knowledge about sustainable development goals (SDGs)? *Micro, small, and medium enterprise cross tabulation

Construct		*Micro, small, and medium enterprise*			*Total*
		Micro enterprise	*Small enterprise*	*Medium enterprise*	
Have You any knowledge about sustainable development goals (SDGs)?	Yes	25	23	8	56
	No	102	69	16	187
Total		127	92	24	243

Findings

SMEs are perceived as a promising sector to address the unemployment crisis by creating new jobs. By fostering and growing corporate activity, this sector has played a vital role in attaining economic growth and earning foreign money. According to the analysis of the study, it was found that SMEs create job opportunities and help to eliminate poverty a lot. But the SMEs in our country face some problems. The significance of SMEs is increasing day by day to achieve the SDGs. According to SME owners, this sector plays a vital role in reducing poverty and making a strong economy and claims that it is inevitable for sustainable economic growth and to touch the SDGs.

This study also found that most SME owners have no or little knowledge about SDG goals. Most of them perform their business because they want to be their own boss and don't like to work for others. Inexperience in business, lack of technical knowledge, poor managerial skills, lack of proper planning skills, market research skills, and technological changes are the major challenges in their present enterprise.

The most important problems are financial problems, high-interest rates, poor infrastructure, credit information gap, marketing problems of SME products, high employee turnover rate, product quality assurance, lack of quality manpower, etc. Direct competitors, indirect competitors, partner competitors, and perceived competitors are their main competitors. The COVID-19 pandemic hit their businesses hard. The SME owners claimed that the performance of their enterprises was not at a satisfactory level (according to annual turnover). The current SMEs have a small part to play in achieving a small part of the SDG goals. Because agro-based SMEs, fisheries SMEs, and some small manufacturing and service industry SMEs were taken into consideration as respondents. The main limitations are that some SME sectors, like green and ocean sectors, are not taken into consideration.

Conclusion and Recommendations

SMEs are now considered as important sectors in the economic development of the country. This sector has incredible potential to expand in rural and urban areas to boost up economic growth by tackling the problem of unemployment in Bangladesh. Sadly, SMEs are not expanding in rural Bangladesh due to limited access to real finance, lack of capital and skilled manpower, poor training facilities, underdeveloped sales channels, and low-level financial inclusion. Sustainable development can be seen as an integrated agenda and fundamental policy that seeks to address economic, social, and environmental challenges. Most of the SME owners have no or little knowledge about SDGs and they are not so aware about the contribution of SMEs on SDGs.

The government should take initiatives to solve the problems of SMEs, provide help to overcome the major challenges. Crate is an awareness program for SME owners related to SMEs and SDGs. A clear institutional structure with management and planning capacity, a participatory process, and regular financial discussions between the government and SME owners are crucial for prioritizing the SDGs and setting work plans accordingly. Innovative and holistic strategies need to be encouraged to expand SMEs and provide a conducive environment for income-generating and employment opportunities for the poor, as well as help to link sustainable economic development.

This chapter made many findings and recommendations and opens up a new avenue for long-term researchers to investigate. The SMEs and SDGs study might create a new research field for future researchers by collecting additional data from a larger sample size and allowing the government authorities' policymakers, academics, and development economists to implement the findings and strategies for achieving SDGs goals. So, this chapter gave insight to Bangladesh Small and Cottage Industries Corporation (BSCIC), banks and other financial institutions in Bangladesh in promoting the utilization of SMEs for achieving SDGs in South Asian developing countries like Bangladesh.

Appendix: Correlations Between Problems Faced by SME Owners and SDGS

	p	*1*	*2*	*3*	*4*	*5*	*6*	*7*	*8*	*9*
P	1	0.271*	0.090	−0.261	0.058	0.061	−0.116	−0.031	0.122	0.288*
1	0.271*	1	0.215	0.046	0.259	−0.155	−0.176	0.067	0.313*	0.288*
2	0.090	0.215	1	0.356**	0.360**	0.071	0.300*	0.248	−0.019	−0.029
3	−0.261	0.046	0.356**	1	0.436**	0.084	0.246	0.150	−0.161	−0.119
4	0.058	0.259	0.360**	0.436**	1	0.391**	0.416**	0.206	−0.012	0.398**
5	0.061	−0.155	0.071	0.084	0.391**	1	0.450**	0.211	0.175	0.317*
6	−0.116	−0.176	0.300*	0.246	0.416**	0.450**	1	0.495**	−0.160	0.251
7	−0.031	0.067	0.248	0.150	0.206	0.211	0.495**	1	0.050	0.248
8	0.122	0.313*	−0.019	−0.161	−0.012	0.175	−0.160	0.050	1	0.069
9	0.288*	0.288*	−0.029	−0.119	0.398**	0.317*	0.251	0.248	0.069	1
10	0.034	−0.077	−0.188	−0.105	−0.090	0.053	0.139	0.061	0.055	0.358**
11	−0.039	−0.125	0.044	0.028	0.114	0.443**	0.393**	0.327*	0.014	0.216
12	−0.052	−0.179	−0.010	−0.103	0.155	0.378**	0.352**	0.187	0.091	0.113
13	−0.340*	−0.118	0.085	0.282*	0.327*	0.206	0.300*	0.277*	0.128	0.066
14	−0.200	−0.009	0.110	0.250	0.392**	0.177	0.448**	0.329*	0.043	0.196
15	−0.287*	−0.087	0.150	0.415**	0.387**	0.373**	0.444**	0.362**	0.003	0.046
16	−0.094	0.027	−0.062	0.278*	0.236	0.335*	0.136	0.232	0.278*	0.116
17	−0.205	−0.070	−0.001	0.274*	0.200	0.213	0.070	−0.001	−0.157	0.050

	p	*10*	*11*	*12*	*13*	*14*	*15*	*16*	*17*
P	1	0.034	−0.039	−0.052	−0.340*	−0.200	−0.287*	−0.094	−0.205
1	0.271*	−0.077	−0.125	−0.179	−0.118	−0.009	−0.087	0.027	−0.070
2	0.090	−0.188	0.044	−0.010	0.085	0.110	0.150	−0.062	−0.001
3	−0.261	−0.105	0.028	−0.103	0.282*	0.250	0.415**	0.278*	0.274*
4	0.058	−0.090	0.114	0.155	0.327*	0.392**	0.387**	0.236	0.200
5	0.061	0.053	0.443**	0.378**	0.206	0.177	0.373**	0.335*	0.213
6	−0.116	0.139	0.393**	0.352**	0.300*	0.448**	0.444**	0.136	0.070
7	−0.031	0.061	0.327*	0.187	0.277*	0.329*	0.362**	0.232	−0.001
8	0.122	0.055	0.014	0.091	0.128	0.043	0.003	0.278*	−0.157
9	0.288*	0.358**	0.216	0.113	0.066	0.196	0.046	0.116	0.050

(continued)

(continued)

	p	*10*	*11*	*12*	*13*	*14*	*15*	*16*	*17*
10	0.034	1	0.186	0.184	0.084	0.166	0.071	−0.036	−0.017
11	−0.039	0.186	1	0.404**	0.179	0.193	0.374**	0.195	0.061
12	−0.052	0.184	0.404**	1	0.309*	0.319*	0.269*	0.035	0.145
13	−0.340*	0.084	0.179	0.309*	1	0.606**	0.518**	0.266*	0.254
14	−0.200	0.166	0.193	0.319*	0.606**	1	0.596**	0.272*	0.150
15	−0.287*	0.071	0.374**	0.269*	0.518**	0.596**	1	0.379**	0.258
16	−0.094	−0.036	0.195	0.035	0.266*	0.272*	0.379**	1	0.271*
17	−0.205	−0.017	0.061	0.145	0.254	0.150	0.258	0.271*	1

Note P = Problems, 1 = SDG No. 1; 2 = SDG No. 2; 3 = SDG No. 3; 4 = SDG No. 4; 5 = SDG No. 5; 6 = SDG No. 6; 7 = SDG No. 7; 8 = SDG No. 8; 9 = SDG No. 9; 10 = SDG No. 10; 11 = SDG No. 11; 12 = SDG No. 12; 13 = SDG No. 13; 14 = SDG No. 14; 15 = SDG No. 15; 16 = SDG No. 16; 17 = SDG No. 17. (** Correlation is significant at the 0.01 level (2-tailed).* Correlation is significant at the 0.05 level (2-tailed))

Appendix shows the relationship between SME problems and SDG goals. The correlation value highlighted that SDG No. 1, 2, 4, 5, 8, 9, and 10 are positively correlated with SME problems to achieve SDGs and SDG No. 3, 6, 7, 11, 12, 13, 14, 15, 16, and 17 are not positively correlated with SME problems to gain SDGs. Because in this research, agro-based SMEs, fisheries SMEs, and some small manufacturing and service industry SMEs took consideration as respondents. The main limitations are that some SME sectors, like green and ocean sectors, are not taken into consideration. But this sector has a great impact to achieve the SDGs.

References

Abdin, M. J. (2019, August 18). MSMEs—Both a choice and a reality for Bangladesh. *The Financial Express*. https://thefinancialexpress.com.bd/views/views/msmes-both-a-choice-and-a-reality-for-bangladesh-1566055028

Acma, M. Q. (2015). Productivity and performance evaluation of SME sector in Bangladesh: Evidence from the historical data. *Journal of Islamic Finance and Business Research, 3*(1), 14–22.

Ahmed, K., & Chowdhury, T. A. (2009). Performance evaluation of SMEs of Bangladesh. *International Journal of Business and Management, 4*(7), 126–133.

Ahmed, M. U., Mannan, M., Razzaque, A., & Sinha, A. (2004). *Taking stock and charting a Path for SMEs in Bangladesh*. Bangladesh Enterprise Institute.

Alauddin, M., & Chowdhury, M. M. (2015). Small and medium enterprise in Bangladesh-Prospects and challenges. *Global Journal of Management and Business Research, 15*(C7), 1–9.

Alauddin, M., Rahman, M. Z., & Rahman, M. (2015). Investigating the performance of SME sector in Bangladesh: An evaluative study. *International Journal of Small Business and Entrepreneurship Research, 3*(6), 14–28.

Ali, M. R., & Islam, A. (2018). Present status of SMEs and SME financing in Bangladesh: An overview. *Journal of Science and Technology, 8*(1), 55–71.

Andalib, T. W., & Halim, H. A. (2019). Convergence of conceptual innovation model to reduce challenges faced by the small and medium sized enterprises'(SMEs) in Bangladesh. *Journal of Open Innovation: Technology, Market, and Complexity, 5*(3), 63.

Androniceanu, A. (2019). Social responsibility, an essential strategic option for a sustainable development in the field of bio-economy. *Amfiteatru Economic, 21*(52), 503–519.

Angulo-Guerrero, M. J., Pérez-Moreno, S., & Abad-Guerrero, I. M. (2017). How economic freedom affects opportunity and necessity entrepreneurship in the OECD countries. *Journal of Business Research, 73*, 30–37.

Avasilicai, S. (2009). Entrepreneurship: applied research. *Cluj-Napoca: Tedosco*, 123–145.

Bădulescu, D. (2010). SMEs financing: The extent of need and the responses of different credit structures. *Theoretical & Applied Economics, 17*(7), 25–36.

Baumol, W. J. (2014). *The free-market innovation machine*. Princeton University Press.

Beck, T., Demirguc-Kunt, A., & Levine, R. (2005). SMEs, growth, and poverty: Cross-country evidence. *Journal of Economic Growth, 10*(3), 199–229.

Biryukov, E., Elina, O., Lyandau, Y., & Mrochkovskiy, N. (2021). *Russian SMEs in achieving sustainable development goals*. Paper presented at the E3S Web of Conferences.

Bosri, R. (2016). SME financing practices in Bangladesh: Scenario and challenges evaluation. *World, 6*(2), 39–50.

Calvino, F., & Virgillito, M. E. (2018). The innovation-employment nexus: A critical survey of theory and empirics. *Journal of Economic Surveys, 32*(1), 83–117.

CAFOD. (2013). *Big Ideas from small entrepreneurs* (p. 10). http://www.cafod.org.uk/Media/Files/Resources/Policy/Thinking-small

Cherroun, R., & Cherroun, H. (2022). Contribution of SMEs in realizing sustainable development goals. *Journal of Economic Growth and Entrepreneurship JEGE, 5*(1), 1–10.

Chowdhury, M. S. A., Azam, M. K. G., & Islam, S. (2013). Problems and prospects of SME financing in Bangladesh. *Asian Business Review, 2*(2), 109–116.

CPSD. (2021, June). *Unleashing the private sector to sustain development success, country private sector diagnostic creating markets in Bangladesh*. World Bank Group.

Cunningham, L. X., & Rowley, C. (2008). The development of Chinese small and medium enterprises and human resource management: A review. *Asia Pacific Journal of Human Resources, 46*(3), 353–379.

Dobreva, J., & Ilieva-Koleva, D. (2015). *Managing sustainable enterprises and promoting open innovation in Bulgaria*. Paper presented at the The 8th International Conference for Entrepreneurship, Innovation and Regional Development.

Endris, E., & Kassegn, A. (2022). The role of micro, small and medium enterprises (MSMEs) to the sustainable development of sub-Saharan Africa and its challenges: A systematic review of evidence from Ethiopia. *Journal of Innovation and Entrepreneurship, 11*(1), 1–18.

Fayomi, O., Okokpujie, I., & Udo, M. (2018). *The role of research in attaining sustainable development goals.* Paper presented at the IOP Conference Series: Materials Science and Engineering.

Grochowski, E. M., Warschat, J., & Dasher, R. (2016). *The impact of collaboration with big companies on entrepreneurial technology innovation.* Paper presented at the 2016 Portland International Conference on Management of Engineering and Technology (PICMET).

Haider, M. Z., Dutta, C. B., & Siddiqui, O. I. (2015). Role of Bangladesh small and cottage industries corporation in poverty reduction. *South Asia Economic Journal, 16*(1), 103–121.

Hamburg, I. (2020). Facilitating lifelong learning in SMEs towards SDG4. *Advances in Social Sciences Research Journal, 7*, 273–282.

Haque, M. M., & Ahlan, A. R. (2018). Can ICT reduce poverty and unemployment and elevate development integrating micro, small, and medium enterprises in Bangladesh?-Sharing Asian experiences. *International Journal of Engineering and Management Research (IJEMR), 8*(4), 168–174.

Harel, R. (2021). The impact of COVID-19 on small businesses' performance and innovation. *Global Business Review.* https://doi.org/10.1177/09721509211039145

Haryati, R., Yasri, Y., Aimon, H., & Darwin, M. (2021). Development of small, micro enterprises based (SMEs) on innovation and environmental sustainable development in West Sumatera. *International Journal of Entrepreneurship, 25*(4), 1–13.

Hasan, F., & Jamil, G. M. H. (2014). Financing small and medium enterprises in Bangladesh-issues and challenges. *The Asian Journal of Technology Management, 7*(1), 45.

Hasan, R. A., & Islam, K. Z. (2008). Bridging the gap between microfinance and SME financing in Bangladesh: Unlocking the potentials. *Daffodil International University Journal of Business and Economic, 3*(1), 41–57.

Helmsing, A. (2003). Local economic development: New generations of actors, policies and instruments for Africa. *Public Administration and Development: The International Journal of Management Research and Practice, 23*(1), 67–76.

Hossain, M. R., Akhter, F., & Sultana, M. M. (2022). SMEs in COVID-19 crisis and combating strategies: A systematic literature review (SLR) and A case from emerging economy. *Operations Research Perspectives, 9*, 100222.

ICAB. (2021, June 27). Institute of Chartered Accountants of Bangladesh (ICAB) organized the Global Webinar on 'Future-ready SMEs and SMPs: Vital for sustainable economy' on Sunday, 27 June 2021 to mark International SMEs/SMPs day 2021. https://www.icab.org.bd/news/sme%E2%80%99s-contribution-to-gdp-is-very-low-in-comparison-to-other-emerging-economies

Imran, M., Haq, S. U., & Ozcatalbas, O. (2022). Role of microcredit in sustainable rural development. In *Sustainable rural development perspective and global challenges*. IntechOpen.

Islam, R., Khan, S. R., & Siddiqua, K. A. (2013). Present scenario and stratagem to SMEs development in Bangladesh. *Jagannath University Journal of Business Studies, 3*(1–2), 1–13.

Jahur, M. S. (2018). SMEs in Bangladesh—Prospects and challenges. *Journal of Business and Society (JBS), 9*, 16–43.

Jiménez, E., de la Cuesta-González, M., & Boronat-Navarro, M. (2021). How small and medium-sized enterprises can uptake the sustainable development goals through a cluster management organization: A case study. *Sustainability, 13*(11), 5939.

Karpak, B., & Topcu, I. (2010). Small medium manufacturing enterprises in Turkey: An analytic network process framework for prioritizing factors affecting success. *International Journal of Production Economics, 125*(1), 60–70.

Khatun, F., Bhattacharya, D., Rahman, M., Moazzem, K. G., Khan, T. I., Sabbih, M. A., & Saadat, S. Y. (2020). *Four years of SDGs in Bangladesh: Measuring progress and charting the path forward*. Centre for Policy Dialogue.

Khatun, F., & Saadat, S. Y. (2021). *Implementation of the SDGs in Bangladesh: Domestic challenges and regional consideration.*

Le Blanc, D. (2015). Towards integration at last? The sustainable development goals as a network of targets. *Sustainable Development, 23*(3), 176–187.

Manzoor, F., Wei, L., & Sahito, N. (2021). The role of SMEs in rural development: Access of SMEs to finance as a mediator. *PLoS ONE, 16*(3), e0247598.

Massaro, M., Rubens, A., Bardy, R., & Bagnoli, C. (2017). Antecedents to export performance and how Italian and Slovanian SME's innovate during times of crisis. *Journal of Eastern European and Central Asian Research, 4*(1), 22.

Masuduzzaman, M., Akther, M., & Ara, S. (2018). Efficiency of SMEs financed under conventional and Islamic banks: Bangladesh perspective. *Journal of Islamic Economics, Banking and Finance, 113*(6219), 1–10.

Meyer, N., & de Jongh, J. (2018). The importance of entrepreneurship as a contributing factor to economic growth and development: The case of selected European countries. *Journal of Economics and Behavioral Studies, 10*(4(J)), 287–299.

MoF. (2022). *Bangladesh Economic Review*. Ministry of Finance.

Mukole, K. (2010). Job creation versus job shedding and the role of SMEs in economic development. *African Journal of Business Management, 4*(11), 2288–2295.

Nilsson, M., Griggs, D., & Visbeck, M. (2016). Policy: Map the interactions between sustainable development goals. *Nature, 534*(7607), 320–322.

Noranarttakun, P., & Pharino, C. (2021). Strategic implementation to enhance green industry practices in SMEs: Lesson learned from Thailand. *EnvironmentAsia, 14*(1), 93–105.

Oldsman, E. & Hallberg, K. (2002). "Framework for Evaluating the Impact of Small Enterprise Initiatives," SED Working Paper No. 3. Bern, Swiss Agency for Development and Cooperation.

Patwary, A. (2020). Private sector SMEs in realizing SDG 1-'no poverty' in the context of Bangladesh. *International Journal of Humanities, Arts and Social Sciences, 6*(6), 274–288.

Pervin, M. T., Sarker, B. K., & Begum, R. (2020). Challenges and benefits of SMEs in Bangladesh to adopt open innovation: A quantitative study. *GSJ, 8*(4), 1465–1477.

Rahman, M. M., & Khondkar, M. (2020). Small and medium enterprises (SME) development and economic growth of Bangladesh: A narrative of the glorious 50 years. *Barishal University Journal of Business Studies, 7*(1), 9–24.

Read, R., & Staines, A. (2004). *The emerging role of small firms in overcoming economic vulnerability.* Paper presented at the Taiwan: University of Lancaster, UK Presented at Islands of the World, VIII International Conference.

Reeg, C. (2013). *Micro, small and medium enterprise upgrading in low-and middle-income countries: A literature review* (Discussion Paper).

Reim, W., Yli-Viitala, P., Arrasvuori, J., & Parida, V. (2022). Tackling business model challenges in SME internationalization through digitalization. *Journal of Innovation & Knowledge, 7*(3), 100199.

Roy, I. (2016). The socio-economic contribution of women entrepreneurs in Bangladesh. *International Journal of Economics, Commerce and Management, 4*(6), 1–12.

Schmidheiny, S., & Stigson, B. (2000). *Eco-efficiency: creating more value with less impact*. World Business Council for Sustainable Development.

Šebestová, J., & Sroka, W. (2020). Sustainable development goals and SMEs decisions: Czech Republic vs. Poland. *Journal of Eastern European and Central Asian Research (JEECAR), 7*(1), 39–50.

Siekelova, A., Kovacova, M., Adamko, P., & Stehel, V. (2019). Profit management as an instrument for SMEs developing: The case for Slovakia. *Marketing and Management of Innovations, 3*, 285–296.

Shoma, C. D. (2019). Financing female entrepreneurs in cottage, micro, small, and medium enterprises: Evidence from the financial sector in Bangladesh 2010–2018. *Asia & the Pacific Policy Studies, 6*(3), 397–416.

Tonis, R. (2015). SMEs role in achieving sustainable development. *Journal of Economic Development, Environment and People, 4*(1), 41–50.

Uddin, M. T. (2014). A study on financing of SMEs in Bangladesh. *Journal of Economics and Sustainable Development, 5*(11), 161–168.

United Nations. (2015). *Transforming our world: The 2030 agenda for sustainable development* (No. A/RES/70/1). United Nations.

Urbano, D., Aparicio, S., & Audretsch, D. (2019). Twenty-five years of research on institutions, entrepreneurship, and economic growth: What has been learned? *Small Business Economics, 53*(1), 21–49.

Vasilescu, M. D., Dimian, G. C., & Gradinaru, G. I. (2022). Green entrepreneurship in challenging times: A quantitative approach for European countries. *Economic Research-Ekonomska Istraživanja, 36*(1), 1828–1847.

Velinov, E., & Ponomarev, M. (2016). Organizational development of Russian SMEs: current trends. *Journal of Eastern European and Central Asian Research, 3*(2), 10.

Verma, T. L. (2019, April 1). *Role of micro, small and medium enterprises (MSMES) in achieving sustainable development goals*. Small And Medium Enterprises (MSMEs) In Achieving Sustainable Development Goals.

Yousaf, Z., Radulescu, M., Sinisi, C. I., Serbanescu, L., & Păunescu, L. M. (2021). Towards sustainable digital innovation of SMEs from the developing countries in the context of the digital economy and frugal environment. *Sustainability, 13*(10), 5715.

Zindiye, S., Chiliya, N., & Masocha, R. (2012). The impact of government and other institutions' support on the performance of small and medium enterprises in the manufacturing sector in Harare, Zimbabwe. *International Journal of Business Management & Economic Research, 3*(6), 655–667.

PART II

Particular areas of the SDGs

CHAPTER 6

Africa's Performance with Regard to SDGs and the Important Role of MSMEs

Sazir Mayanja Nsubuga

Introduction

SDGs are a culmination of efforts, especially at the global level, to ensure that humankind's well-being is attained through growth and development, without undue destruction of the ecosystem. Most activities in producing goods and providing services are achieved through MSMEs (micro, small and medium-scale enterprises). They are the backbone of the global production system, accounting for 99% of firms, 70% of jobs and over 50% of value added worldwide (James & Massimo, 2021). A decade ago there were about 365–445 million MSMEs in emerging markets. Of these 25–30 million were formal MSMEs, 55–70 million were formal micro, and 285–345 million were informal enterprises (Kushnir et al., 2010). Development must be sustainable, without mortgaging the future. Production must be for sustaining human

S.M. Nsubuga (✉)
Faculty of Business and Management, St. Peter's University, Kampala, Uganda
e-mail: saznsumay@gmail.com

H. Dasaraju and T.T.H. Tambunan (eds.), *Role of Micro, Small and Medium Enterprises in Achieving SDGs*, Sustainable Development Goals Series, https://doi.org/10.1007/978-981-99-4829-1_6

consumption characterized by patterns today that do not undermine human development tomorrow (UNDP, 1998).

This chapter discusses the evolution of the sustainability concept, the emergence of Millennium Development Goals (MDGs) and how they evolved into the SDGs. Africa's performance regarding some of the SDGs is discussed as is the role of MSMEs and microfinance on which many of them rely for survival and growth.

Evolution of Sustainable Development Goals

The expression "Sustainable Development" is often attributed to Barbara Ward Jackson, who presided over a special meeting organized by UNCTAD and UNEP, held on October 8, 1974, in Cocoyoc, Mexico. It dwelt on matters related to technological and economic development in member states of the United Nations Organisation. It became increasingly used globally in 1987 after the World Commission on Environment and Development or Brundtland Report was authored. However, its origin is traced to the professional terminology in forestry of 'sustained yield' translated from the German term 'nachhaltiger Ertrag' used about three centuries ago. At that time foresighted leaders of the Kingdom of Saxony wanted to conserve forests, which they considered valuable resources, but whose existence they deemed extremely threatened by human activity at that time. They wanted future generations to enjoy the benefits of these natural resources like them. This would only be possible if the ecosystem was preserved. The challenges faced then and those anticipated in the future caused a network of experts from different European nations to make effort to find solutions. In response, Hanns Carl von Carlowitz (1645–1714), a German nobleman, wrote the earliest and most comprehensive book on the sustainability of the environment, the 'Sylvicultura Oeconomica' in 1713. The moral of the 400-page book was that getting by with less despite wanting more should be the guiding principle of life. As drivers of growth and development, MSMEs dominate economic activity, no matter the level of development.

How MDGs Transformed to SDGs

In 2000 Member States of the UN General Assembly issued the Millennium Declaration, according to which they undertook collective responsibility to uphold the principles of human dignity, equality and equity at the global level and to eradicate the world's most extreme and deplorable conditions, including poverty and destitution. There were eight goals (MDGs) in the declaration. Their adoption in 2002 marked a historic effort to mobilize resources to achieve a set of common societal priorities. Fifteen years later, the likelihood of a child dying before age five had decreased by half and global maternal mortality ratio dropped by 45%, among many other achievements (UNO, 2015). Nevertheless, many countries had not registered sufficient progress with regard to most of the eight goals. It was also the consensus of many members that the scope of the MDGs was narrow. Accordingly in 2015, member states of the United Nations Organisation adopted the Sustainable Development Goals (SDGs), also known as the Global Goals, to replace the eight MDGs, to be achieved by 2030. The goals were more than doubled to seventeen. At the center of SDGs is focus on the people, planet, peace, prosperity and partnerships. Sustainable development is essentially anchored on three pillars, namely the environment, economy and society (Mensah, 2019). Many natural disasters occurred between 2000 and 2105 which raised concerns among member states. Three SDGs directly relate to environmental protection. SDG13 is about taking urgent action to combat climate change and its impacts. SDG 14 emphasizes conserving and sustainably using the oceans, seas and marine resources for sustainable development. SDG 15 is about protecting, restoring, and promoting sustainable use of terrestrial ecosystems, sustainably managing forests, combating desertification, and halting and reversing land degradation, and halting biodiversity loss.

Africa's Performance with Regard to SDGs

The United Nations Development Program (UNDP) introduced the concept of "human development" and its first annual Human Development Report (HDR) was published in 1990. Its concept of human development was presented as progress toward greater human well-being. The UNDP has named the metric to measure well-being as the Human Development Index (HDI). It tries to establish average achievement in three basic dimensions, namely, standard of living, education and health, which are assessed by three indicators—Gross National income (GNI) per capita (expressed as purchasing power parity in $), expected years of schooling and life expectancy. An important assertion made by the UNDP from the start and one which it has maintained all along is that: *People are the real wealth of a nation. The basic objective of development is to create an enabling environment for people to enjoy long, healthy and creative lives. This may appear to be a simple truth. But it is often forgotten in the immediate concern with the accumulation of commodities and financial wealth* (UNDP, 1990: 9).

In Sub-Saharan Africa, on average, social safety protection covers only 17.8% of the population. In West Africa, it can be as low as 10% in some countries. In South Africa, it is about 48%. Other countries doing fairly well are Botswana, Cape Verde, Lesotho, Mauritius, Namibia, Seychelles, Eswatini and Zanzibar (United Republic of Tanzania). The situation remains dire for many of the countries on the continent (ILO, 2017). The last decade registered an increase in the expenditure committed to education. Africa, the Caribbean, Latin America and the Middle East spent a modest 3% in relation to their GDP as an investment in education. This is lower than expenditure by most of the emerging economies where investment averaged 4% of GDP. The Western Hemisphere shows a relatively high level of public spending per student in tertiary education, but together with South Asia, it lags behind other regions regarding primary and secondary education. Regarding expected years of schooling, countries with high and very high HDI reached the level of 14–16 years, while the countries with medium and low HDI—10 to 12 years (UNDP, 2016). Overall, the performance of African countries with regard to SDGs has been less than satisfactory since the beginning. Below are several tables summarizing HDI performance by region and some countries.

Table 6.1 reveals that Africa's performance with regard to selected, and fairly representative SDG indices, is dismal. For life expectancy at birth, expected years of schooling, mean years of schooling and GNI per capita, the Organisation for Economic Cooperation and Development (OECD) registered 0.899, 79.0, 16.5, 9.1, and $45,087 respectively. For Sub-Saharan Africa, the corresponding figures are 0.547, 67.9, 10.3, 6.0, and $ 3,669. It should be noted that higher life expectancy is the most sought for goal for human kind. The general quality of education of a nation has much bearing on the attainment of all the other SDGs. For example, it increases opportunities of being employed, fighting poverty, conserving the environment, increasing gender equity.

Table 6.2 disaggregates the summary indicating the performance of what have been referred to as very high human development countries (with regard to HDI). They almost score a perfect 1. Their GNI per

Table 6.1 Global performance summary with regard to selected SDGs 2021

Category	*Human development index HDI 2021*	*SDG 3 Life expectancy at birth*	*SDG 4.3 Expected years of schooling*	*SDG 4.4 Mean years of schooling*	*SDG 8.5 Gross net income (GNI) per capita (at 2017 PPP $*
Very high human development	0.896	78.5	16.5	12.3	43,752
High human development	0.754	74.7	14.2	8.3	15,167
Medium human development	0.636	67.4	11.9	6.9	6353
Low human development	0.518	61.3	9.5	4.9	3009
Developing countries	0.685	69.9	12.3	7.5	10,704
Region					
Arab states	0.708		12.4	8.0	13,501
East Asia and Pacific countries	0.749	70.9	13.8	7.8	15,580
Europe and Central Asia	0.796	75.6	15.4	10.6	19,352
Latin America and the Caribbean	0.754	72.9	14.8	9.0	14,521
South Asia	0.632	72.1	11.6	6.7	6481
Sub-Saharan	**0.547**	**67.9**	**10.3**	**6.0**	**3669**
Least Developed countries	0.540	64.2	10.2	9.0	2881
Small island developing countries	0.730	70.3	12.4	5.2	16,872
OECD	**0.899**	**79.0**	**16.5**	**9.1**	**45,087**
World	0.732	71.4	12.8	8.61	1752

Source Human Development Report 2021

capita ranges from about \$ 50,000 to \$ 66,000. Their life expectancy averages above eighty years.

Table 6.3 gives a summary of the worst performing countries. Out of the 33 countries, it is only Pakistan, Haiti and Afghanistan that are not from Africa.

Table 6.2 Very high human development countries

Country	*Human development index HDI 2021*	*SDG 3 Life expectancy at birth*	*SDG 4.3 Expected years of schooling*	*SDG 4.4 Mean years of schooling*	*SDG 8.5 Gross net income (GNI) per capita (at 2017 PPP $*
Switzerland	0.962	84.0	16.5	12.3	66,933
Norway	0.954	83.2	18.2	8.3	64,660
Iceland	0.959	82.7	13.9	6.9	55,782
Hong Kong, China SAR	0.952	85.5	17.3	12.2	62,607
Australia	0.951	84.5	21.1	12.7	49,238

Source Human Development Report 2021

MSMEs in the Achievement of SDGs in Africa

MSMEs provide the means through which economic growth and development are attained, in all countries. In this section, some aspects of the seventeen SDGs are discussed, namely, poverty alleviation, employment and innovation, especially with regard to their contribution to the achievement of SDGs in the continent. The three are interrelated in that poverty alleviation requires that the population has what to do (employment). To survive and grow in very competitive markets, MSMEs must be innovative.

Poverty Alleviation (SDG 1)

The first SDG is the reduction of poverty. The World Bank defines poverty as the lack or insufficiency of money to meet basic needs, including food, clothing and shelter. However, poverty is much more than the mere lack of money. It is also about deprivation in other important areas of well-being such as education, health, water and housing. Extreme ***poverty*** is measured as the number of people living on less than $1.90 per day (World Bank, 2016b). Different business scales offer various implications for poverty reduction. Micro and small enterprise play a bigger role in alleviating poverty than medium-sized ones. The former reduces not only the percentage of poor people but also the Poverty Gap and Severity Index (Nursini, 2020). Research in 76 countries, established that there is

Table 6.3 Low human development

Category	*Human development index HDI 2021*	*SDG 3 Life expectancy at birth*	*SDG 4.3 Expected years of schooling*	*SDG 4.4 Mean years of schooling*	*SDG 8.5 Gross Net Income (GNI) per capita (at 2017 PPP $*
Tanzania	0.549	66.2	9.2	6.4	2264
Pakistan	0.544	66.1	8.7	4.5	4624
Togo	0.539	61.6	13.0	5.0	2167
Haiti	0.535	63.2	9.7	5.6	2848
Nigeria	0.535	52.7	10.1	7.2	4790
Rwanda	0.534	66.1	11.2	4.4	2210
Benin	0.525	59.8	10.8	4.3	3409
Uganda	0.525	62.7	10.1	5.7	2131
Lesotho	0.514	53.1	12.0	6.0	2700
Malawi	0.512	62.9	12.7	4.5	1466
Senegal	0.511	67.1	9.0	2.9	3344
Djibouti	0.509	62.3	7.4	4.1	5025
Sudan	0.508	65.3	7.9	3.8	3575
Madagascar	0.501	64.5	10.1	5.1	1484
Gambia	0.500	62.1	9.4	4.6	2172
Ethiopia	0.498	65.0	9.7	3.2	2361
Eritrea	0.492	66.5	8.1	4.9	1729
Guinea-Bissau	0482	59.7	10.6	3.6	1908
Liberia	0.481	60.7	10.4	5.1	1908
Congo (DRC)	0.479	59.2	9.8	7.0	1289
Afghanistan	0.472	62.0	10.3	3.0	1824
Sierra Leone	0.477	60.9	9.6	4.6	1622
Guinea	0.465	58.9	9.8	2.2	2401
Yemen	0.455	63.8	9.1	3.2	1314
Burkina Faso	0.449	59.3	9.1	2.1	2118
Mozambique	0.446	59.3	10.2	3.2	1198
Mali	0.428	58.9	7.4	2.3	2133
Burundi	0.426	61.7	10.7	3.1	732
Central African Republic	0.404	53.9	8.0	4.3	966
Niger	0.400	61.6	7.0	2.1	1240
Chad	0.394	52.5	8.0	2.6	1364
South Sudan					

Source Human Development Report 2021

a strong association between the activities of MSMEs and GDP per capita growth. It was concluded that there is a relationship between the relative size of the MSME sector, economic growth and poverty using a new database on the share of SME labor in the total manufacturing labor force (Beck et al., 2003).

Most of Africa is mainly agricultural. In Uganda, like in most African countries, micro and small farm holdings dominate the agricultural sector. The country witnessed a drop in extreme poverty from 53.2% in 2006 to 34.6% in 2013. Agriculture has the potential to create lucrative livelihoods and lift thousands of Ugandans out of poverty, especially with the adoption of modern techniques and better-quality inputs (World Bank, 2016a).

MSMEs and Unemployment (SDG 8)

Poverty and unemployment have a predictable relationship. When the majority of the population have what makes them gainfully engaged such that they have purchasing power or even discretionary income, they are able to fulfil the basic necessities of life. By definition, they are not poor. Having something to do and making an earning out of it is a privilege that unemployed people lack (Toran et al, 2020).

In Africa, MSMEs account for approximately 80% of jobs. There are over 45 million MSMEs in the continent and they are a significant mechanism for socioeconomic growth (CSIS, 2021). MSMEs are key enablers of job creation and livelihood generation. More than 90% of all businesses globally are in this category, contributing more than half of all employment and more than a third of the GDP. Findings in empirical studies have shown that MSMEs contribute to over 55% of GDP and over 65% of total employment in high-income countries. Formal and informal MSMEs account for over 60% of GDP and over 70% of total employment in low-income countries. They contribute about 70% of the GDP and 95% of total employment in middle-income countries (Kongolo, 2010).

A big percentage of MSMEs are in the informal (nonfarm sector), especially in the least developed economies. Nevertheless, their contribution to GDP and poverty reduction has been substantial (AFDB, 2019; ILO, 2015). Although unemployment is a global challenge, it is more prominent in developing countries than the developed ones. This is partly because of the high rate of growth in the working-age population in these countries. For example in Africa, it is increasing by 265.8% in Africa compared to the global average of 28.3%. The rate is expected to increase by 306.6% in sub-Saharan Africa, by 2030 (Bhorat & Oosthuizen, 2020). The continent is thus in a precarious position regarding unemployment, thus the need for an enabling environment, including policies, legal and regulatory framework as well as political stability for MSMEs to thrive and offer employment opportunities (Dey, 2012). Vibrant entrepreneurship in the MSME will go a long way in providing solutions to the challenges of unemployment and economic growth among others. Many of the benefits which are envisaged as sustainable development goals will be attained as a result of deliberate efforts in this direction (Devine & Kiggundu, 2016). Being labor-intensive, MSMEs are crucial in solving the dilemmas paused by unemployment by supporting decent job creation and entrepreneurship (SDG 8.3 and 9.3) (Brixiova et al., 2020). In order for the MSMEs to play a key role in job creation, policymakers must ensure that local infrastructure is conducive, for example, public utilities such as good road network and availability of reliable sources of energy, an appropriate legal and regulatory framework and access to finance (EDFI, 2016). In many cases, entrepreneurial abilities in low-income countries are spurred by lack of employment and typically in such cases the enterprises are not highly productive. This is because individuals are forced into starting their own businesses and lack creativity (Doran et al., 2018).

Inclusive Sustainable Economic Growth (SDG 8).

The promotion of sustained, inclusive, and sustainable economic growth, full and productive employment and decent work for all requires a sustainable environment.

Small-scale food producers make up 90% of farmers worldwide (Edelman, 2014). Agribusinesses and smallholder farmers have a critical role to play in conserving land-based ecosystems. Both of these are rapidly depleting and degrading the world's soil, water and biodiversity. Environmental damages caused by industrial and agricultural activities cost the world US $3 trillion each year, according to the United Nations. It is, therefore, necessary to adopt agroecological methods through capacity building of small-scale producers that can foster biodiversity, natural soil fertility, water conservation and biological control of insects and shift food production to a sustainable path (UNO, 2019).

The interrelated nature of the human activity and the environment is aptly explained by Kate Raworth model. It defines parameters of the environment and social well-being, and in this way elaborated on the optimum size and mix of the three circles. It is argued that just as there is an environmental ceiling, beyond which lies unacceptable environmental degradation, so too there is a social foundation, below which lies unacceptable human deprivation. The arguments are summarized in Fig. 6.1. Given that MSMEs dominate human activity, they present a great challenge in managing environmental sustainability (Raworth, 2012).

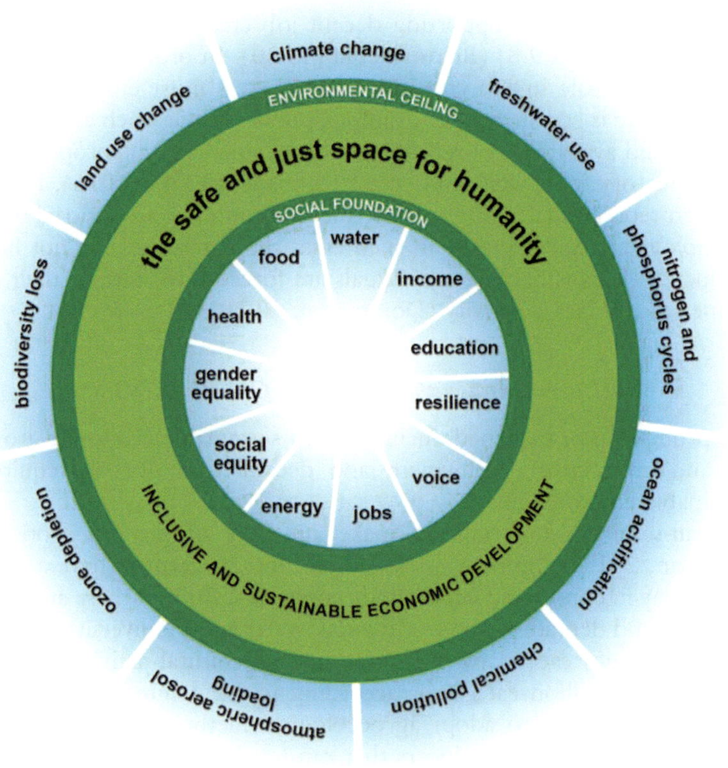

Fig. 6.1 Planetary and social boundaries (*Source* Raworth, 2012, Planetary and social boundaries: a safe and just space for humanity)

Decoupling growth from environmental degradation, and shifting to low-carbon, climate-resilient development pathways are key concerns in The 2030 Agenda for Sustainable Development and Sustainable Development Goals (SDGs), together with the Paris Agreement on Climate Change and the Addis Ababa Action Agenda on Financing for Development. Given their immense contribution to GDP and activities geared toward transforming the livelihoods of human kind, there is need for policies that help MSMEs become more resource-efficient and investment in R&D, technology upgrading, and employee training to ensure that the digital transformation benefits the low-carbon transition. This needs a support framework requiring broader emphasis on investment, trade, MSME and environmental policies that support the business case for green entrepreneurship and lower the additional entry costs associated with ensuring compliance with green standards and production processes. The concept of "born green via digital" refers to those entrepreneurs and MSMEs that invest and trade in green products and services and integrate responsible business conduct into their business model (James & Massimo, 2021).

Formulation and implementation of policies and programs that support MSMEs in becoming greener will help to ensure long-term economic growth achieved in a health of its environment: (OECD, 2014).

There are two broad categories of green MSMEs, namely: green innovators, which are developing new products, technologies and approaches that can have transformational impacts, and green performers, the vast majority of SMEs, which can take steps to make their operations more resource-efficient and environmentally friendly (UNEP, 2017). Governments have an important role to play, not just in providing tools to SMEs for becoming greener but also in helping to expand the market for green products in order to help enterprises make a better business case for investing in greener practices (OECD, 2018).

Promote Inclusive and Sustainable Industrialization and Foster Innovation (SDG 9)

(i) The innovativeness concept is about improving or doing things better to achieve a competitive edge or positioning among competitors. It may relate to inputs, activities, capabilities and outputs (Carvalho et al, 2020). MSMEs need to be innovative to survive, make a profit and grow in a highly competitive environment.

MSMEs' competitiveness is related to sustainability, information knowledge and network-open innovation. Deliberate efforts to innovate in business models can significantly modify products, operations processes or processes, distribution methods, marketing methods, management methods as well as all types concurrently (Müller et al., 2018). In Africa, MSMEs are essentially enabling invention, innovation, and the creation of new ideas and technologies. They are providing for pre-incubation, incubation, introduction and commercialization of innovation and technology into the market. They are thus important in providing a platform for creating and testing new products before they can be upscaled and disseminated into the larger industry through the macroeconomic systems. It is therefore not surprising that in countries like Uganda, Kenya and Ethiopia, up to 90% of the working population are employed within MSMEs (Facundo & Sergio, 2017).

(ii) MSMEs and Innovation for Financial Inclusion

A major technological innovation that has transformed the performance of MSMEs is mobile money technology. Its history can be traced back to M-PESA in Kenya. It was developed by Vodafone, a telecommunications service provider, and first deployed by its Kenyan affiliate Safaricom. By end of 2017, approximately 50% of informal business owners globally had a mobile money account or an account at a bank or other formal financial institution. In some countries, progress has been phenomenal. For example in China and Kenya, nearly 100% of informal business owners poses a mobile phone (Klapper et al., 2019). Innovation has not only played an important role in easing the transaction of MSMEs but also enhanced financial inclusion.

(iii) Mobile Money Technology Innovation

The Group Managed Service Accounts (GSMA) is a body that represents the interests of mobile operators worldwide, uniting more than 750 operators with almost 400 companies. The majority are MSMEs. In its broader mobile ecosystem are handset and device makers, software companies, equipment providers and internet companies, as well as ancillary industry sectors. These are versatile innovators and most are MSMEs. In its GSMA "State of the Industry Report on Mobile Money 2021", there are important revelations. It states that there are 5.2 billion mobile users worldwide. These offer great potential to the mobile industry to be a key player in consolidating efforts and to deepen financial inclusion and build more equitable societies. The GSMA reiterates its commitment to achieving the UN Sustainable Development Goals (SDGs), and leveraging its Innovation Funds. This is intended to help digital pioneers deliver clean energy solutions, vital health information, life-changing tools for farmers and to enhance opportunities for women to participate as equal partners in a digital society. In the report it is revealed that the rate of growth in mobile money activities is enormous. From just $15 billion in 2015, the value of transactions between mobile money platforms and banks grew fourfold, reaching $68 billion in 2020. During the year monthly transactions averaged over $1 billion being sent and received. GSMA reported for the year-on-year growth for 2019 was 34%, with transactions per unique customer growing by 13% (GMSA, 2021).

(iv) Innovation for Economic Prosperity

Both innovation and entrepreneurship are increasingly seen as essential ingredients for economic and social prosperity. A key requirement for the achievement of all SDGs is the adoption of efficient technological solutions (Independent Evaluation Group, 2013). Innovation is a major source of productivity, growth and employment. It introduces new ways of doing things and is geared toward efficiency, effectiveness and competitiveness (Oughton et al., 2002). Given the importance of innovation for MSMEs governments should enhance regulatory frameworks and policies that promote it. They should identify and assist university-MSME cooperation networks as well as procure goods and services from innovative MSMEs (WB, 2016; IMC World Wide 2016). In order to be competitive in the market, entrepreneurs are increasingly adopting frugal and incremental innovation which has enabled them to redesign products and business models which enhance cost reduction (*The Economist*, 2010). During the early stages of development firms adopt incremental innovation characterized by the adoption of foreign technology but increasingly turn to that generated locally. As the firm grows, high-technology R&D-based innovation matters more (OECD, 2011).

(v) Technical Assistance for Innovation

Recognizing the importance of leveraging technological innovation and that what might work in developed countries may not necessarily work in developing countries, interventions like TRANSFORM (as a result of efforts between Unilever, the UK's Department for International Development (DFID) on the one hand and Microsoft 4Afrika were initiated and have assisted millions of MSMEs. Their interventions are intended to benefit entrepreneurs by enhancing locally relevant technology and skills development. In Kenya, for example, they have worked with local companies, Mawingu Networks, and Shuujaaz Inc, to enable youths and women to access WIFI at prices 60% lower than before. Microsoft together with a local company, Sustain Technology Platform, developed a platform that provides relevant software solutions which enable partner organizations to deal with each other securely using their Azure Cloud technology. MSMEs that embrace technological innovations create new opportunities and they increase their chances and potential to expand. In particular, high mobile phone penetration has greatly improved economic activities of MSMEs in rural and urban areas of Africa (Amankwah-Amoah et al., 2018).

(vi) Microfinance as a Tool for financing MSMEs

The idea of extending small loans to underprivileged communities is traced back to the fifteenth century. The forms differed according to times and locations. There were community-oriented pawn shops and financial services were provided to those in need. Centuries later, in the early parts of the eighteenth century, Jonathan Swift, an Irish nationalist and writer, conceived the idea of how to extend financial services to the rural and urban poor. These had been ignored by the traditional and profit-oriented commercial financial institutions for long. This was in a form different from the arrangements of the thirteenth century. This saw the beginning of loan funds which extended small credit amounts to the poor, and actual money was exchanged. By the 1840s there were over 300 of them throughout Ireland. At some point in time over 20% of households in Ireland were accessing these funds. Borrowing from Swift's ideas, financial institutions by descriptions such as 'people's bank', 'workers bank', credit union and SACCOs sprung up throughout Europe. Around the same time the German altruist, Friedrich Wilhelm Raiffeisen urged the rural population to stop borrowing from moneylenders if they were to improve on their welfare, thus the Raiffeisen model, which revolved around the provision of subsidized agricultural credit through cooperatives, with emphasis on mobilizing poor people in rural areas and motivating them to save.

Modern Microfinance Becomes a Matter of Global Concern

However, the expression microfinance as currently used can be traced to 1970 with two pioneers, namely, Akther Hameed and Muhammad Yunus in Bangladesh. For his efforts, the latter was awarded a Nobel Prize. The Consultative Group for the Alleviation of Poverty (CGAP) is among the leading exponents for the advancement of microfinance and by 2004 had outlined some principles summarizing over a century-and-half period of microfinance practices. These principles were endorsed by the G8 Summit on June 10th, 2014 (Table 6.4).

Table 6.4 Key issues or elements of these principles of microfinance

Item	*Issue*
1	Poor people need not just loans, but also savings, insurance and money transfer services
2	Microfinance must be useful to poor households as it helps them raise income, build up assets and or cushion themselves against external shocks
3	Microfinance can pay for itself and subsidies from donors and government are scarce and uncertain and so, to reach large numbers of the poor people, microfinance must pay for itself (operational self-sufficiency and institutional financial self-sufficiency
4	Microfinance means building permanent local institutions
5	Microfinance requires integration of the financial needs of poor people into a country's mainstream financial system
6	The job of government is to enable financial services, not to provide them
7	Interest rate ceilings hurt the poor people by preventing microfinance institutions from covering their costs, which chokes off the supply of credit
8	The key bottleneck is the shortage of strong institutions and competent managers and therefore donors should focus on capacity building
9	Microfinance institutions should measure and disclose their performance—both financially and socially
10	Donor funds should complement private capital, not compete with it

Source CGAP (2014)

Again in 2004 a study of '*alternative financial institutions*' had indicated that 665 million client accounts existed at over 3,000 financial institutions serving people who cannot access services of commercial banks. Of these accounts 120 million were held with institutions principally practicing the provision of microfinance services. Increasingly, more MSME entrepreneurs find it easy to effect person-to-person (P2P) transfers. Adoption of digital technologies provides an important solution to financial exclusion (GPFI, 2018).

Liquidity as a Constraint to the Growth of MSMEs and the Role of Microfinance

Despite the important roles MSMEs play in economies, a major constraint to their growth is a lack of finance and poor liquidity management practices. For example, many do not conform to the matching principle. If they did, assets like plants and equipment would be purchased using funds that are payable in installments over the long term. The purchase of raw materials and other consumables would be financed using funds repayable in the near future such as overdrafts. It is inconceivable that a business enterprise can have, as its key objective, to survive without growing and not making a profit, at least in the medium and long term. Most mission statements of business enterprises are the justification of why they seek existence and residence in a social and economic environment (Sazir & Shakira, 2020).

Microfinance, Green Microfinance MSMEs and SDGs

Microfinance has evolved to serve the economically active poor. Thus it has long served both social and economic transformation missions, which have subsequently been prominently captured in the Millennium Development Goals and eventually in the Sustainable Development Goals. This is particularly so with regard to poverty alleviation and employment. MSMEs have been noted to be key drivers in the achievement of many of other SDGs. Given the nature of activities that the micro-entrepreneurs participate in, especially agriculture, most of these people exploit the environment, which results into devastating effect on the environment. For example, they cook using firewood in the absence of other sources of energy. The relationship between microfinance and MSMEs is enhanced by the difficulty they have in accessing finance from mainstream financiers. Microfinance has been credited with unlocking entrepreneurial ambitions of the economically active poor and sustaining their income-generating activities (Lakuma et al., 2019). The seamless interaction between SDGs, MSMEs and microfinance makes it impossible to discuss any without the others.

From the foregoing, there are several distinguishing features between conventional banking and microfinance. First, the amounts lent out by microfinance institutions are typically small. Ironically there exists the thinking that the interest charged thereon is most times high. Secondly, the repayment periods and frequency of these repayments are characteristically short, and in some cases daily.

Third, providers of microfinance have much preference for women. The arguments for this are that they have displayed relatively better loan repayment experience compared to men. This is attributed to what has been referred to as their triple burden. Triple burden simply refers to the very crucial roles women play in (especially poor) communities. First is the reproductive role, as mothers. Second is the productive role as tillers of gardens or trading in markets to provide both subsistence and finance for the household. Finally they provide crucial community linkages like when you find them representing the community in greater proportions and with more active participation in social activities. The triple burden perhaps explains why as breadwinners and mothers women are more honest than men and why they easily fall in with the collateral substitute requirements which characterize the lending technology of microfinance.

Conclusion

Sustainability as a concept has a long history. It evolved arising out of the need to protect future generations from the result of reckless exploitation of resources by those before them. The center of focus for SDGs are people, planet, peace, prosperity and partnerships. Unfortunately the performance with regard to SDGs has been dismal for Africa. This is despite some landmark achievement as in the case of mobile money technology. MSMEs have been critical in the modest performance by the continent for example by providing employment and poverty alleviation. MSMEs are resilient as a result of flexibility and innovation. A major challenge for MSMEs is lack of financing. Conventional banking principles and practices make it difficult to provide liquidity for MSMEs. This necessitated having alternatives. The most prominent is microfinance. It substitutes traditional security requirements with collateral substitutes.

References

AFDB. (2019). *Creating decent jobs strategies, policies, and instruments: Policy research document 2.* African Development Bank, Abidjan, Ivory Coast

Amankwah-Amoah, J., Osabutey, E. L. C., & Egbetokun, A. (2018). Contemporary challenges and opportunities of doing business in Africa: The emerging roles and effects of technologies. *Technological Forecasting and Social Change, 131*, 171–174. https://doi.org/10.1016/j.techfore.2018.01.003

Beck, T., Demirgüç-Kunt, A., & Levine, R. (2003). *Small and medium enterprises, growth, and poverty: Cross-country evidence* (Policy Research Working Paper; No. 3178). © World Bank.

Bhorat, H., & Oosthuizen, M. (2020). *Jobs, economic growth, and capacity development for youth in Africa* (Working Paper Series N° 336). African Development Bank, Abidjan, Ivory Coast.

Brixiova, Z., Kangoye, T., Yogo, U. T. (2020). *Access to finance among small and medium-sized enterprises and job creation in Africa* (IZA Discussion Papers, No. 13708). Institute of Labor Economics (IZA), Bonn.

Carvalho, G. D. G., Cruz, J. A. W., Carvalho, H. G., Duclos, L. C., & Correa, R. O. (2020). Innovativeness and coopetition in tourism SMEs: Comparing two cooperative networks in Brazil. *Journal of Hospitality and Tourism Insights, 3*(4), 469–488. https://doi.org/10.1108/jhti-12-2019-0134

CCIS Briefs. (2021, July 7). *Supporting small and medium enterprises in sub-Saharan Africa through blended finance.* https://www.csis.org/analysis/supporting-small-and-medium-enterprises-sub-saharan-africa-through-blended-finance

CGAP. (2014). Advancing financial inclusion to improve lives of the poor. *Annual report, 1818.* Retrieved from 2006. https://www.cgap@worldbank.org

Devine, R. A., & Kiggundu, M. N. (2016). Entrepreneurship in Africa: Identifying the frontier of impactful research. *Africa Journal of Management, 2*(3), 349–380. https://doi.org/10.1080/23322373.2016.1206802

Dey, P. (2012). Rapid incubation model for the development of micro and small enterprises in sub-Saharan Africa. *Global Journal of Management and Business Research, 12*(10), 78–85.

Doran, J., McCarthy, N., & O'Connor, M. (2018). The role of entrepreneurship in stimulating economic growth in developed and developing countries.

Cogent Economics & Finance, 6(1), 1–14. https://doi.org/10.1080/23322039.2018.1442093

Edelman, M. (2014). Food sovereignty: Forgotten genealogies and future regulatory challenges. *The Journal of Peasant Studies, 41*(6), 959–978.

EDFI. (2016). *Investing to create jobs, boost growth and fight poverty* (Flagship Report). Association of the European Development Finance Institution.

Facundo, A., & Sergio, L. S. (2017). Small and medium enterprises (SMEs) finance, improving SMEs' access to finance and finding innovative solutions to unlock sources of capital. https://www.worldbank.org/en/topic/smefinance

GMSA. (2021). *The state of the industry report on mobile money 2021.* Nairobi, Kenya.

GPFI. (2018). *G20 policy guide—Digitisation and informality: Harnessing digital financial inclusion for individuals and MSMEs in the informal economy.* Los Angeles. www.gsma.com/mobilemoney

ILO. (2015). *Small and medium-sized enterprises and decent and productive employment creation* (Report IV, Issue), Geneva, Switzerland.

ILO. (2017). *World social protection report 2017–19: Universal social protection to achieve the Sustainable Development Goals* (p. 119). Geneva.

IMC. (2016). *World wide connect to grow: Supporting SME growth through innovation and partnership—A review of the landscape.* Retrieved from the Practitioner Hub for Inclusive Business website. http://www.inclusivebusinesshub.org/wp-content/uploads/2016/02/Connect-Baseline-FINAL_public-July-2016.pdf

Independent Evaluation Group. (2013). *World Bank Group support for innovation and entrepreneurship: An independent evaluation.* World Bank Group, © World Bank. https://openknowledge.worldbank.org/handle/10986/16665. License: CC BY 3.0 IGO.

James, Z., & Massimo, M. (2021). Born green via digital: Increasing MSME's competitiveness and sustainability—A facilitators' reflection on the G20 Policy Toolkit. UN Journal on Transnational Corporations, Geneva.

Kushnir, K., Mirmulstein, M. L., & Ramalho, R. (2010). *Micro, small, and medium enterprises around the world: How many are there, and what affects the count?* IFC, the World Bank Group.

Klapper, L., et al. (2019). *Leveraging digital financial solutions to promote formal business participation.* Report International Bank for Reconstruction and Development/The World Bank, 1818 H Street NW Washington, DC 20433.

Kongolo, M. (2010). *Job creation versus job shedding and the role of SMEs in economic development.* http://citeseerx.ist.psu.edu/viewdoc/download?doi=10.1.1.188.5761&rep=rep1&type=pdf

Lakuma, C. P., Marty, R., & Muhumuza, F. (2019). Financial inclusion and micro, small, and medium enterprises (MSMEs) growth in Uganda. *Journal of Innovation and Entrepreneurship, 8*(15), 1–20. https://doi.org/10.1186/s13731-019-0110-2

Mensah, J. (2019). Sustainable development: Meaning, history, principles, pillars, and implications for human action: Literature review. *Cogent Social Sciences, 5*(1), 1653531. https://doi.org/10.1080/23311886.2019.1653531

Müller, J. M., Buliga, O., & Voigt, K. I. (2018). Fortune favors the prepared: How SMEs approach business model innovations in industry 4.0. *Technological Forecasting and Social Change, 132*, 2–17. https://doi.org/10.1016/j.techfore

Nursini, N. (2020). Micro small and medium enterprises (MSMEs) and poverty reduction: Empirical evidence from Indonesia. *Journal of Development Studies Research, 7*(1), 153–166.

OECD. (2011). *Studies on SMEs and entrepreneurship Thailand: Key issues and policies*. OECD Publishing.

OECD. (2014). Mainstreaming green growth into national development planning. In *Towards green growth in Southeast Asia*. OECD Publishing. https://doi.org/10.1787/9789264224100-5-en

OECD/Economic Research Institute for ASEAN and East Asia. (2018). Environmental policies and SMEs. In *SME policy index: ASEAN 2018: Boosting competitiveness and inclusive growth*. OECD Publishing; Economic Research Institute for ASEAN and East Asia. https://doi.org/10.1787/9789264305328-9-en

Oughton, C., Landabaso, M., & Morgan, K. (2002). The regional innovation paradox: Innovation policy and industrial policy. *The Journal of Technology Transfer, 27*(1), 97–110.

Raworth, K. (2012). *A safe and just space for humanity*. Oxfam. https://www.oxfam.org/sites/www.oxfam.org/files/dp-a-safe-and-just-space-for-humanity-130212-en.pdf

Sazir, N. M., & Shakira, N. M. (2020). Relationship between liquidity management and growth of MSMEs in Africa: A case study of selected districts of Uganda. *American Journal of Finance, 5*(1), 24–42. ISSN 2520-0445 (Online).

The Economist. (2010, April 23). For an analysis of innovation in emerging markets.

Toran, L. V., Nema, D. K., & Rahul, P. (2020). Role of MSMEs in poverty alleviation and rural development in India. *International Journal of Multidisciplinary Research and Development, 7*(9), 61–65.

UNDP. (1990). *Human development report*. UNDP.

UNDP. (1998). *Human development report*. Oxford University Press.

UNDP. (2016). *UN human development report 2016*. Human Development for Everyone, Technical Notes.

UNEP. (2017). *Mobilizing sustainable finance for small and medium sized enterprises*. United Nations Environment Programme. http://unepinquiry.org/wpcontent/Uploads/2017/06/Mobilizing_Sustainable_Finance_for_Small_and_Medium_Sized_Enterprises.pdf

UNO. (2015). *Millennium development goals report*. UNO.

UNO. (2019). *Global sustainable development report 2019: The future is now—Science for achieving sustainable development*. United Nations Department of Economic and Social Affairs.

WB. (2016). *World Bank and Ministry of Planning and Investment of Vietnam. Vietnam 2035: Toward prosperity, creativity, equity, and democracy*. Word Bank.

World Bank. (2016a). *Agriculture: A driver of growth and poverty reduction*. © World Bank.

World Bank. (2016b). *Uganda poverty assessment 2016b: Fact sheet*. © World Bank.

CHAPTER 7

The Role Played by Research and Development in Uganda's MSME Toward Promoting SDGs

Tukundane Benson and Emuron Lydia

Introduction

Humankind has forever striven to improve the quality of life lived. Among the efforts beginning in 1990 was the formulation of the Millennium Development Goals (MDGs) to provide a united front among the member states of the United Nations Organisation (UNO). Leaders of 189 member countries of UNO gathered at the United Nations headquarters and signed the Millennium Declaration in September 2000. They committed to the achievement of eight goals, with several targets for each by 2015. These were the eradication of poverty and hunger, achievement of universal primary education, the promotion of gender equality

T. Benson · E. Lydia (✉)
Kampala International University, Kampala, Uganda
e-mail: lydiaemuron@kiu.ac.ug

T. Benson
e-mail: benson.tukundane@kiu.ac.ug

H. Dasaraju and T.T.H. Tambunan (eds.), *Role of Micro, Small and Medium Enterprises in Achieving SDGs*, Sustainable Development Goals Series, https://doi.org/10.1007/978-981-99-4829-1_7

and empowerment of women, reduction of infant mortality, improvement of maternal health, combating HIV/AIDS and malaria among other diseases, and finally, ensuring environmental sustainability. By 2015, many of the goals had been achieved to some extent.

Member states of the United Nations agreed that there was a need not only to continue pursuing the goals albeit in a modified format and of related targets but also to widen the scope. Thus, seventeen Sustainable Development Goals (SDGs) were agreed upon. Leaders of 192 member states of UNO met in 2015 to agree on the SDG framework. These are summarized in Table 7.1.

Table 7.1 Summary of the seventeen sustainable development goals (SDGs)

Goal	*Summary*
1	End poverty in all its forms everywhere
2	End hunger, achieve food security and improved nutrition and promote sustainable agriculture
3	Ensure healthy lives and promote well-being for all at all ages
4	Ensure inclusive and equitable quality education and promote lifelong learning opportunities for all
5	Achieve gender equality and empower all women and girls
6	Ensure availability and sustainable management of water and sanitation for all
7	Ensure access to affordable, reliable, sustainable and modern energy for all
8	Promote sustained, inclusive and sustainable economic growth, full and productive employment and decent work for all
9	Build resilient infrastructure, promote inclusive and sustainable industrialization and foster innovation
10	Reduce inequality within and among countries
11	Make cities and human settlements inclusive, safe, resilient, and sustainable
12	Ensure sustainable consumption and production patterns
13	Take urgent action to combat climate change and its impacts
14	Conserve and sustainably use the oceans, seas, and marine resources for sustainable development
15	Protect, restore, and promote sustainable use of terrestrial ecosystems, sustainably manage forests, combat desertification, and halt and reverse land degradation and halt biodiversity loss
16	Promote peaceful and inclusive societies for sustainable development, provide access to justice for all and build effective, accountable, and inclusive institutions at all levels
17	Strengthen the means of implementation and revitalize the global partnership for sustainable development

Source UN (2017)

The MDGs had eight goals and eighteen targets. On the other hand, the seventeen goals had 169 underlying targets. The new framework, the SDGs, first of all, emphasizes comprehensiveness. Secondly, it requires interconnectedness of the 5P's, namely, People, Planet (sustainable development), Prosperity (elimination of poverty), Peace (peace required for development), and Partnerships (multi-stakeholder approaches).

Micro, Small, and Medium Enterprises and Their Relevance to the Achievement of SDGs

The concept of micro, small, and medium enterprises (MSMEs) dates back to 4000 years ago during the ancient cultures when business flourished among Arabs, Egyptians, Romans, Jews, and Greeks among many others. During that time, the cottage industries thrived and the workman in craft produced goods at his home and would sell the finished products by himself to the market. As time passed by, there was a major resurgence in small-scale industries throughout the world (Fiseha & Oyelana, 2015; Rotar et al., 2019). Throughout the whole world, MSMEs have been found to address impediments like poverty, unemployment, and inequality. Despite their importance in all economies, developing, emergent or developed MSMEs, there is no universal definition of what they are. In Uganda, however, a qualitative approach describes what they are. Micro enterprises are those businesses employing not more than 5 people and their total assets do not exceed ten million Uganda shillings (approximately $2500). Small enterprises employ between 5 and 49 people and total assets are between ten million Uganda shillings (approximately $2500) but not exceeding one hundred million Uganda shillings (approximately $25,000). Medium enterprises are those employing between 50 and 100 people with total assets of more than one hundred million Uganda shillings (approximately $25,000) but not exceeding three hundred and sixty million Uganda shillings (approximately $90,000) (Sazir, 2020).

Despite the lack of a universal definition, the different descriptions of MSMEs try to establish a relationship between the size and turnover of an enterprise. MSMEs are defined by their size. In the European Union, a Small and Medium enterprise has no more than 250 employees, not more than 50 million Euro turnover and not more than 25% of the shares of such a company are owned by another company (Hauser, 2005). MSMEs are at the center of achieving SDG goals because of their immense contribution to sustainable, inclusive, and resilient economic growth, shared prosperity, and decent work for all. They account for 90% of businesses, 60–70% of employment, and 50% of GDP worldwide. In recognition of this, The United Nations General Assembly designated 27 June as Micro-, Small, and Medium-sized Enterprises Day (UN Secretary-General & UNCTAD, 2020).

Some specific cases corroborate the global importance of MSMEs. In 2017 they contributed to 53% and 86% of employment in Organization for Economic Co-operation and Development Countries (OECD) such as the UK and Greece respectively. Over the same year, they constituted 98% of private enterprises in Peru, contributing to 42% of GDP and accounting for 60% of employment. Likewise, MSMEs provide about 50% and 80% of employment in Cambodia and Kenya (OECD, 2017b).

Economic Contributions of MSMEs

MSMEs are particularly important as avenues for employment in developing countries. For example, in Africa, growth in the working-age population is expected to be very rapid, increasing by 265.8% in Africa overall and by 306.6% in sub-Saharan Africa, compared to 28.3% globally (Bhorat & Oosthuizen, 2020). Globally, countries under the OECD, are facing the challenge of low growth, weak trade and investment, and persistently rising inequality among micro, small, and medium enterprises. MSMEs continue to struggle with the prolonged impact of the 2007–2008 global crisis. MSMEs are key players in the economy and the wider ecosystem of firms. Enabling them to adapt and thrive in a more open environment is essential in boosting economic growth and development (OECD, 2017a).

MSMEs largely contribute to the economies of developed countries. They act as a source of employment, provider of subsistence, ensuing equitable income distribution, increase in financial capital gains, and importantly contribute to the gross domestic product (GDP) of emerging economies like India, Brazil, China, and Malaysia. Governments of developing countries must invest substantially in the growth, development, and sustainability of MSMEs through the provision of essential infrastructures, human resources, finance, security, and adequate policies (Motilewa et al., 2015). Determinants of productivity and growth have been documented for industrialized countries with research and development regarded as the key to growth. Firms invest in research and development to develop new products and processes in particular. However, in most developing countries, the majority of the firms are still operating far below the technological level with lower levels of human capital and older vintage of machinery. Several initiatives have been undertaken to develop measures of innovation activities in developing countries; they include human resources, linkages with the human capital development of firms, adoption of ICT, quality control systems, acquisition of embodied technology, and Research and Development (Goedhuys, 2007).

In Uganda, MSMEs are faced with a number of challenges that lead to business failure. These causes of failure are quite diverse in nature. They have resulted in more than fifty per cent of businesses failing in the first five years of their existence and fighting a difficult battle from the start (Tushabomwe-Kazooba, 2006). Furthermore, the World Bank in its survey (2014) identified impediments to MSMEs growth and competitiveness which included the informal sector with many enterprises only lasting five years or less. In line with Vision 2040, MSMEs were identified as major sources of technological innovations and new products as they constitute an important sector in manufacturing, services, and agribusiness engaged in local, regional, and international business. The government of the Republic of Uganda, its development partners, and the private sector have come up with a number of initiatives to promote the MSMEs sector but this has to date not yielded to their expectations and still performs poorly (Ishengoma & Kappel, 2007; MTIC, 2015). Despite the fact that MSMEs have been found worthy in strengthening financial sectors in economies by creating jobs, they are still faced with challenges that threaten their existence, performance, and growth Hussein Kakembo et al. (2021). In light of the above, this study sought to investigate whether research and development has an effect on the growth of micro, small, and medium enterprises in the Greater Kampala Metropolitan Area, Uganda.

Innovation by MSMEs for Achievement of SDGs

SDG 9 "build resilient infrastructure, promote inclusive and sustainable industrialization and foster innovation" is very relevant in MSME survival, profitability, and growth. With these achieved, MSMEs augment several of the SDGs. Most specifically, end poverty in all its forms everywhere (SDG1), End hunger, achieve food security and improved nutrition, and promote sustainable agriculture (SDG2), promoting sustained, inclusive, and sustainable economic growth, full and productive employment, and decent work for all (SDG8), ensuring healthy lives and promote well-being for all at all age (SDG3), reducing inequality within and among countries (SDG10) and all goals directly related to personal health, survival, and development. Regarding matters of environmental conservation, all activities by MSMEs have an impact on them.

Research and Development (R&D) and Growth of MSMEs

Research and development (R&D) is the work that businesses conduct toward innovation, introduction, and improvement of their products and procedures. It covers activities from basic scientific research performed in universities and laboratories all the way to testing and refining products before commercial sale or use. Research and Development is generally connected to the notion of corporate or governmental innovation. Known as research and technical development in Europe (Hall, 1988).

Globally, it is thought that pharmaceutical and technology companies when they hear "R&D," but other firms, including those that produce consumer products, invest time and resources into R&D as well. For example, a spaghetti sauce brand's many variations on the original product—"Chunky Garden," "Four Cheese," and "Tomato Basil Garlic"—are the results of extensive R&D. It takes place in companies of all sizes. Any business that creates and sells a product or service, whether it's software or spark plugs, invests in some level of R&D (Business Encyclopedia).

Research and Development is a systematic activity joining together basic and applied research, which aims at realizing solutions to problems or creating new goods and knowledge. R&D may result in ownership of intellectual property such as patents. In accounting for R&D costs, the development costs may be carried forward but the basic and applied research costs are often written off as incurred (Investopedia, 2023).

Empirical evidence on the relationship between R&D and the survival of firms varies and sometimes conflicting. This may be attributed to overlooking the scale of effects of R&D and the complementarity between R&D intensity and market concentration. R&D and market concentration complement each other and as a result, firms that have vibrant R&D systems have longer survival times. This shows that the firm's ability to grow depends on strong R&D systems. Smaller firms that put their focus on growth as the only means of enhancing their market have a high risk of exiting the market due to a gap that exists between the growth ambitions and the growth management ability (Ugur et al., 2016).

Various theories of growth concentrate on technological innovations which come as a result of Research and Development activities to explain productivity growth. In a study to identify the relationship business R&D expenditure and growth of productivity in the European manufacturing sector, findings revealed a strong causality from increasing firm R&D intensity to expanding high and medium high manufacturing. The study concluded that manufacturing firms should put in place R&D policies that will provide transformation in the manufacturing sector into high-tech specifications coupled with continued growth (Karahan, 2015).

In the MSMEs Development Strategy of Georgia, as an example, one of its priority actions is that government assistance is necessary to support the development of R&D, this is because the private sector is not investing in R&D but, instead purchasing their products from abroad. There is a need to avail finances to private businesses for research and development in the form of a research grants program. Due to the failure to commercialize R&D, its use is hindered in the economy. The government pledged to work with research institutions and private firms to commercialize R&D which has been ranked as key to the growth of firms if well implemented (OECD, 2016).

In a study to explore the influence which MSMEs internal funding capacity and credit ratings have, is based on whether they conduct R&D and how much they spend on it, major findings revealed that micro, small, and medium enterprises with high-profit margins spent 14% more on research and development than the businesses which have medium profit margins. MSMEs heavily depend on R&D activity and on enterprise internal funding capacity due to specific characteristics like the uncertainty of business success which hamper financing even for MSMEs with good credit ratings. It was recommended from that study that, more funding should be directed to MSMEs R&D systems to enhance their investment and growth capacity (Zimmermann, 2017).

To address the role of R&D strategies in sales and growth of MSMEs, data from a database of large MSMEs revealed that R&D offshore sourcing positively affects the growth of sales both directly and indirectly. This shows that the more MSMEs invest in R&D activities, the more the sales growth and vice versa which in turn lead to the growth of the firm (Rodríguez & Nieto, 2016).

The Study

Methodology

Research paradigm: based on the preceding discussions and literature reviewed, this study was guided predominantly by a positivism paradigm. Positivism paradigm holds the view that, causes probably determine effects and outcomes as seen in experiments (Creswell, 2014). Deductive theorist and quantitative methodology approaches were applied to develop a quantitative survey questionnaire instrument. According to (Guba & Lincolin, 1994), positivism allows the researcher to establish how things are and how things really work.

Research Approach

Quantitative research methods and statistics were used to measure reality or views with a focus on the measurement of relationships between variables with a view to building models that can predict outcomes (Creswell, 2008).

Research Design

When conducting a good study, a good researcher is expected at the inception after determining the area of research to come up with an outline of the strategy to be followed which is called the research design. It is composed of the methods to be followed, concepts that will be measured, and the method chosen by the researcher which will adequately perform the task (Dikko, 2016). Particularly, Descriptive, Correlational, survey, and cross-sectional designs were employed. Descriptive designs involved the use of means, frequencies, standard deviation, and percentages to describe and get meaningful interpretation of the study variables; Correlational designs were used on variables where it was found to be necessary as per the study's objective to establish the relationship between two variables; Survey design was used because the data was collected from a large number of respondents at a certain point of time. Since the data was collected at once to achieve this study's objective, it made the study Cross-sectional thus justifying the use of cross-sectional design as well.

Target Population

The target population of this study was all registered MSMEs in the Greater Kampala Metropolitan Area. From each of the MSMEs, the researcher targeted one top administrator (owner, manager, director, CEO, or supervisor). These enterprises that were considered should have been in operation for a duration of not less than three years. The researcher chose to carry out the study on MSMEs in GKMA because over a third of MSMEs are located in the central region with most businesses in Kampala, Mukono, and Wakiso (FSD Africa et al., 2015; Kuteesa et al., 2017).

Sample Size

The study used Krejcie and Morgan (1970) table to determine the sample of enterprises and respondents to be involved in the study and a total of 226 firms was selected from 571 MSMEs that met the requirements to participate in this study (researcher's computations according to UMA website, 2018, and Uganda MSMEs Policy (MTIC, 2015)).

Data Collection Instruments

According to Amin (2005), in a survey research design exists main procedures for collecting data from the respondents. These methods include questionnaires and interview administration. This study adopted a survey research design and thus used self-administered questionnaires to collect data from the respondents of the study. Structured questionnaires were used. The study was dominated by structured questions in order to control the study's problems, get rid of error, and take care of the respondent's confidentiality by allowing the questionnaires to be filled without any interference. The questionnaire had several sections with related items where the respondents were asked to give their views by circling or ticking the best choice from the provided alternatives.

Validity and Reliability of the Instruments

Often, validity is referred to as the extent to which a tool measures what it purports to measure (Kimberlin & Winterstein, 2008). Validity of the instrument was measured according to face validity, content validity, and construct validity (Sarantakos, 1998). Kaiser–Meyer–Olkin (KMO) Test was also used to measure how the data was suited for factor analysis (Cerny & Kaiser, 1977). Using the KMO test, it was revealed that all the values for KMO were above 0.60 the acceptable threshold according to Cerny and Kaiser (1977). Growth of MSMEs scored 0.666 ranked as mediocre while the independent variable (research and development) scored an average of 0.721 which ranked as middling. Accordingly, KMO values above 0.60 threshold indicated that the data was a good suit for factor analysis.

The reliability of an instrument can be achieved if it can consistently and with no bias measure the concepts it is supposed to measure. This can be achieved through test-re-test reliability which is referred to as the measure of reliability that can be obtained by administering the same test twice over a period of time to a group of individuals but you obtain the same or parallel results (Dikko, 2016). The reliability of the instrument was also tested by using Cronbach's Alpha (α) test to test for the consistency of the instrument. According to Amin (2005), Cronbach's Alpha (α) test is one of the most commonly used measures of the reliability of data collected because it is based on the internal consistency of the test. This is done to minimize the errors and biases in data. A reliability of at least 0.70 is acceptable (Gable & Wolf, 1993). This study used SPSS (Statistical Package for Social Scientist), Cronbach alpha (α) test was conducted on constructs to see if the benchmark of 0.70 was adhered to, Cronbach's Alpha value was 0.886 which was way above the benchmark of 0.70. These results indicated that the instrument was reliable and consistent.

Findings

To be able to establish the effect of R&D on growth of MSMEs in the Greater Kampala Metropolitan Area, the researcher collected data on the level of research and development which was correlated with the level of growth of MSMEs in GKMA. Results generally rated research and development as high as per this study's findings. This was revealed by the overall mean of 3.07 (Table 7.2) from the 9 constructs that were used to measure research and development. Of the nine constructs, business introducing new products, services and processes was ranked number one with a mean value of 3.2 and a standard deviation of 0.83; whereas business is always the first to bring new products on the market was ranked lowest with a mean value of 2.73. These results suggest that as much as the business firms introduce new products, services and procuresses, they are not always the first to bring them to the market; businesses have creative modes of operation, businesses establish current ways of reaching their customers and businesses frequently trying out new ideas were also rated high with means of 3.20, 3.17 and 3.12 respectively. These were followed by businesses investing in research and development of new products, services, and processes, businesses using new marketing methods, and businesses trying out and using new technologies like mobile banking with means 3.10, 3.08, and 3.04 respectively. Management pursuing to try out new ways was ranked second to last with a mean value of 2.96 which also was rated high. These relatively high rankings on research and development indicated that businesses in the Greater Kampala Metropolitan Area were performing their businesses in line with research development.

Table 7.2 Descriptive statistics on R&D

Indicators of Research and Development	*Mean*	*Std. Deviation*	*Interpretation*
Our business introduces new products, services, and processes	3.21	0.8282	High
Our business has creative modes of operation	3.20	0.8823	High
Our business establishes current ways of reaching their customers	3.17	0.8647	High
Our business frequently tries out new ideas	3.12	0.8258	High
Our business invests in the research and development of new products, services and processes	3.10	0.9024	High
Our business uses new marketing methods	3.08	0.9045	High
Our business tries out and used new technologies like mobile banking	3.04	0.9330	High
Management pursues to try out new ways	2.96	0.8466	High
Our business is always the first to bring new products on the market	2.73	1.0600	High
Average mean	3.07	0.8942	High

Note

Mean range	*Response range*
3.26–4.00	Very High
2.51–3.25	High
1.76–2.50	Low
1.00–1.75	Very Low

Source Computed from Primary Data

Confirmatory Factor Analysis for R&D in GKMA

For this study to be able to ascertain whether items used to measure R&D were valid and significant measures of the construct of research and development, confirmatory factor analysis was done. Results for "total variance explained for R&D in MSMEs in the greater Kampala metropolitan area" (Table 7.3) revealed that of the 9 items that were used to measure R&D, only the first 2 components recorded Eigenvalues which were above 1 (3.199 and 1.152). These two components all together explained 48% of the variance as indicated in the cumulative % column above.

The factor loadings for these 9 items on research and development (Table 7.4) revealed that all items measuring research and development were valid and relevant since each of the items loaded highly that is above 0.5 on any of the 2 components. All nine items were, therefore, valid measures for research and development which implied that the data from these questions can be used to perform further parametric analyses.

Table 7.3 Total variance explained for R&D in MSMEs in the Greater Kampala Metropolitan Area

Component	*Initial eigenvalues*			*Extraction sums of squared loadings*		
	Total	*% of variance*	*Cumulative %*	*Total*	*% of variance*	*Cumulative %*
1	3.199	35.544	35.544	3.199	35.544	35.544
2	1.152	12.798	48.343	1.152	12.798	48.343
3	0.889	9.875	58.218			
4	0.777	8.634	66.851			
5	0.748	8.316	75.167			
6	0.671	7.457	82.624			
7	0.620	6.886	89.510			
8	0.523	5.812	95.322			
9	0.421	4.678	100.000			

Extraction Method: Principal Component Analysis
Source Computed from Primary Data

Table 7.4 Component matrix a for R&D in MSMEs in the Greater Kampala Metropolitan Area

	Component	
	1	*2*
Our business establishes current ways of reaching their customers	0.731	
Our business has creative modes of operation	0.636	
Our business uses new marketing methods	0.618	
Our business frequently tries out new ideas	0.610	
Our business introduces new products, services, and processes	0.603	
Our business invests in the research and development of new products, services, and processes	0.596	
Management pursues trying out new ways	0.582	
Our business tries out and used new technologies like mobile banking	0.547	
Our business is always the first to bring new products to the market		0.620

Extraction Method: Principal Component Analysis
a. 2 components extracted
Source Computed from Primary Data

Pearson's Linear Correlation for R&D and Growth of MSMEs in the Greater Kampala Metropolitan Area

Further, for this study's objective to be achieved, Pearson's linear coefficient correlation test was conducted on R&D and growth of MSMEs. Results revealed that there was a positive and significant relationship between research and development and the Growth of micro, small, and medium enterprises in the Greater Kampala Metropolitan Area with $r = 0.432$, $p = 0.000$ (Table 7.5). In line with Cohen's (1998) findings on determining the strength of a relationship, these results indicated that an improvement in research and development is likely to increase the growth of MSMEs. The direction of the relationship was found to be medium and positive, this implies that R&D and growth of MSMEs in the Greater Kampala Metropolitan Area, tend to increase together with the magnitude of the relationship approximately medium.

Effect of R&D on the Growth of MSMEs in the Greater Kampala Metropolitan Area

After thorough preliminary analyses were performed to ensure no violation of the assumptions, the simple linear regression analysis was applied

Table 7.5 Pearson's linear correlation on R&D and growth of MSMEs in the Greater Kampala Metropolitan Area

		R&D	*Growth of MSMEs*
R&D	Pearson Correlation	1	0.432**
	Sig. (2-tailed)		0.000
	N	189	189
Growth of MSMEs	Pearson Correlation	0.432**	1
	Sig. (2-tailed)	0.000	
	N	189	189

** Correlation is significant at the 0.01 level (2-tailed)
Source Computed from Primary Data

to establish the strength effect of R&D on the growth of MSMEs in the Greater Kampala Metropolitan Area. Results revealed that 18.7% (Table 7.6) of growth of MSMEs in the Greater Kampala Metropolitan Area was explained by variations in R&D. Under coefficients, it was further confirmed that R&D contributes positively and significantly to the growth of MSMEs in the Greater Kampala Metropolitan Area ($\beta = 0.353$, $p = 0.000$). Beta value of 0.353 indicated that an increase in R&D by one unit leads to a 0.353 increase in growth of MSMEs. The constant value of 2.028 indicates that when R&D is zero, the growth of these enterprises is equal to 2.028, which is also significant at the 0.01 level.

Discussion

It reveals from this study's findings that R&D is a key factor in the growth of firms if it was well implemented. This calls for the need to avail finances to private businesses for R&D. Equally, Zimmermann (2017), in his study explores the influence which MSMEs internal funding capacity and credit ratings have based on whether they conduct R&D and how much they spend on it, also agrees with this study's findings, it was revealed that MSMEs with high-profit margins spent 14% more on R&D than the businesses which have medium profit margins.

Centrally to this study's findings, according to Ugur et al. (2016), the relationship between R&D and the survival of firms varies and sometimes is conflicting. This was attributed to overlooking the scale of effects of R&D and complementarity between R&D intensity and market concentration.

Table 7.6 Regression analysis for R&D and growth of MSMEs in the Greater Kampala Metropolitan Area

Model summary[b]

Model	R	R^2	*Adjusted* R^2	*Std. error of the estimate*
1	0.432[a]	0.187	0.182	0.39026

[a]Predictors: (Constant), Research and Development
[b]Dependent Variable: Growth of Micro, Small, and Medium enterprises

ANOVA[a]

Model		*Sum of squares*	*Df*	*Mean square*	*F*	*Sig*
1	Regression	6.540	1	6.540	42.943	0.000[b]
	Residual	28.480	187	0.152		
	Total	35.021	188			

[a]Dependent Variable: Growth of Micro, Small, and Medium enterprises
[b]Predictors: (Constant), Research and Development

Coefficients[a]

Model		*Unstandardized coefficients*		*Standardized coefficients*	t	*Sig*	*Decision on* H_{O1}
		B	*Std. Error*	*Beta*			
1	(Constant)	2.028	0.168		12.079	0.000	
	Research and Development	0.353	0.054	0.432	6.553	0.000	Rejected

[a]Dependent Variable: Growth of Micro, Small, and Medium enterprises

Again, in a study that used mixed methods design, factors such as age and size of the firm, organizational culture and experience of the employees were found to significantly influence innovation practices. In the same study, it was revealed that R&D does not have a significant influence on the growth of MSMEs. Thus, MSMEs managers and decision-makers are advised to cautiously choose innovation practices that suit their firm's needs as at that time (Hungund, 2018).

Conclusions

This study indicates that R&D has a strong positive relationship with growth of MSMEs. It is well known that MSMEs, especially micro and small enterprises (MSEs) in developing countries, including Uganda, experience many obstacles to grow or to increase their production and be sustainable. Whereas sustainable production growth is one of the requirements for MSMEs to be able to play their role optimally in achieving the SDGs. So, the empowerment of MSMEs is very necessary, and, as shown in this study, the ability of MSMEs to conduct their own R&D should be considered as one of the efforts to empower MSMEs.

Recommendations

Since this study has established that R&D has an effect on the growth of MSMEs, it recommends to the government of the Republic of Uganda, MSMES business owners, and all other stakeholders in the MSMEs sector, to embrace, support, and adopt R&D practices, frameworks and policies that support growth of MSMEs.

MSMEs should invest heavily in promoting R&D activities which include but are not limited to the introduction of new products, services, and processes; creative modes of operation; establishment of current ways and means of reaching their customers; frequently trying out new ideas; adoption of new technologies like mobile banking; and ensuring that they are always the first to bring new products on the market and to the customers.

Further, this study recommends to all MSMEs allocate some of their income to finance their R&D activities. Whereas, on the side of the government of Uganda, grants should be introduced to support R&D activities and mechanisms in MSMEs.

References

Amin, M. E. (2005). *Social science research: Conception methodology and analysis*. Makerere University Printery.

Bhorat, H., & Oosthuizen, M. (2020). *Jobs, economic growth, and capacity development for youth in Africa* (Working Paper Series N° 336). Abidjan: African Development Bank.

Cerny, B. A., & Kaiser, H. F. (1977). A study of a measure of sampling adequacy for factor-analytic correlation matrices. *Multivariate Behavioral Research, 12*(1), 43–47. https://doi.org/10.1207/s15327906mbr1201_3

Cohen, J. W. (1998). *Statistical power analysis for the behavioral sciences* (2nd ed.). Lawrence Erlbaum Associates.

Creswell, J. W. (2008). *Research design: Qualitative, quantitative and mixed methodsapproaches* (3rd ed.). Sage.

Creswell, J. W. (2014). *Research design: Qualitative, quantitative and mixed methods approaches* (4th ed.). Sage.

Dikko, M. (2016). Establishing construct validity and reliability: Pilot testing of a qualitative interview for research in Takaful (Islamic Insurance). *The Qualitative Report, 21*(3), 521–528. https://doi.org/10.46743/2160-3715/2016.2243

FSD Africa, Nathan Associates London Ltd., & TNS East Africa. (2015). *National small business survey 2015.* Kampala.

Gable, R. K., & Wolf, M. B.. (1993). *Instrument Developing in the Effective domain.* Boston: Kluwer.

Fiseha, G. G., & Oyelana, A. A. (2015). An assessment of the roles of small and medium enterprises (SMEs) in the local economic development (LED) in South Africa. *Journal of Economics, 6*(3), 280–290. https://doi.org/10.1080/09765239.2015.11917617

Goedhuys, M. (2007). *The impact of innovation activities on productivity and firm growth: Evidence from Brazil* (MERIT Working Papers 2007-002). United Nations University—Maastricht Economic and Social Research Institute on Innovation and Technology (MERIT).

Guba, E. G., & Lincolin, Y. S. (1994). *Competing paradigms in qualitative research.* Sage.

Hall, B. H. (1988). Research and development. *International Encyclopedia of the Social Sciences, 19*(1), 58–59. https://doi.org/10.1016/S0026-2692(88)80209-X

Hauser, H.-E. (2005, November 3–4). *A qualitative definition of SMEs*. OECD, Statistics Directorate.

Hungund, S. (2018, July). Innovation approaches: Practices and firm performance among select software product SMEs: A case of Bangalore firms. *DHARANA—Bhavan's International Journal of Business, 12*(2). https://www.informaticsjournals.com/index.php/dbijb/search/institutions

Hussein Kakembo, S., Abduh, M., & Pg Hj Md Salleh, P. M. H. A. (2021). Adopting Islamic microfinance as a mechanism of financing small and medium enterprises in Uganda. *Journal of Small Business and Enterprise Development, 28*(4), 537–552. https://doi.org/10.1108/JSBED-04-2019-0126/FULL/XML

Investopedia. (2023). Research and Development (R&D) Definition. https://www.investopedia.com/terms/r/randd.asp#ixzz5M6LfHugP

Ishengoma, E. K., & Kappel, R. (2007). Business constraints and growth potential of micro. *Business Management Review Journal, 11*(1), 1–29.

Karahan, Ö. (2015). Intensity of business enterprise R&D expenditure and high-tech specification in European manufacturing sector. *Procedia—Social and Behavioral Sciences, 195*, 806–813. https://doi.org/10.1016/j.sbspro.2015.06.180

Kimberlin, C. L., & Winterstein, A. G. (2008, December 1). Research fundamentals: Validity and reliability of measurement instruments used in researc. *American Journal of Health-System Pharmacy, 65*(23). https://doi.org/10.2146/ajhp070364

Krejcie, R. V., & Morgan, D. W. (1970). Determining sample size for research activities. *Educational and Psychological Measurement, 30*(3), 607–610.

Kuteesa, A., Lakuma, P., Gupta, R., & Kasirye, I. (2017). *Improving the growth and transition of small and medium enterprises in the Greater Kampala Metropolitan Area*. Economic Policy Research Centre (EPRC). https://doi.org/10.22004/AG.ECON.256744

MTIC. (2015, June). *Uganda micro, small and medium enterprise (MSME) policy* (Final Draft). Ministry of Trade, Industry and Cooperatives.

Motilewa, B. D., Ogbari, M., & Aka, D. O. (2015). A review of the impacts of SMEs as social agents of economic liberations in developing economies. *Journal of International Review of Management and Business Research, 4*(3), 903–914.

OECD. (2016). *Supporting SME competitiveness in the eastern partner countries: Recommendations for Georgia's SME development strategy 2016–2020.*

OECD. (2017a, June 7–8). *Enhancing the contributions of SMEs in a global and digitalised economy*. Meeting of the OECD Council at Ministerial Level, Paris.

OECD. (2017b). *Entrepreneurship at a glance 2017*. OECD Publishing. https://doi.org/10.1787/entrepreneur_aag-2017-en

Rodríguez, A., & Nieto, M. J. (2016). Does R&D offshoring lead to sme growth? Different governance modes and the mediating role of innovation. *Strategic Management Journal, 37*(8), 1734–1753. https://doi.org/10.1002/smj

Rotar, L. J., Pamić, R. K., & Bojnec, Š. (2019). Contributions of small and medium enterprises to employment in the European Union countries. *Economic Research-Ekonomska Istraživanja, 32*(1), 3296–3308. https://doi.org/10.1080/1331677X.2019.1658532

Sarantakos, S. (1998). *Social research* (2nd ed.). Macmillan Education Australia.

Sazir, N. M. (2020). Assessing the status of growth of MSMEs in a developing country: Uganda as a case study. *Science Journal of Business and Management*,

Special Issue: *Business Policy& Strategic Management*, *8*(2), 83–89. https://doi.org/10.11648/j.sjbm.20200802.15

Tushabomwe-Kazooba, C. (2006). Causes of small business failure in Uganda: A case study from Bushenyi and Mbarara towns. *African Studies Quarterly*, *8*(4), 27–35.

Ugur, M., Trushin, E., & Solomon, E. (2016). Inverted-U relationship between R&D intensity and survival: Evidence on scale and complementarity effects in UK data. *Research Policy*, *45*(7), 1474–1492. https://doi.org/10.1016/j.respol.2016.04.007

UN. (2017). *Sustainable development goals report 2017*. United Nations.

UN Secretary-General & UNCTAD. (2020). Entrepreneurship for sustainable development. Report of the Secretary-General, Geneva and New York.

World Bank Group. (2014). *Doing business economy profile 2015: Uganda*. https://hdl.handle.net/10986/21005. License: CC BY 3.0 IGO.

Zimmermann, V. (2017, December). *Research and development (R&D) in SMEs : Internal funding capacity determines scope of R&D expenditure* (KfW Research: Focus on Economics, No. 190). https://www.kfw.de/PDF/Download-Center/Konzernthemen/Research/PDF-Dokumente-Fokus-Volkswirtschaft/Fokus-englische-Dateien/Fokus-2017-EN/Focus-No.-190-December-2017-R-D-in-SMEs.pdf

CHAPTER 8

MSMEs as a Driving Force for Achieving SDGs in Developing Economies: Special Focus on Gender Equality in India

Himachalam Dasaraju, Sudarsan Kasa, Shibaprasad Parhi, and Nithya US

Introduction

The MSMEs are key drivers of the development of the economy and the backbone of many economies—both developed and developing, accounting for 90% of the business, 60–70% of employment, and 50% of GDP worldwide (UN, 2022a). The MSMEs delineate 95% of all the existing companies across the world, and more than 90% of all enterprises can be considered as MSMEs, of which a large number of enterprises

H. Dasaraju (✉)
School of Commerce and Management, Sri Venkateswara University, Tirupati, India
e-mail: dhchalam@gmail.com; dhchalam@yahoo.com

S. Kasa
Sreenivasa Institute of Technology and Management Studies, Chittoor, India
e-mail: drsuntpt@gmail.com

H. Dasaraju and T.T.H. Tambunan (eds.), *Role of Micro, Small and Medium Enterprises in Achieving SDGs*, Sustainable Development Goals Series, https://doi.org/10.1007/978-981-99-4829-1_8

can be classified as micro firms, with lesser than ten employees in many countries. They may be small individually, but put together the micro and small enterprises with their workers account for a staggering 70% of employment worldwide as per revised ILO data (ILO, 2019). The SMEs/ MSMEs have a lion's share in terms of the number of enterprises and job creation among G20 countries. This sector plays a significant role in economic development and employment generation in all countries. "SMEs sector (firms with either less than 300 employees or 300 million yen in the capital in general) account for 99.7% of firms, employing approximately 70% of the workforce in Japan. They produce about half of the value added in the manufacturing sector. The legal or statistical definitions of SMEs vary across countries and also across sectors" (ICSB, 2020).

The MSMEs in the formal sector contribute up to 40% of the GDP in most emerging countries. When informal SMEs are included, these numbers are significantly higher. According to the estimates of the World Bank, around 600 million jobs will be required by 2030 to absorb the ever-increasing workforce globally, which makes SMEs a top priority sector for promotion and nurturing across the world. Most of the jobs in the formal sector, around 7 out of 10, will be generated by SMEs in emerging markets. But this sector is not thriving well due to myriad problems encountered, in which getting access to finance is an obstacle to the development of SMEs in emerging markets and developing countries as well (World Bank, 2020). The MSMEs play a critical role in most economies in general and developing economies in particular. In emerging economies, the MSMEs contribute around 45% of total employment in the formal sector and up to 33% of GDP. When enterprises in the informal sector are included, these numbers would be much higher significantly. They have a huge potential for employment generation mostly in developing and underdeveloped countries. In the formal

S. Parhi
Adani Institute of Infrastructure Management, Ahmedabad, India
e-mail: bkshibprasad@gmail.com

Nithya US
Sree Narayana College for Women, Kollam, India
e-mail: nithyaus05@yahoo.co.in

sector, 4 out of 5 new positions were created by MSMEs, which is about 90% of total employment in emerging markets. As reported by the World Bank, a huge number of jobs will be needed in the next 15 years to absorb the ever-growing global workforce, mainly in Asia and sub-Saharan Africa (World Bank, 2018a). Most MSMEs largely rely on foreign trade for their business activities, as they export their products through various channels, and they import their inputs to manufacture the goods for sale in the domestic market. The MSMEs are the employers of huge unemployed mostly women and young people, and they are the new innovators of new things (WTO, 2020).

MSMEs: World Scenario

The MSMEs are the strongest propellers of economic development, innovation, and employment generation. According to Asian Development Bank (ADB) statistics, in the Asian region specifically, MSMEs account for 96% of total enterprises, employ 62% of the labor workforce, and contribute an average of 42% of gross domestic product (GDP) (OECD, 2017). In most developing countries, the MSMEs constitute over 90% of total enterprises, are generators of huge employment growth, and contribute a major share of industrial production and exports. The MSMEs are recognized globally for their contributions to innovation and enhancing economic conditions and employment generation. The MSMEs Economic Indicators Database 2019 indicated the region-wise number of formally registered MSMEs across 176 economies, as mentioned in the following table. The numbers in the parenthesis indicate the rank in terms of the number of MSMEs in various regions.

Table 8.1 indicates that the East Asia and Pacific region is having a higher number of MSMEs followed by South Asia, Sub-Saharan Africa, Europe, Central Asia, North America, etc., The MSMEs employment generation would directly help the poor and vulnerable sections, more so youth and women, thereby increasing economic status and reducing poverty of the people significantly (Fig. 8.1).

The entrepreneurs, of the MSMEs sector, are the lifeline of the economy in the process of effective utilization of untapped local resources such as men, money, and material resources. According to SME Finance Forum, the MSMEs are the major employment generators, mostly for 9/10 of businesses and 2/3 of jobs worldwide. By the end of 2030, MSMEs requires to meet more employment needs by building resilience to absorb

Table 8.1 Region-wise MSMEs (numbers)

S. No.	*Region*	*MSMEs per Region Nos.*	*%*	*Rank*
1	East Asia and the Pacific	102,289,112	31.72	(1)
2	Europe and Central Asia	35,027,833	10.86	(4)
3	Latin America and the Caribbean	17,217,911	5.34	(6)
4	Middle East and North Africa	10,066,801	3.12	(7)
5	North America	33,729,168	10.46	(5)
6	South Asia	75,924,258	23.55	(2)
7	Sub-Saharan Africa	48,185,712	14.94	(3)
Total		322,440,795	100.00	

Source MSME Economic Indicators (MSME-EI, 2019) https://www.smefinanceforum.org. Data Sites (/data-sites)

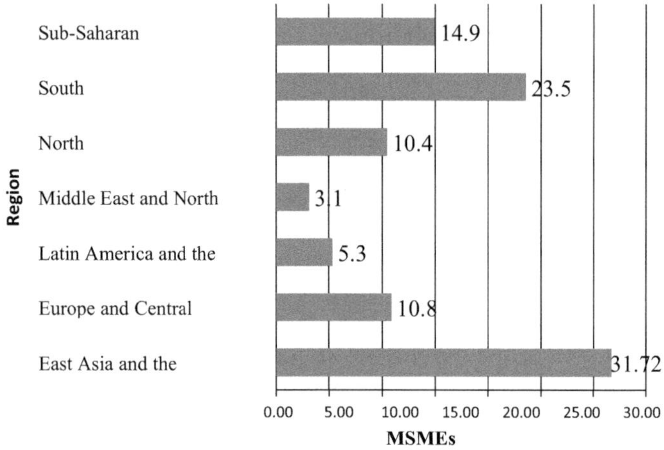

Fig. 8.1 Region-wise percentage of MSMEs (*Source* MSME Economic Indicators [MSME-EI, 2019] https://www.smefinanceforum.org). Data Sites (/data-sites)

the ever-increasing workforce across the world, both skilled and unskilled in formal and informal sectors. The International Council for Small Business (ICSB) has prioritized the top 10 trends for MSMEs in 2019 (ICSB, 2019a) as detailed below (Fig. 8.2).

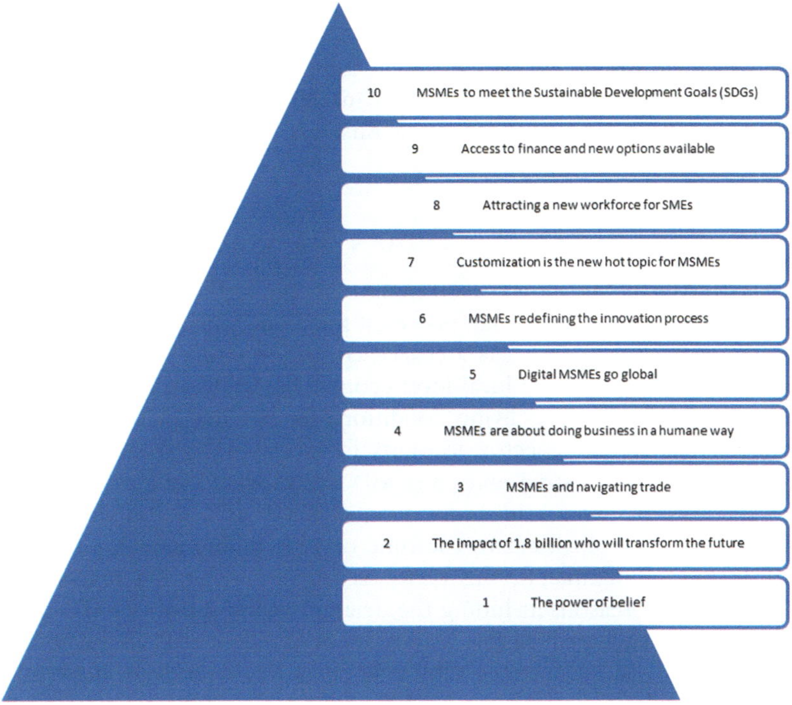

Fig. 8.2 Top ten trends for MSMEs predicted by International Council for Small Business (ICSB) (*Source* ICSB [2019]. ICSB Top 10 MSMEs Trends for 2019, December 27, 2018, Washington, DC, USA. https://icsb.org/icsb-top-10-msme-trends-2019/)

International MSMEs Day, 27th June

Keeping in view the vital role of MSMEs in achieving SDGs, the UN General Assembly has decided to observe the 27th June of every year as "International Day for MSMEs" to spread public awareness of their contributions to sustainable development and the global economy. The delegation of Argentina, the 193-member body, has invited all stakeholders including the Member States, UN entities, and civil society organizations to observe the day and create public awareness of their contributions to the 2030 agenda for Sustainable Development with different themes. The topic for UN International MSMEs Day, 27th June

2022 was "Resilience and Rebuilding: MSMEs for Sustainable Development" (UN, MSMEs Day, 2022b). "These MSMEs can become the engines that sustain growth for long-term development in developing countries," the representative said, thanking the contribution made by the International Council for Small Business (ICSB) to the creation of the Day" (Power2SMEs, 2017).

MSMEs: Indian Scenario

The MSMEs in India employ 60 million people in the rural areas mostly and are recognized as a huge potential for employment generation next to the agriculture sector. As a crucial contributor to economic growth, the MSMEs develop the local level economies, by creating employment opportunities and good living conditions for the low-income group of people. The MSMEs sector accounts for about 45% of manufacturing output, 95% of industrial units, and 40% of exports to various countries.

The MSMEs sector in India comprises 63 million enterprises and created huge employment to the tune of 111 million jobs as of August 2021. But unfortunately, this sector is lagging in getting registrations due to various reasons including the deleterious impact of COVID-19. If MSMEs get registered, all the establishments qualify to get all the government benefits in the form of subsidies, quick loan approvals, discounted electricity bills, and access to MSME clusters for skill upgradation and technology development. As of December 2019, the registered MSMEs are only 8.3 million, accounting for 13% of all MSMEs. These registered MSMEs are generating around 46.6 million jobs accounting for 42% of jobs in India. Others, 87% are unregistered and their potential for job creation is 58% of total employment generated by MSMEs in India. The top 80% of registered are widely spread in various sectors as detailed in Figure 8.3. Assessing the huge employment-creating potential of MSMEs and their significant contributions to the exports and GDP, the Indian government has been promoting registering more establishments as MSMEs. Towards this end, the government of India has simplified in registration process further and established a new portal "Udyam Registrations" to integrate the Income tax and databases of GST on a regular basis (MSME Report, 2020).

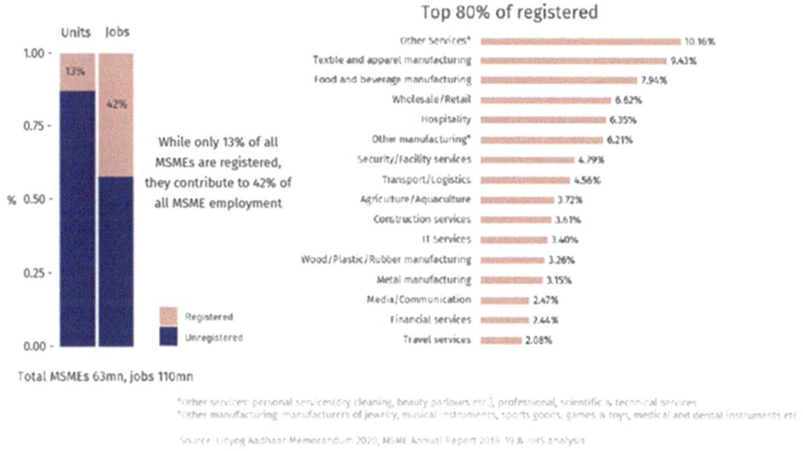

Fig. 8.3 MSMEs as large employment generators in India (*Source* Udyog Aadhar Memorandum [2020], MSME Annual Reports 2019–20; Adapted from: Viola Levis [2021], India Spend—MSMEs as Larger Employment Creators, IIHS Urban Informatics Lab, Bangalore, India)

Major Issues

Access to and meeting financial needs is a big challenge and hinders the growth of MSMEs in all regions of the world. MSMEs need more financial support to expand and flourish. Around 131 million (41% of formal MSMEs) in developing economies are suffering a lot for want of required financing needs. The unmet financial demand is estimated to be around $4.5 trillion. The women-owned businesses account for 23% of MSMEs and 32% of the financing gap (IFC, 2019). As per estimates of the IFC around 65 million firms or 40% of formal MSMEs have an unmet financing need of $5.2 trillion every year, which is equivalent to 1.4 times the current level of global MSME lending in developing countries. Both East Asia and the Pacific region alone account for the highest share around 46% of the gap in total global finance, followed by Latin America and the Caribbean (23%), Europe and Central Asia (15%), and this gap of volume changing region to region considerably. The Latin America and the Caribbean and the Middle East and North Africa regions are having more coherence in the finance gap in particular and then look at the potential demand, accounting for 87% and 88%, respectively. Around 50% of SMEs in the formal sector have limited access to finance for their business operations. The financing gap is even larger when micro and informal enterprises are taken into its ambit (World Bank, 2020).

According to the information published by the WTO Secretariat, the MSMEs are miserably suffering due to the COVID-19 pandemic, which impacted the supply chain badly. Micro businesses are the hardest hit by the crisis. Further, WTO reported that disruptions in the supply chain had a very bad impact on MSMEs and a big challenge for micro and small firms with scanty capital and supply options in the market. The report focused on comprehensive measures to remedy the problems by the governments to help MSMEs, like measures to address financial issues, widen trade opportunities, and make them more vibrant and successful enterprises. To preserve jobs and productivity as well, support systems for MSMEs are very crucial (Dasaraju et al., 2022). The WTO members and observers have felt that the supporting measures are vital and introduced stimulus packages and measures to address the cash flow issues confronted by MSMEs. According to WTO, 152 members had introduced immediate stimulus and backstop measures for MSMEs by the end of July 2020, which are meant for payment deferrals and reductions, state loans, state-backed loan guarantees, and wage financial support (WTO, 2020) (Fig. 8.4).

MSMEs are confronting a lot of issues and running a small business can be challenging even in favorable conditions. This sector is suffering from myriad problems to run its business operations successfully. From the employers' perception, this sector is facing major challenges like a non-conducive regulatory environment, low access to finance, shortage of required skills and access to infrastructure, whereas, from the worker's

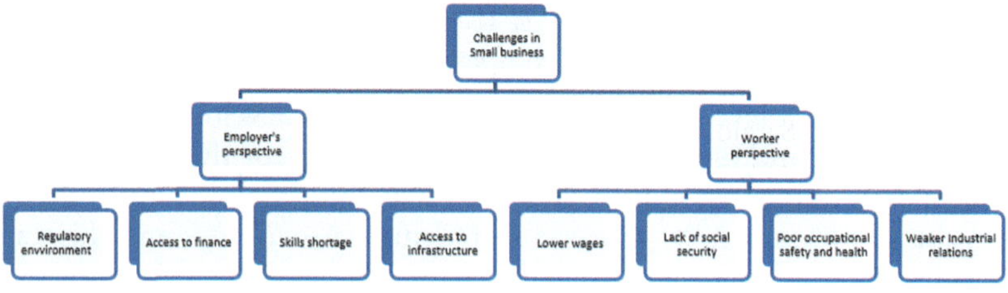

Fig. 8.4 Major challenges of MSMEs

perspective, low wages, lack of job and social security, poor occupational safety, health, and poor industrial relations are some prominent issues.

It's evident in many developing countries, including Indonesia, that the development of many MSMEs is hampered by several constraints. These problems may vary from region to region and between different enterprises within the sector. However, certain problems are common to all MSMEs, which include limited access to working capital, lack of business acumen and required information, lack of technology and skilled workers, and the like. The marketing issues, non-availability of required raw materials, managerial deficiencies, and government policies are often detrimental to the MSMEs' growth and development (Tambunan, 2019). The experts believe that new technology will be the "new normal post COVID-19 for MSMEs." All businesses are now looking forward to new kinds of technology and to building a robust digital-driven delivery capability. The drive towards digital transformation and greater use of disruptive technologies may affect physical employment opportunities. Emerging disruptive technologies may replace the human workforce. Further, it needs to focus more on upskilling the workforce to cope with the rapid pace of change and supporting technology in MSMEs (Dasaraju et al., 2020).

MSMEs and SDGs

United Nations General Assembly in September 2015 framed 17 SD Goals to transform the world for the well-being of the people, which are ratified and were presented as a universal call to action that is targeted to be achieved by 2030. The SDGs are the core of the 2030 agenda for sustainable development adopted by 191 member states of the United Nations in 2015 and it requires massive financing from public and private sources as opined by the Asian Development Bank (ADB). The ADB works to help realize the 17 UN SDGs by all means that aim to eradicate poverty, protect the planet and ensure prosperity, and peace for all people across the world (ADB, 2022). The SDGs are the pathway to accomplishing a more sustainable future for the well-being of people across the world. Table 8.2 provides linkages of SDGs with 5 Ps, Viz., People, Prosperity, Planet, Peace, and Partnership and with various sectors like Social, Economic, Environmental, Fostering Peace, and Partnerships.

In most developing countries, the MSMEs play a very critical role in the stimulation of economic activities. MSMEs constitute the major

Table 8.2 SDGs related to 5Ps and industrial sectors

5Ps		*SD Goals*	*Sectors*
End poverty and hunger in all forms and ensure dignity and equality	People	Goal 1: No Poverty Goal 2: Zero Hunger Goal 3: Good Health and Well-being Goal 4: Quality Education Goal 5: Gender Equality Goal 6: Clean Water and Sanitation	Social
Ensure prosperous and fulfilling lives in harmony with nature	Prosperity	Goal 7: Affordable and Clean Energy Goal 8: Decent Work and Economic Growth Goal 9: Industry, Innovation and Infrastructure Goal 10: Reduced Inequality Goal 11: Sustainable Cities and Communities	Economic
Protect our planet's natural resources and climate for future generations	Planet	Goal 12: Responsible Consumption & Production Goal 13: Climate Action Goal 14: Life below Water Goal 15: Life on Land	Environmental
Foster peaceful, just and inclusive societies	Peace	Goal 16: Peace and Justice, Strong Institutions	Fostering Peace and Partnership
Implement agenda through a solid global partnership	Partnership	Goal 17: Partnerships to achieve Goal	

Source Rajras (2021), Rajasthan SDG Index, Rajasthan, India. https://www.rajras.in/rajasthan-sdg-index/

segment of businesses and are prime contributors to employment generation and economic development across the world. SMEs represent around 90% of businesses and create more than 50% of total employment worldwide. MSMEs in the formal sector alone contribute 40% of national income (GDP), particularly in emerging economies, and when SMEs in the informal sector are included, these numbers are considerably higher.

According to the estimates of the World Bank, by the end of 2030, more than 6 hundred million jobs will be needed to accommodate the ever-increasing workforce, which assigned prominent priority to SMEs as a high potential employment generator for many governments. Most of the formal jobs are generated by SMEs, which create 7 out of 10 jobs in emerging markets. But these are not free from problems and confront multifarious challenges. However, access to required finance is the main hurdle to SMEs' growth and scanty financial resources rank second in the most cited obstacles facing SMEs to grow their businesses in emerging markets and developing countries (World Bank, 2020). According to the Business and Sustainable Development Commission, sustainable business models could open economic opportunities worth $12 trillion and create 380 million jobs by 2030, with more than 50% being located in developing countries. Such opportunities for MSMEs have been identified under each goal. MSMEs possess immense potential for the achievement of all SDGs directly or indirectly. Individual MSMEs have the potential to adopt actions in their business practice to contribute to the goal. New business models and solutions of MSMEs will help to achieve the SDGs (Dasaraju et al., 2020).

Role of MSMEs in Achieving SDGs

The MSMEs are vital contributors to economic development in both developed and developing economies across the world. The MSMEs have been assigned an important role in developed and developing countries as well, due to their huge potential for solving socio-economic challenges. Providing a congenial environment and ample opportunities for MSMEs in emerging markets is a pathway to economic development and curbing poverty levels (IFC, 2020). It indicates the pivotal role to be played by MSMEs in emerging economies where poverty is persistent and lack of opportunity to work for their livelihood is a serious problem, hindering the social status of the people. Therefore, the MSMEs are vital for poverty eradication in women and other underprivileged people. Keeping in view its critical contribution to the economy, the UN has assigned a great role to MSMEs in achieving the economic-related SDGs, fostering inclusive and sustainable economic growth, enhancing decent job opportunities, and appropriate work for the poor people, promoting innovation and sustainable industrialization, and begetting a congenial environment for good quality of life, good education and good health for the people across the world (OECD, 2017).

MSMEs' employment generation capacity would directly help the people who are poor and vulnerable in terms of income levels, particularly among women and youth, thereby alleviating poverty, increasing income levels, and enhancing household investments in education and health sectors. Nurturing and promoting MSMEs has the potential for impacting the SDGs positively across the world, including "SDG 1 (end poverty), SDG 2 (zero hunger), SDG 3 (good health and well-being), SDG 5 (gender equality), SDG 8 (promote inclusive and sustainable economic growth, employment and decent work), and SDG 9 (improve sustainable industrialization and fostering innovation)" (Dasaraju et al., 2022).

The MSMEs are having potential for huge employment generation, enhanced operational efficiency, social responsibility activities, and ethical laws and bylaws. Higher employment generation meets SD "Goal 1: Poverty Alleviation, Goal 5: Gender Equality, Goal 8: Decent Work and Economic Growth, Goal 9: Industry Innovation and Infrastructure, Goal 10: Reduced Inequalities," Sectorial Operational Efficiency meets "Goal 2: Zero hunger, Goal 9: Industry Innovation and Infrastructure, Goal 7: Affordable and Clean Energy, Goal 12: Responsible Consumption and Production, Goal 17: Partnership for Goals," Social responsibility activities meet "Goal 3: Good Health and Well-being, Goal 4: Quality Education, Goal 6: Clean Water and Sanitation, Goal 11: Sustainable Cities and Communities," and Ethical laws and bylaws achieve "Goal 13: Climate Action, Goal 14: Life below Water, Goal 15: Life on Land, and Goal 16: Peace Justice and Strong Institutions."

New business models/solutions of MSMEs will help to achieve the SDGs. Further strengthening the capacity of MSMEs through greater access to finance is identified as a key target in specific SDGs. MSMEs involved in certain areas could help in attaining most of the SD goals (Blue Orchard, 2017). The details of its linkages and associations are presented in Fig. 8.5.

The UN targets of SDGs are very ambitious and it requires a lot of effort to transform business activities-both public and private, toward the prosperity and well-being of the people and planet. This requires the adoption of new business models, innovation and technology, prudent management of resources, and sustainable business with ethical values

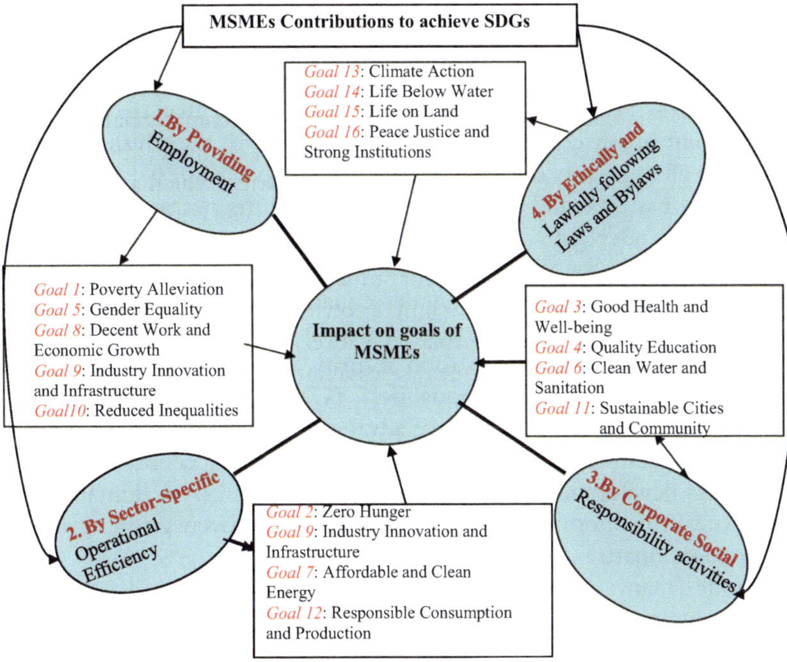

Fig. 8.5 MSMEs contributions and their impact on achieving SDGs (*Source* Blue Orchard Academy, SMEs, and SDGs, 2017 and Role of MSMEs in Achieving Sustainable Development Goals, 2019. https://doi.org/10.18231/2454-9150.2019.0189)

(Lessidrenska, 2019). According to the Business and Sustainable Development Commission, Sustainable business models could open up new economic opportunities worth $12 trillion and create 380 million jobs by 2030, mostly by the units located in developing countries. These new opportunities for the individual MSMEs units have the potential to contribute immensely to every SDG for the well-being of the people.

Women's Empowerment: World Scenario

The concept of "Women's Empowerment has gained a lot of importance owing to its inherent strengths of women worldwide, as the global decisions of various countries as a part of global partnership for development, centering on population development, more so on women empowerment. The task of promoting women's empowerment is included in the Millennium Development Goal and ICPD Programme of Action, Principle-4" (ICPD, 1994). The UN report has also emphasized that the economic development of any country is closely associated with the advancement of women in all segments of the economy (UN Report, 2020).

The UN has reported as follows: "Women's empowerment and minimizing the gender gaps in the world of work are key to achieving the 2030 UN Agenda for Sustainable Development and achieving the SDGs, particularly Goal 5, to achieve gender equality, and Goal 8, to promote full and productive employment and decent work for all; also Goal 1 on ending poverty, Goal 2 on food security, Goal 3 on ensuring health and Goal 10 on reducing inequalities" (UN SDGs, 2020). In developed economies where women have advanced in terms of socio-economic and financial status, undoubtedly there will be steady economic growth and higher development. In contrast, where women have limitations to participate in economic activity, the economy has been grinding halted (Dasaraju & Bhattrai, 2019).

Despite many international agreements on human development affirming women's rights, women are still lagging behind men as they are poor and illiterate across developing and underdeveloped economies. The differences in gender in laws affect both developing and developed countries and women across various regions. The World Bank has reported clearly that "Over 2.7 billion women globally are legally restricted from having the same choice of jobs as men. Of 189 countries assessed in 2018, 104 economies still have laws preventing women from working in specific jobs, 59 economies have no laws on sexual harassment in the workplace, and 18 economies where husbands can legally prevent their wives from working" (World Bank, 2018a, 2018b).

Women are having limited access to all resources and facilities than men usually in developing economies. They are far less likely than men to be politically and economically active and far more likely to be victims of domestic violence. When women access more work and participate actively in the economic activity in any country, that economy would flourish and grow faster. Economic empowerment of women boosts productivity and increases economic diversification and income equality in addition to other positive development outcomes (IMF, 2018). "For example, increasing the female employment rates in OECD countries to match that of Sweden could boost GDP by over USD 6 trillion."

Nevertheless, the growth does not automatically lead to a reduction in gender-based inequality. Conversely, it is estimated that gender gaps cost the economy some 15% of GDP (UN, 2018).

Gender Equality and Woman Empowerment

Gender equality is a fundamental human right, and it's a necessary foundation for a healthy, peaceful, prosperous, and sustainable world. The empowerment of women is a critical aspect of achieving gender equality. Though gender equality is well recognized as a human right, in reality, the entire world is confronting a persistent gap in getting access to all resources—physical, financial, and natural, and opportunities and decision-making power between women and men. Women are left with lesser opportunities than men in getting income, participating in economic activities, trivial access to basic needs in higher education, health and safety, and less political representation across the world (Tambunan et al., 2022). Ensuring the rights of women and giving them opportunities to reach their full potential is critical not only for attaining gender equality but also for meeting a wide range of international development goals. Empowered women and girls contribute to the health and productivity of their families, communities, and countries, creating a ripple effect that benefits everyone across the world (Peace Corps, 2019).

Women empowerment and gender equality are recognized globally as vital elements to achieving progress in all areas. The policies and programs that enhance and broaden the scope of participation of women in all areas of life as equal partners improve their access to all resources to the fullest extent possible and exercise their fundamental rights wherever they can. The Beijing Conference (1995) was the first effort and significant milestone toward ensuring gender equality and women's empowerment in the journey of gender equality and women empowerment. The salient features and prime objectives of the Conference were (UN Women, 2020): (i) Equal sharing of power; (ii) Ensuring full access to the resources of development; (iii) Mitigating poverty levels; (iv) Peace promotion and protecting the basic rights of women; and (v) Educating and motivating the young women to work closely together for getting equality.

Table 8.3 Top 10 and India Gender Gap Index at global and regional in 2022

Global rank	*Country*	*Regional rank*	*Gender gap score*
1	Iceland	1	0.908
2	Finland	2	0.86
3	Norway	3	0.845
4	New Zealand	1	0.841
5	Sweden	4	0.822
6	Rwanda	1	0.811
7	Nicaragua	1	0.81
8	Namibia	2	0.807
9	Ireland	5	0.804
10	Germany	6	0.801
135	*India*	*6*	*0.629*

Source World Economic Forum (WEF), Global Gender Gap Index (2022)

Gender-Related Development Index (GDI)

The Gender-related Development Index (GDI) and the Gender Empowerment Measure (GEM) were introduced in 1995 in the Human Development Report of the UNDP. The following table and figure present the Gender Gap Index at Global and Regional in 2022.

In terms of the Global rank, as detailed in Table 8.3 and Fig. 8.6, Iceland tops the list (No.1), followed by Finland (2), Norway (3), New Zealand (4), and Sweden (5). Whereas, India ranks 135 global and 6 regional ranks in 2022.

The following Table 8.4 and Fig. 8.7 present the Gender Gap by Region in 2022. It furnishes details about the gender gap in various regions.

Next, Table 8.5 and Fig. 8.8 provide the details of the GGG Index which indicates the extent of closeness to the gender gap in the top 10 countries, as reported by the WEF report of GGG Index, July 2022.

Global Gender Gap Index 2022 in South Asia (Out of 146 Countries)

The following Table 8.6 and Fig. 8.9 present the country-wise gender parity in South Asia (percentage) and their respective global rank. In terms of regional rank, Bangladesh tops the list (1), followed by Nepal

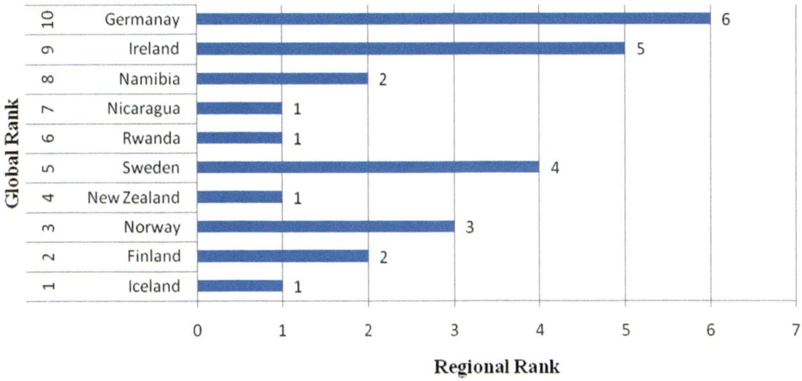

Fig. 8.6 Top 10 Gender Gap Index at global and regional in 2022

Table 8.4 Gender gap by region in 2022

S. No.	*Region*	*Gender gap (%)*
1	North America	76.90
2	Europe	76.60
3	America and the Caribbean	72.60
4	Central Asia	69.10
5	East Asia and the Pacific	69.00
6	Sub-Saharan Africa	67.80
7	The Middle East and North Africa	63.40
8	South Asia	62.40

Source World Economic Forum (WEF), Global Gender Gap Index (2022)

(2), Sri Lanka (3), Maldives (4), Bhutan (5), India (6), and so on. India got 135th global and 6th regional rank.

It is observed from the above diagram of the GGG Index 2022 of WEF, that India got 135th rank out of 146 countries, and 62.9% of the gender gap in the gender parity in South Asia. "India has slipped 28 places to rank 135 among 146 countries as reported by the Global Gender Gap Report 2022, by World Economic Forum. In 2020, India ranked 112th among 153 countries. Iceland has topped the index as the most gender-equal country in the world for the 12th time."

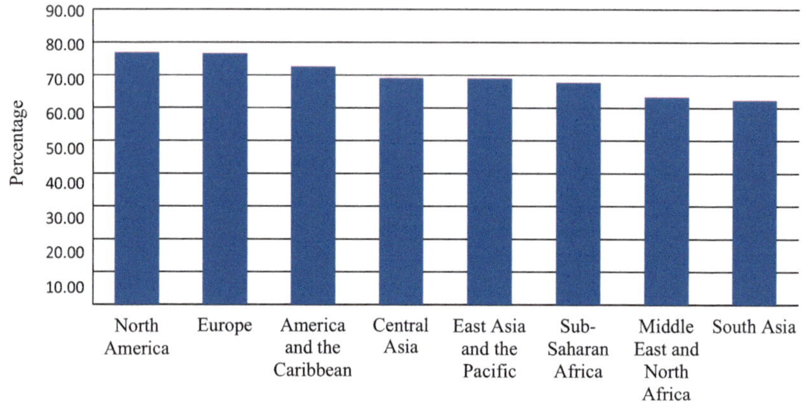

Fig. 8.7 Gender gap by region in 2022

Table 8.5 How close are countries to closing the gender gap?

S. No.	*Country*	*Global Gender Gap Index*
1	Iceland	0.908
2	Rwanda	0.811
3	Germany	0.801
4	United States	0.769
5	Poland	0.709
6	Japan	0.65
7	Saudi Arabia	0.636
8	*India*	*0.629*
9	Iran	0.576
10	Afghanistan	0.435

Source World Economic Forum (WEF), Global Gender Gap Index (2022)

Table 8.7 and Fig. 8.10 furnish the details of the country-wise HD Index. India ranks 131 in 2019, decreasing by 2 from 2018 due to various reasons. It requires improving the HD Index to enhance the country's overall performance in different variables.

Interestingly, India dropped a place in the overall Human Development Index (HDI) ranking though improved in its absolute value of HDI to 0.645 in 2019 from 0.642 in 2018, a change of −2 from 2018. We have miles to go to reach the considerable global HDI,

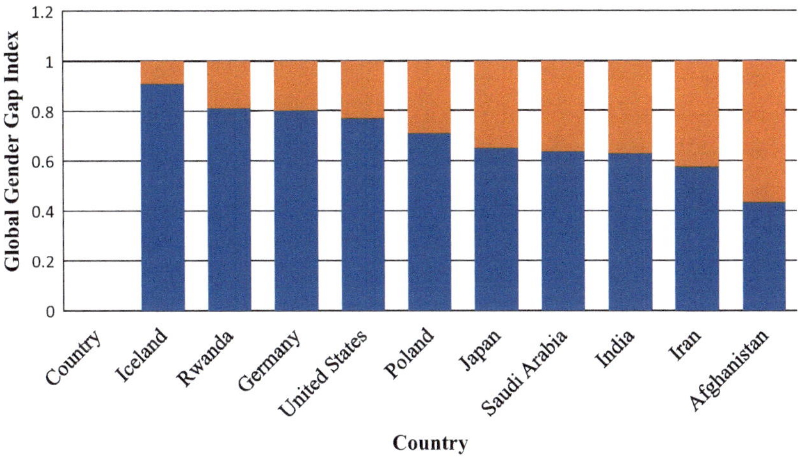

Fig. 8.8 How close are countries to closing the gender gap?

Table 8.6 Global and regional rank Gender Gap Index 2022 in South Asia

S. No.	*Country*	*Regional rank*	*Global rank*
1	Bangladesh	1	71
2	Nepal	2	96
3	Sri Lanka	3	110
4	Maldives	4	117
5	Bhutan	5	126
6	*India*	*6*	*135*
7	Iran, the Islamic Republic of	7	143
8	Pakistan	8	145
9	Afghanistan	9	146

Source World Economic Forum (WEF), Global Gender Gap Index (2022)

followed by Bangladesh, and Nepal. "The Global Gender Gap Report is annually published by the World Economic Forum (WEF). It benchmarks countries on their progress toward gender parity in four dimensions: Economic Participation and Opportunity, Educational Attainment, Health and Survival, and Political Empowerment" (WEF, 2021).

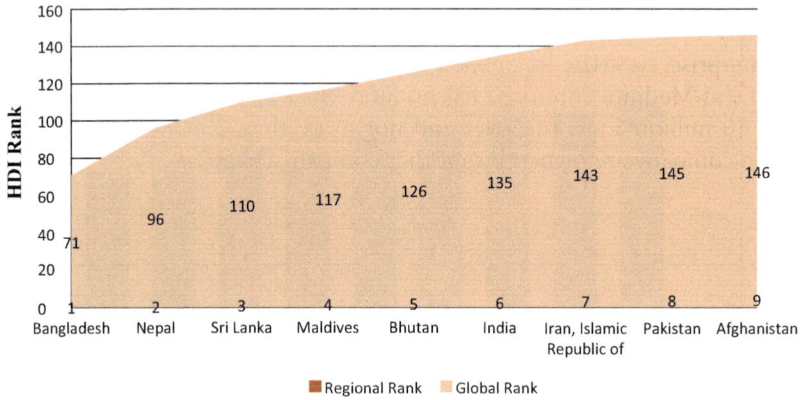

Fig. 8.9 Global Gender Gap Index in South Asia region 2022

Table 8.7 Country-wise Human Development Index rank 2019

Country	*HDI rank (2019)*	*Change from 2018*
Russia	52	−3
Sri Lanka	72	1
Brazil	84	0
China	85	2
South Africa	114	1
India	*131*	*−2*
Bangladesh	133	1
Nepal	142	1
Pakistan	154	0

Source World Economic Forum (WEF), Global Gender Gap Index (2021)

Comparison of Global/Sub-Index of India in 2006 and 2022

The following Table 8.8 and Fig. 8.11 present the Comparison of the Global/Sub-index of India in 2006 and 2022. Table 8.8 presents the factor-wise India's index and score in 2006 and 2022. It can be observed from the table that the global gender gap index score rose from 0.601 in 2006 to 0.625 in 2022 and the rank rose from 98 in 2006 to 140 in 2022. It indicates that the global rank slipped 42 ranks from 2006 to 2022.

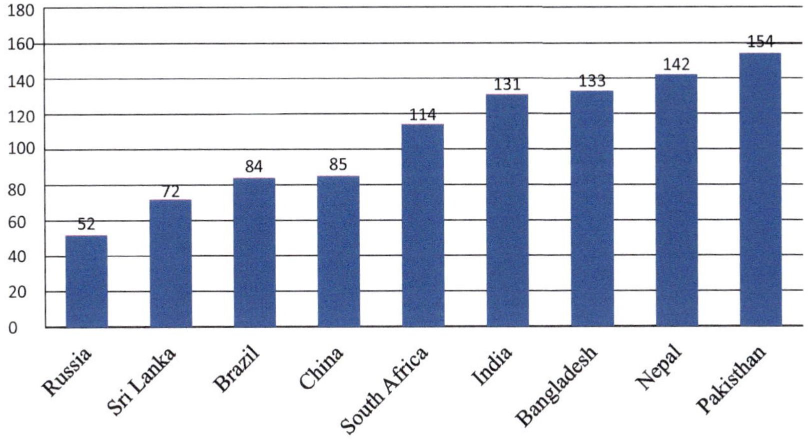

Fig. 8.10 HDI rank 2019

Likewise, all other factors like economic opportunity and participation, educational attainment, health and survival, and political empowerment have declined from 2006 to 2022.

It is interesting to note that, as reported in the global gender gap index 2021, the women's share in administrative positions like senior and managerial positions remains low: 14.6% of these positions are held by women only and there are only 8.9% of firms are having women as top managers. "The estimated income level of women in India is only one-fifth of that of men, which puts the country among the bottom 10 globally on this indicator. Serious discrimination against women is also reflected in the health and survival sub-index statistics. With 93.7%

Table 8.8 Comparison of Global/Sub-index of India in 2006 and 2022

Index/Sub-index	*Rank (2006)*	*Score*	*Rank (2022)*	*Score*
Global Gender Gap Index	98	0.601	135	0.629
Economic participation and opportunity	110	0.397	143	0.35
Educational attainment	102	0.819	107	0.961
Health and survival	103	0.962	146	0.937
Political empowerment	20	0.227	48	0.267

Source World Economic Forum (WEF), Global Gender Gap Index (2006) and (2022)

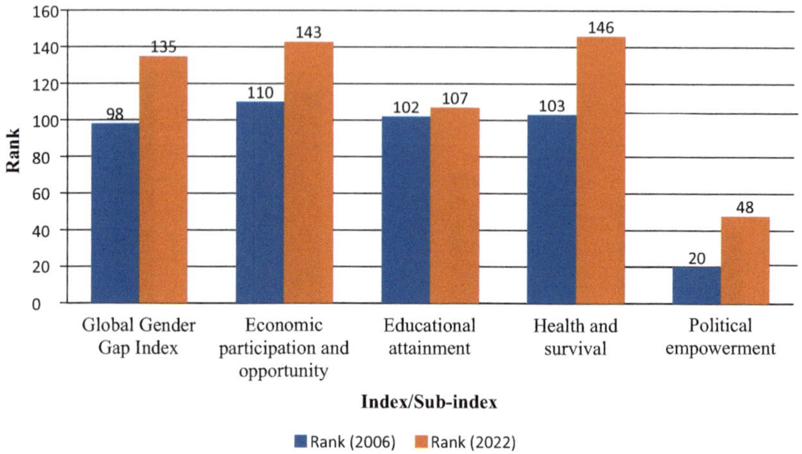

Fig. 8.11 Comparison of Global/Sub-index of India in 2006 and 2022

of this gap closed to date, India ranks among the bottom five countries in this sub-index." As per the report, more than one out of four women has faced brutal violence in her lifetime. "Conversely, 96.2% of the educational attainment sub-index gender gap has been closed, with parity achieved in primary, secondary, and tertiary education. Yet, gender gaps persist in terms of literacy: one-third of women are illiterate (34.2%) compared to 17.6% of men," the report revealed (WEF, 2021).

Gender equality is not a fundamental human right alone, but also a basic necessary foundation for a sustainable world with a sense of peace and prosperity. It is observed that there exists a considerable development in the last decades: "School going girls are increased, the early marriage rate of girls decreased, the number of women serving in parliament is increased and increased positions of leadership, women related laws are being reformed to improve gender equality rate. Despite these gains, many serious problems and challenges are persisting like discriminatory laws and social norms that remain pervasive. Women are still underrepresented at various political leadership levels and other administrative positions. 1 in 5 women and girls between the ages of 15 and 49 are reported experiencing physical or sexual violence by an intimate partner within 12 months" (UN SDGs, 2020).

The ill effects of the "COVID-19 pandemic could reverse the limited progress that has been made on gender equality and women's rights. The coronavirus outbreak exacerbates existing inequalities for women and girls across every sphere of activity." Women play a greater role in combating COVID-19, serving as frontline healthcare workers and caretakers at home and office. The unpaid care work of women has multiplied due to the closure of schools and offices, and the needs of older people at home (UN, 2021).

Economic Empowerment of Women

The economic empowerment of women is crucial for all economies to flourish further. Economic empowerment of women is vital to realize the rights of women and gender equality as well. It includes the ability of women to participate equally in existing business economic activity, their access to productive resources, access to decent work, control over their own time, lives, and resources; and increased voice, and active participation in economic decision-making at all levels from the household to international institutions. Women's economic independence is good for the business and economic development of the country. Fostering the capacity of women entrepreneurs and business owners has a significant potential to reduce poverty and inequality globally. According to UN Women, the global economy could see as much as US$28 trillion growth by 2025, if women participate equally as men in entrepreneurship (UN Women, 2017).

The great challenge that women are facing usually at work is not being properly recognized and no equal distribution of wealth and income. Women have a great contribution to the economy and society in the agriculture and food production sectors in developing countries. This is in addition to their role as homemakers which are unmeasured in economic terms. As reported by the UN, "53% of work in developing countries is done by women, and out of $16 trillion in global output which is invisible, $11 trillion is contributed by women." Even after the significant contribution of women, no proper recognition and acknowledgment are given to women in every sphere of life in all economies (UN, 2021). The issue is much larger than what we are seeing and it is not just about the economic empowerment of women, but it is about the deep-rooted women's discrimination and the bias that is existing in society for a long of all nations with a degree of variance in the severity.

(a) **Low-income generation and flow of capital in the women**

Women need more income levels and capital for investment independently. The more the women get empowered economically and generate income the more the chance of getting progress and development. Moreover, a steady income stream brings more confidence among women to take more appropriate economic decisions in their business. The status of women in economic and financial is sound, which enhances their social status and progress in their life.

(b) **Low women empowerment status**

According to the UN report, less than 50% of working-age women are in the labor market. More unpaid domestic and care work is shouldered mostly by women, restraining their economic potential as the COVID-19 pandemic additionally affects women's jobs and livelihoods, as per the UN report. "The World's Women, 2020: Trends and Statistics have compiled 100 data stories that provide a snapshot of the gender equality status across the world. The report analyses gender equality in six critical areas: population and families; health; education; economic empowerment and asset ownership; power and decision-making; and violence against women and the girl child as well as the impact of COVID-19" (UN, 2020). The unequal distribution of unpaid care work is holding back progress in women's economic environment. The development community is committed to addressing the issues in sustainable development goals.

(c) **Glass ceiling**

According to UN Report, no country has achieved gender equality and violence against women and girls remains rampant globally. Women job market participation stagnating at less than 50% for the last 25 years. "Women held only 28% of managerial positions globally in 2019—almost the same proportion as in 1995 in terms of power decision-making and only 18% of enterprises had a female Chief Executive Officer in 2020. As per the UN women's report, 2020, only 7.4%, or 37 Chief Executive Officers, were women among Fortune 500 corporations. Women's representation in parliament has more than doubled globally; it has still not crossed the barrier of 25% of parliamentary seats in 2020. Women's representation among cabinet ministers has quadrupled over the last 25 years, yet remains well below parity at 22%" (UN Women, 2020).

(d) **The rise in women's participation in education**

There is a substantial rise in women's and girls' participation in primary education across the world. Even in tertiary education, the enrolment of women is increasing faster than men. But women are underrepresented in the science, technology, engineering, and mathematics domains representing only slightly more than 35% of the world's Science, Technology, Engineering, and Mathematics (STEM) graduates. Women's economic empowerment has important linkages with gender equality. As a result, India has witnessed several schemes and progressive legislation—both at the central and state levels, that aim to empower women and increase their participation in the economy. However, despite the government's targeted efforts, India has had limited success. Several reasons can explain why these policies in India may not have had the anticipated impact.

The WEF, 2022 report released recently in July 2022, reported slightly different on various aspects. India ranks 135 out of 146 countries, an improvement over its 140th rank out of 156 countries in 2021, recording a sharp decline from the 105th rank out of 135 countries in 2012. When the GGG Index report was first introduced in 2006, India ranked 98th position among 115 countries. According to the WEF, 2021 report, India has declined by 13.5% points on the political empowerment index and recorded a decline to 9.1% in 2021 from 23.1% in 2019 in the number of women ministers. But India has continued to perform relatively well than other countries, ranking at 51 in women's participation in politics. India has been ranked 114in the educational attainment index.

The World Economic Forum's Global Gender Gap Report 2022, indicates the following in the Indian context (WEF, 2022).

1. India ranked 135 among 146 countries (India ranked 140 among 156 countries in 2021) in the overall global rankings. It is noticed that India has fallen 28 places in the overall ranking globally.
2. India's ranking in 4 sub-indexes is

    (a) Economic Participation and opportunity-143,
    (b) Educational Attainment-107,
    (c) Health and Survival-146 and

(d) Political Empowerment-48.

3. In South Asia, India lagged behind neighboring countries like Bangladesh, Nepal, Srilanka, Maldives, and Bhutan.
4. India has fared badly in Health and Survival, which includes sex ratio and economic participation in national and regional economic activity.
5. In India, the income level of women estimated is only one-fifth of men, which puts the country among the bottom 10 globally on this indicator.

The violence against women is yet another indicator of the extent of gender inequality. Over one-third of adult women have experienced sexual or physical violence. Domestic violence is one of the most common forms of violence faced by women in both urban and rural sectors. The high levels of tolerance for domestic violence show the need for continued efforts to raise awareness of women's rights. Constraints to women's economic participation, social and political recognition, and women's entrepreneurship are persisting which is limiting the scope for attaining sustainable development across the world.

Conclusion

The MSME sector is the most vital sector with huge potential to impact most of the UN SDGs across the world. It gained immense importance in achieving SDGs in developing countries, and their most vibrant and flexible nature and capacity make them a key player in achieving these goals. The huge potential for the employment generation of MSMEs will benefit the poor and vulnerable section of the people. It helps young women and youngsters in reducing their poverty, reducing the severity of hunger, enhancing living standards, and nurturing entrepreneurship. It makes a substantial contribution to achieving the SDGs, which are directed toward better prospects for people and the planet as well. Keeping in view its vital role in achieving inclusive growth and sustainable development, the UN has assigned top priority and a key role to play in this prodigious task.

Women's empowerment and gender equality are the key elements to achieving economic progress across the world without any exception. Gender equality is a major task globally and it is one of the eight-millennium goals and seventeen sustainable development goals set for the wellbeing of the people. It is a prime priority to enable women to realize their full energy, social recognition, and economic stability in life. It is a prioritized task to ameliorate women's status in developing economies like India, as women are deprived of many socio-economic benefits and are backward in many ways. But in reality, the situation is not conducive as expected globally. No country has achieved gender equality and violence against women and girls remains rampant globally as emphasized by the UN report. Women alone would change the living conditions and empower themselves by nurturing and promoting women's entrepreneurship in MSMEs as well. Certainly, this would change the Indian economic and entrepreneurial scenario in general and women's empowerment and gender equality in particular. As the MSMEs sector is suffering from multifarious challenges globally, this sector requires more stimuli to rejuvenate them further to serve better. It requires creating a congenial atmosphere for the MSMEs to prosper and access the new technology and global markets for their survival. It is a dire need to revitalize and make MSMEs resilient to play a crucial role to contribute to accomplishing SD Goals as targeted by the UN by 2030.

References

ADB. (2022). *Asian Development Bank & SD goals*. ADB's Strategy 2030-Supports 2030 Agenda, Philippines.

Blue Orchard. (2017). *SMEs and SDGs: Supporting small and medium enterprises to achieve sustainable development goals*. https://www.blueorchard.com/wp-content/uploads/Blue-Orchard-pages-11-12.pdf

Dasaraju, H., Kishore, S., & Sreenivasa Murthy, K. (2020, July). MSMEs in developing economies and their role in achieving sustainable development goals in the context of Covid19: A theoretical exposition. *International Journal of Small and Medium Enterprises and Business Sustainability, Indonesia, 5*(2), 93–120.

Dasaraju, H., Abraham, S., Kishore, S., & Aparna, S. (2022). Reflections of COVID-19 on MSMEs in developing economies: A case of India. In *COVID-19 and its reflections on SMEs in Developing Countries* (pp. 41–58). Nova Publishers. ISBN: 978-1-68507-454-8.

Dasaraju, H., & Bhattrai, H. (2019). Nurturing entrepreneurship for inclusive economic development in emerging economies: Present status and global challenges (a case of India and Nepal). *International Journal of Management and Development Economics (JMDE) Global College of Management, Kathmandu, Nepal, 9*(1), 114–127.

ICPD. (1994, September 5–13). *Report of international conference on population and development*. Cairo, Egypt.
ICSB. (2019a). *ICSB top 10 MSMEs trends for 2019*. December 27, 2018, Washington, DC, USA. https://icsb.org/icsb-top-10-msme-trends-2019/
ICSB. (2019b, June). *International council for small business, annual global micro-, small and medium-sized enterprises report*. Washington, DC, USA. ISBN: 978-1-7324980-4-4.
ICSB. (2020). *Annual global micro-, small, and medium-sized enterprise report*. The George Washington University, USA.
IFC. (2019). *SME finance forum Annual Report, 2019* (p. 8). Published by International Finance Corporation (IFC), World Bank Group, Washington, DC, USA. https://www.smefinanceforum.org/sites/default/files/SMEFF%20Annual-Report-2019Final.pdf
IFC. (2020). *MSME finance gap*. International Finance Corporation, World Bank Group, Washington, DC, USA. https://www.smefinanceforum.org/data-sites/msme-finance-ga
ILO. (2019). *Info stories, The power of small: Unlocking the potential of SMEs*. Geneva, Switzerland. https://www.ilo.org/infostories/en-GB/Stories/Employment/SMEs#power-of-small
IMF. (2018). *World Bank Annual Report 2018*. World Bank, Washington DC, USA.
Lessidrenska, T. (2019). *SMEs and SDGs: Challenges and opportunities*. OECD Development Matters, OECD, Paris, France. https://oecd-development-matters.org/2019/04/23/smes-and-sdgs-challenges-and-opportunities/
MSMEs Report. (2020). *MSME Annual Report, 2019–20*. Ministry of MSMEs, Govt. of India.
OECD. (2017). *Unlock the potential of SMEs for the SDGs, 2017*. OECD, Paris, France. https://oecd-development-matters.org/2017/04/03/unlocking-the-potential-of-smes-for-the-SDGs/
Power2SMEs. (2017). *Power2SMEs Report, 2017*. New Delhi, India.
Peace Corps. (2019). *Global issues: Gender equality and women's empowerment*. Peace Corps, Washington, DC, USA.
Rajras. (2021). *Rajasthan SDG Index*. Rajasthan, India. https://www.rajras.in/rajasthan-sdg-index/
Tambunan, T. T. H. (2019). Recent evidence of the development of MSMEs in Indonesia" Journal of Global Entrepreneurship Research. *Journal of Global Entrepreneurship Research, Springer Open, 9*(18), 11.
Tambunan, T. T. H., Dasaraju, H., & Kavita, S. (2022, April). Accelerating gender equity for promoting empowerment of women and women's entrepreneurship in developing countries: A case of Indonesia and India. Policy Discussion Paper series, Sustainable Development Program, No. 7, Universitas Trisakthi, Indonesia.

The Beijing Conference. (1995). UN women, world conference on women, Beijing. https://www.unwomen.org
Udyog Aadhar Memorandum. (2020). *MSME Annual Reports 2019–20*. Adapted from: Levis, V. (2021). *India spend—MSMEs as larger employment creators*. IIHS Urban Informatics Lab, Bangalore, India. https://www.indiaspend.com/data-viz/msme-sector-employment-creators-766041
UN. (2018). *UN Report—The world's women 2018*. Trends and Statistics, New York, USA.
UN. (2020). *UN Report—The world's women 2020*. Trends and Statistics, New York, USA
UN. (2021). *Report on the UN Women's global response to Covid19, 2021*. New York, USA, https://www.un.org
UN. (2022a). *UN observances*. United Nations, New York, USA. https://www.un.org/en/observances
UN. (2022b, June). *United Nations MSMEs day*. United Nations, New York, USA. www.un.org/observances. https://www.un.org/en/observances/micro-small-medium-businesses-day
UN Report. (2020). *The sustainable development goals report, 2020*. United Nations, New York, USA.
UN SDGs. (2020). *Do you know all 17 SDGs?* United Nations, New York, USA. www.sdgs.un.org/goals
UN Women. (2017, July 17). *Women entrepreneurs can drive economic growth*. Lakshmi Puri Speech. http://www.unwomen.org/en/news/stories/2017/7/speech-ded-puri-she-era
UN Women. (2020). *The world's women report 2020: Trends and statistics*. New York, USA.
WEF. (2021, April). *Global gender gap report*. World Economic Forum, Geneva, Switzerland.
WEF. (2022, July). *Global gender gap report*. World Economic Forum, Geneva, Switzerland.
World Bank. (2018a). *The world development report 2018*. World Bank, Washington, DC, USA.
World Bank. (2018b). *The World Bank SMEs finance*. The World Bank Group, Washington, DC, USA. www.worldbank.org/en/topic/smefinance
World Bank. (2020). *SMEs finance forum*. The World Bank Group, Washington, DC, USA.
World Economic Forum (WEF). (2006). *Global gender gap index*, Geneva, Switzerland. https://www.weforum.org
WTO. (2020, June 3). *Helping MSMEs navigate the COVID-19 crisis*. Geneva, Switzerland. https://www.wto.org/english/tratop_e/covid19_e/msmes_report_e.pdf

CHAPTER 9

Role of MSMEs for Achieving SDGs in Developing Economies with Special Reference to India and Bangladesh

Uma Shankar Yadav, Ravinder Rena, and Ravindra Tripathi

U.S. Yadav · R. Tripathi
Department of Humanities and Social Sciences, Motilal Nehru National Institute of Technology, Prayagraj, India
e-mail: usyadav@mnnit.ac.in

R. Tripathi
e-mail: ravindra@mnnit.ac.in

R. Rena (✉)
DUT Business School, Faculty of Management Sciences, Durban University of Technology, Durban, Republic of South Africa
e-mail: RavinderR@dut.ac.za

H. Dasaraju and T.T.H. Tambunan (eds.), *Role of Micro, Small and Medium Enterprises in Achieving SDGs*, Sustainable Development Goals Series, https://doi.org/10.1007/978-981-99-4829-1_9

Introduction

We know that most acceptable definition of sustainable development, that has defined by WSED Which is sort form of world economic council of sustainable development) has defined this in 1987, which defines it as "Development that meets the needs of the present without compromising the ability of future generations to meet their own needs." Several other factual definitions have been given by some individuals, famous commission working on sustainable development, and autonomous organization and several Countries share a concern for sustainable development. However, incorporating sustainable development into governmental programs has never been an easy undertaking (Sharma, 2009).

The concept of sustainable development encompasses a variety of factors, including ecological considerations as well as economic, social, and environmental considerations. The social component covers things like community development, equal opportunity, human rights, and respect for individuals. While the key economic factors are steady and profitable growth, risk management, shareholder return, etc. Permits and licenses are included in the environmental aspects. Compliance, management of biological diversity, control of air pollution, regulation of discharges and water chemical use, etc. In addition to these, there are numerous socio-economic, eco-economic, and socio-environmental problems that are a part of development that is stable and eco-friendly. The goal of sustainable development is to create a society in which people and nature coexist peacefully. Economic growth is encouraged, but not at the expense of environmental deterioration. The chapter argues that the SDGs, in which a total of 17 important goals focus on global, policy, and socio-economic welfare, and may unleash a step change in growth and productivity, with an investment boom in sustainable infrastructure as a vital driver. Not only small industries and the handicraft sector worldwide, but all firms can promote the SDGs. The UN Global Compact urges businesses to conduct themselves ethically by utilizing commercial innovation and teamwork to address societal challenges. To construct their company for the society they envision—one that is sustainable on all fronts—economic, social, and environmental, a report from business.com—each MSME owner must now take responsibility for this.

MSMEs are seen as a crucial pillar in accomplishing sustainable development objectives because of their wide range of operational skills. The

United Nations General Assembly has designated June 27 as International MSME Day in recognition of the influence MSMEs have on the world stage. The conference's subject, "MSMEs: Small Businesses, Big Impact," emphasizes the significance of MSMEs in attaining sustainable development.

MSMEs are viewed as incubators for innovation and entrepreneurship. Developing countries' MSMEs are widely scattered throughout the nation and manufacture more than 9000 different goods and products to suit the local and international market, supply, and demands, thereby integrating into national and international supply chains. MSMEs are at the center of developing economies including India's industrial types of equipment. Nevertheless, there are various definitions of MSMEs in different countries on the basis of turnover, investment, demand, and capital used in machinery. In India for the first time the systematic definition was given in 2006, but in 2020 the definition was changed and now on the basis of turnover and investment, for example, the definition of MSME in two countries is given below. The division is as follows.

MSMEs Are Categorized in Accordance with the MSMED Act of 2020 in India

The revamped definition of MSMEs may provide a uniform description for everything connected to, investment, taxation, and more, Union Minister Nitin Gad Kari stated in October 2019. An amendment that would for the medium-sized business environment for Indian firms was to be used to apply the modified definition. The modification to switch the MSMEs classification criteria from "investment in plant and machinery" to "annual turnover" had been accepted by the Union Cabinet. Nirmala Sitharaman, the finance minister, added the extra term of turnover along with the investment on May 13, 2020, as given in Table 9.1 (Yadav et al., 2020).

MSMEs are the instruments that have made policies like Digital India, Mudra scheme, stand-up India, start-up India, Make in India, etc. successful. The expansion and development of MSMEs have a significant impact on the industrial sector's advancement. In various regions of India, there are more than 63 million MSME units in operation. MSMEs make up 30.5% of the services sector and 45% of industrial production. According to data from the Directorate General of Commercial Intelligence and Statistics (DGCIS), MSME-related items in India have a value

Table 9.1 Classification of MSMEs in India according to the Two New Definitions

Modified and improved Classification of MSME in India			
New definition: annual turnover and Investment			
Classification	*Micro*	*Small*	*Medium*
Manufacturing & services	Turnover < Rs. 5 crore Investment < Rs. 1 crore	Turnover < Rs. 50 crore and Investment < Rs. 10 crore	Turnover < Rs. 100 crore Investment < Rs. 20 crore

Source Table has been adopted from the Ministry of MSME 2020

of Rs. 147,390.08 million and made up 48.56% of the nation's exports in 2017–2018 (PIB, 2018). There are approximately 11 crore workers in these MSME businesses. The MSME industry if they receive the necessary assistance and encouragement, has the potential to become a pillar of this economy and a growth engine.

Definition of MSMEs in Bangladesh

MSMEs are defined differently in each nation. To understand the difference between "micro," "small," and "medium" firms in Bangladesh, some better criteria have been used each with a shifting nature. At the moment, it is described as given in Table 9.2.

In Bangladesh, 31 million people work for MSMEs which account for more than 75% of household income, according to Zaman and Islam (2011). Additionally, MSMEs make up over 95% of industrial units in Bangladesh and over 85% of the employment in the industry. Micro enterprises make up over 95% of all industrial units and over 85% of the employment in the industry. Therefore, it is anticipated that the MSME sector would make a greater contribution to the Bangladeshi economy, making research on the MSME sector's international competitiveness more important. For this following research objective has been proposed:

1. To analyze SDG in India and Bangladesh
2. To study the contribution of MSMEs to achieving SDGs in developing economies in south-east Asia, especially India and Bangladesh.
3. To evaluate a definition of MSMEs in India and Bangladesh which have the capacity to achieve the SDG.

Table 9.2 Classification of MSMEs in Bangladesh

Type		*Fixed asset, other than land and building*		*Employed manpower*
Sector		*BDT (in million)*	*U.S Dollar*[a]	
	Manufacturing	1.0–7.5	\$12,294–\$92,213	16–30
Micro	Service	Less than 1.0	Less than \$12,294	15
	Business	Less than 1.0	Less than \$12,294	15
	Manufacturing	7.5–150	\$92, 213–\$18,44,258	31–120
Small	Service	1.0–20	\$12,294–\$245,872	16–50
	Business	1.0–20	\$12,294–\$245,872	16–50
Medium	Manufacturing	150–500	\$18,44,258–\$61,47,527	121–300
	Service	20–300	\$245,872–\$3,687,450	51–120

[a]1 USD = 81.3430 BDT (Note 2)

Review of the Literature

According to Kamal-Chaoui (2017), governments can only accomplish the SDGs if they develop robust MSMEs. The researcher stressed the necessity for policies to establish a positive business climate to increase the productivity of MSMEs while discussing the significance of MSMEs in, gender equality, employment generation, inclusive development, innovation, etc. He expressed worry about the funding of MSMEs, their involvement in, banking sector reforms, global value chains, and novel financing strategies (Yadav et al., 2022). The relationship between innovation, financial performance, and economic growth was examined by Oncioiu et al. (2017). They discovered that financial success, innovation, and economic expansion are all positively correlated. In addition, they noted that to enhance both their economic and environmental performance, Indian MSMEs are putting more emphasis on the environmental aspect of sustainable innovation. They also stated that creativity is the secret to Indian entrepreneurship success. Tambunan (2018) has broadly focused on MSMEs and their impact on the Indonesian economy.

Freimann et al. (2014) conducted extensive research on how to assess both subjective and objective well-being (economic development) (perception of residents about the quality of life). They stated that to gauge sustainable development, objective well-being must be tracked and quantified. They discussed various metrics for gauging sustainability worries about systematic measurement and monitoring advances. They concluded that benchmarks and indicators do, to some extent, measure the efforts and outcomes connected to sustainable development (Yadav et al., 2020) described the role of government in the classification and new definition of MSMEs.

Liu (n.d.) talked about the value of MSMEs in achieving the SDGs and described the opportunities and obstacles that may stand in the way of their expansion. The author also highlighted the contribution of MSMEs in creating jobs but contends that these employment prospects may not always be in line with work requirements. Tambunan (2018) explained new clients or markets in other nations not affected by the crisis or switching to the domestic market, were found to be the most often used crisis-mitigating (CM) measure by MSMEs hit by the 2008–2009 crisis, according to survey results. The most crucial change made by other respondents who altered their staff was to shorten the working day.

Methodology

The study is descriptive in nature and is based on secondary data. The information was gathered by researchers from reputable, published secondary sources, journals, magazines, reports, websites, and other sources are among them. The majority of the data was gathered from the United Nations website (https://sustainabledevelopment.un.org) and the Indian Government's Ministry of MSME website and from Bangladesh. The statistical information from various studies has been mentioned to emphasize the significance of the problems that have been covered. Researchers have characterized MSMEs' duties under four categories of activity attributes, including providing employment, operational efficiency, engaging in CSR activities, and adhering to the law and bylaws in an ethical and legal manner.

Role of MSMEs in Attaining SDGs

MSMEs are crucial to the growth of the developing economy. By providing jobs for the nation's expanding labor force, they aid in the reduction of poverty. They promote economic growth in remote and rural places. They will create jobs, raise the GDPs of countries, improve lifestyles, make people happier, develop children's skills, minimize capital major production, preserve nature, protect them from pollution, and provide everyone with work or decent work. Therefore, we may conclude that women's participation in MSMEs can address the issues of unemployment, establish sustainability, provide work for everyone, keep the environment clean, and banish pollution.

MSMEs dominate both developing and developed countries, and in that situation, MSMEs would find it difficult to decide how this sector should contribute to maintaining sustainability in the modern and digital era and who, in particular women, can solve the problem of some concerns about Job, GDP, and increase happiness. Even though MSMEs directly cannot help the Goal of Sustainable development in the same way, they can indirectly. MSMEs dominate the private sector and economic activity in both developed and developing countries, according to the Organization for Economic Cooperation and Development (OECD). Given the wealth of information available to them and their customary lack of it is clear why MSMEs would find it difficult to decide how they should contribute when compared to larger organizations.

Many MSME core employees, such as women and tribes in the handicraft industry, would likely be shocked to learn just how "great" their company already is and how minor adjustments might make it even better, in the words of Flagler's for sustainable enterprise. Because customers want to support businesses that are genuinely changing the world, being able to share that story can help MSMEs stand out from their rivals. By 2030, sustainable and profitable business models could provide 380 million new employments and $12 trillion in economic opportunities, claims the Business and Sustainable Development Commission (BSDC) UNDP 2020.

And of this share, MSMEs account for 60%, with the full participation of women, children, and tribal people comprising 50% of the population of developing nations. It is crucial to understand how small businesses can advance toward sustainable development goals and why MSMEs are so crucial given their significance in the global economy. This was accomplished by tying SDG 3 to SDG 8 and the sub-goal of SDG 10. Provide opportunities for apprenticeships. Encourage an entrepreneurial culture. And then gently bring up SDG 10 B, working slowly to give entrepreneurship education and training in collaboration with civil society networks.

MSMEs working in particular sectors could contribute to a variety of aims in many different ways (Blue Orchard, n.d.). We have broken down

MSMEs' contributions to achieving those goals into four categories of activity attributes: creating jobs, operating efficiently, engaging in corporate social responsibility (CSR), and abiding by the law and regulations in an ethical and legal manner.

Each action aids in the accomplishment of particular objectives. Although some activity attributes may be useful in achieving several goals, we have taken a straightforward approach and only assigned one goal to each activity element.

MSMEs immediately contribute to the achievement of goals including reducing inequality and alleviating poverty (SDG 1), decent work and economic growth (SDG 8), gender equality (SDG 5) industry innovation and infrastructure (SDG 9), and economic growth (SDG 10). By operating efficiently in their respective sectors, MSMEs help achieve several objectives, such as Partnership for Goals (SDG 12), Responsible Consumption and Production (Goal 7), and Ending Hunger (Goal 2).

MSMEs also contribute to the SDGs by carrying out CSR activities in a variety of ways that support the objectives of promoting good health and well-being (Goal 3), access to clean water and sanitation (Goal 6), and the development of sustainable cities and communities (Goal 11), high-quality education (Goal 4) and other related goals.

The MSMEs play critical roles in achieving the objectives of some aims by upholding the rules and ordinances in an ethical and legal manner and employing the community's resources wisely. Under this, Goals 13 (Climate Action), 14 (Life Below the Water), 15 (Life on Land), and 17 (Peace, Justice, and Strong Institutions) get the majority of funding (Goal 16). The classification of SDGs is given in Table 9.3.

Some Important SDGs and the Role of MSME in Achieving These Goals

(i) SDG 1: Reduction of poverty and MSME

More than 650 million people, or 11% of the world's population, are estimated by the World Bank to be living in poverty. In South Asia and Sub-Saharan Africa, nearly 80% of people are living in poverty. The biggest challenge facing the globe is still the issue of poverty. Despite having one of the economies with the greatest rate of growth, India is home to 20.6% of the world's impoverished. MSMEs make sure that they

Table 9.3 Classification of selected SDGs on the basis of the activity attribute

1	On the basis of innovation with technology	Reduced Inequalities (Goal 10), Innovation and Infrastructure (Goal 9)
2	By sector specific operational efficiency	Responsible Consumption and Production (Goal 12), Affordable and Clean Energy (Goal 7), Partnership for goals (Goal 17), Zero hunger (Goal 2), Industry Innovation and Infrastructure (Goal 9.3)
3	By CSR activities	Good Health and Well-being (Goal 3), Quality Education (Goal 4), Sustainable Cities and Communities (Goal 11), Clean Water and Sanitation (Goal 6)
4	By ethically and lawfully following laws and bylaws	Life on Land (Goal 15), Life Below Water (Goal 14), Climate Action (Goal 13), Peace Justice and Strong Institutions (Goal 16)

Source Researchers' own classification

have enough income to survive by hiring the unemployed and under-privileged. This helps to ensure that they have everything they need to survive. Over 60 million people are employed by MSMEs in India across over 11 lakh businesses. Greater sustainable MSMEs could lead to more employment prospects being produced.

(ii) SDG 2: Ending Hunger

The majority of farmers and food producers are MSMEs, which contribute to an agro-based MSMEs, sustainable food production system. Play a significant part in establishing and preserving a sustainable food supply (Blue Orchard Academy, 2017). Government programs like the Kisan Sampada Yojana present, in India and Kishon welfare in Bangladesh MSMEs with an excellent opportunity to produce and process food at a minimal cost.

(iii) SDG 3: Optimal Health and Welfare

In rural areas, health issues including pregnancy mortality, child mortality, etc. still need to be better addressed. MSMEs could assist in setting up, healthcare facilities, and awareness campaigns. Free medical examinations, ambulance services, etc. in the rural areas as part of their CSR programs. MSMEs have taken steps to promote social health, including must use caution when disposing of the waste from their business. They must take care to avoid polluting the areas around them, as this ultimately causes a lot of sickness issues in the community. Only in 2016, did home and outdoor air pollution cause almost 7 million fatalities (Division for Sustainable Development Goals [DSDG], n.d.).

(iv) SDG 4: Quality Education

Education is more than just attending classes. It's about gaining new knowledge and skills at any age. MSMEs train young people in rural areas by giving them skills and training, this is the time of the Digital age so SME workers are transforming them toward the digital use of the machine and connecting with sustainability. Everyone who works for a microenterprise must learn new things and develop new abilities in order to do their jobs effectively. Workers are now being trained to better their skills through employee training programs administered by Skill India. Additionally, MSMEs are supporting the opening of schools in rural and underdeveloped areas as part of their social responsibility requirements.

(v) SDG 5: Gender equality

MSMEs could further the cause of gender equality and women's empowerment by giving women entrepreneurial chances (Kamal-Chaoui, 2017). Women are in charge of a sizable number of micro-businesses in India. Around 123 lakh proprietary businesses are held by female entrepreneurs, according to the MSME Annual Report. Women are now independent and are successful in their businesses. Additionally, they have received training in their talents through various training programs, putting females on par with men in terms of employment.

(vi) SDG 7: Inexpensive and Clean Energy

MSMEs have been able to reduce operating costs and have improved energy efficiency thanks to their creative approaches. When it comes to access to electricity, the government is supporting MSMEs. The MSME ministry has released guidelines and invited proposals for the establishment of industrial facilities, automatic voltage stabilizers, using LED-based lighting systems, electric transmission line hardware, electric solar coolers, distribution transformers, solar panels, etc. Numerous MSMEs are attempting to produce energy-efficient goods, and thousands of them are receiving accreditation from the ministry. In the end, this will increase the share of renewable energy in the overall energy mix. India's A fantastic illustration of delivering affordable electricity is Unnat Jyoti by Affordable LED for All (UJALA). MSMEs are playing a crucial part in such a project.

(vii) SDG 8: Decent Employment and Economic Development

Promote full and productive employment, sustained, inclusive, and sustainable economic growth, and decent work for everyone 12 targets make up this goal. They cover a variety of topics, such as promoting decent work, entrepreneurship, financial inclusion, and international cooperation programs like aid for trade, as well as a global strategy on young employment. The goal is to give everyone complete and fruitful employment. Despite a major decline in unemployment over the years, there are still a lot of unemployed individuals in several emerging nations, including India. There is no denying the MSME sector's contribution to job creation in India. After agriculture, the sector is the second-largest employer. According to the Ministry of MSME (2018), it employs over 110 million people across approximately 60 million businesses, giving a vast number of people access to employment and respectable employment possibilities.

The industry contributes to a rise in labor productivity and a fall in the unemployment rate. As a result, there is a decrease in the amount of rural poor migration. During the period of the global recession and slowdown in the economy, the industry has been able to create and offer work to a sizable population (Rena, 2010). For the next 15 years, India is expected to require 15 million new employees annually (Bhattacharya & Bijapurkar, 2017; Rena, 2010). It is impossible to imagine how much job growth there has been without the assistance of MSMEs. Can MSMEs contribute to Goal 8?

MSMEs support economic expansion and the GDP. MSMEs are the most common type of business in Organization for Economic Co-operation and Development (OECD) nations, and they significantly contribute to economic growth, generating, on average, between 50 to 60% of value added. About 83% of MSMEs can contribute up to 33% of GDP in emerging economies. In most nations, regardless of economic levels, MSMEs contribute to more than half of GDP when taking into consideration the role of informal enterprises.

MSMEs can encourage ethical employment practices and entrepreneurship in addition to creating jobs. MSMEs are related to SDG 8, which is decent work. MSMEs contribute up to 45% of all employment in developing market nations, as was covered under Goal 1. MSMEs have gradually increased employment globally over the past few decades. The overall number of full-time workers in MSMEs increased from 79 million in 2003 to 156 million in 2016, across 132 countries, about 86% of total MSMEs in the thirteen Sub-Saharan African nations (Freimann et al., 2014), there is evidence that MSMEs account for up to 90% of employment. MSMEs are often more labor-intensive-skilled, industries such as light manufacturing and services. However, the jobs offered by MSMEs may pay less and be less secure. Compared to those of larger businesses. Due to their informal nature, informal MSMEs struggle to guarantee that all of their employees have access to a social safety net (Liu et al., n.d.). Therefore, despite accounting for a sizable portion of employment, MSMEs might not provide chances for decent employment under ILO standards. Nevertheless, they have the ability to advance respectable work standards and job prospects when appropriately supported. The formalization and growth of entrepreneurship as envisioned in this goal will necessitate addressing a number of issues MSMEs face, including the creation of regulatory frameworks, more enabling policy, increasing access to market information and finance, developing MSMEs' capacities, and expanding access to the good infrastructure needed for MSME operations. Goal 8 has as a target the financing of MSMEs. Access to financial services is emphasized in Goal 9 and sub-goal 3. This goal emphasizes the significance of providing MSMEs with enough financial resources, but it also emphasizes the importance of enhancing MSMEs' market access and ability to grow their businesses through value chains, moving beyond the purview of basic finance. For formal MSMEs, the present credit gap is estimated to be worth US$1.2 trillion, and for informal MSMEs globally, it is worth US$6 trillion (Bhattacharya & Bijapurkar, 2017).

In emerging economies, almost 70% of MSMEs do not have access to loans. Formal and informal MSMEs are frequently viewed by commercial financing institutions as high-risk borrowers since they have few assets to utilize as collateral, which results in greater expenses in terms of fees and interest rates. Interest charges accrued numerous national governments and international organizations have tried to close the gap in recent years by promoting grant facilities, preferential rates, and loan guarantee programs. To guarantee that small business owners only take on reasonable amounts of risk and steer clear of bad debts, well-managed initiatives must offer support and capacity building. Venture capital and other creative and alternative financing for MSMEs continue to draw a lot of interest and see innovations.

Individual MSMEs may decide to change their business methods in order to further the objective. They can encourage an entrepreneurial culture, provide apprenticeship possibilities, and support/mentor new businesspeople. MSMEs might also enforce a strict policy against discriminatory hiring and support ethical workplace practices.

(viii) Goal 9: Infrastructure, Industry, and Innovation

Building a solid infrastructure, promoting inclusive and sustainable industrialization, and fostering innovation are the goals.

MSMEs are crucial to the industrialization process. Increased industry participation in employment, exports, and GDP have benefited India's development (GDP). In the current environment, the increase in industrial units is providing positive indicators for the economy. Due to a shortage of resources, MSME entrepreneurs constantly come up with new, straightforward ways to execute complex tasks, which ultimately results in low-cost innovations. MSMEs are regarded as innovation centers because of this. According to economist William Baumol, new and small businesses have been a major source of disruptive innovation during the past ten years since they operate outside of established paradigms and have weak ties to existing goods and technologies (Kamal-Chaoui, 2017). The MSME owners contribute to the creation of high-quality infrastructure in the region where they are present. They also work on building roads, hospitals, schools, and other structures as part of their civic obligation. Nine targets make up this aim, the majority of which are focused on three main areas: building robust infrastructure, industrialization, especially the growth of small-scale industrial firms, and fostering research and innovation.

Can MSMEs assist in achieving Goal 9? Small industrial businesses are directly related to the objective. SDG 9 is concerned with expanding the function and potential of small industrial businesses. According to UNIDO, only a few nations have provided statistics on this aim. The data that is currently available indicates that small businesses' manufacturing value added varies greatly among emerging nations. While small-scale industries contributed 21% of the manufacturing in Albania, they only made up 0.4% of the manufacturing in Ecuador. The chapter does, however, point up that notwithstanding its little contribution to industrial output overall, the function of small-scale businesses plays a crucial role in job creation, particularly in emerging nations, because of the high percentage of labor they absorb from traditional industries like agriculture or fishing. MSMEs produce between 50 and 60% of value added on average in OECD nations, making up roughly 60% of manufacturing jobs and 75% of service jobs. To reach the target, MSME financing for small industrial firms would be provided. Despite playing a significant part in industrial development in developing nations, one of the major obstacles small businesses confront is gaining access to financial services. Since it enables business owners to innovate, increase productivity, expand into new markets, and create new employment possibilities, financial access is essential for the expansion of small-scale industries. The Solomon Islands (64%), Nicaragua (59%), and Benin have the greatest percentages of small-scale enterprises having a loan or line of credit in manufacturing, according to the World Bank Enterprise Surveys for the years 2015 to 2017 (52%). In contrast, other nations, including Laos (6.7%), Zimbabwe (5.7%), and Egypt (4.3%), had extremely low shares. MSMEs support invention. While not all MSMEs are creative, new and small businesses are frequently the engine behind the innovations vital to economic growth because they can operate outside of dominant paradigms, capitalize on technological or business opportunities passed up by more established businesses, or facilitate the commercialization of knowledge that would otherwise go untapped. For example, about 20% of patents in Europe's biotechnology-related industries, which are one indicator of innovation, are held by MSMEs. MSMEs also contribute to value creation by distributing new or specialized products as well as utilizing innovation created by others and adapting it to various contexts through small adjustments. Finding the right knowledge partners and networks at the local, national, and international levels as well as developing the skills and management techniques necessary to integrate the knowledge produced by external partners with internal practices and innovation processes present significant challenges for many MSMEs.

Individual MSMEs may decide to change their business methods in order to further the objective. They can set guidelines and encourage rules to make sure initiatives and programs are sustainable and manageable. By offering all stakeholders the chance to propose innovative solutions to sustainability concerns, they could also encourage creativity.

(ix) New MSMEs' Business Strategies or Solutions that will Aid in Achieving the SDGs

The SDGs encourage resource usage efficiency and increased industrial process adoption that is environmentally sustainable. MSMEs are well-suited to reproduce and commercialize technology and add value in a range of fields, such as nanotechnology software, clean technologies, and biotechnology. This shift of enterprises and the economy will open up business prospects. Building capacity, financial resources, technology and knowledge transfer, and an enabling policy environment supportive of improving the ease of doing business are all necessary to assist MSMEs in adopting these new technologies and business domains.

(x) SDG 10: Lessens the Disparity

MSMEs have the potential to mitigate or decrease the disparity within the population of developing economies by creating jobs, making entrepreneurs, by empowering women. MSMEs have helped to lessen income disparities within and between nations. These MSMEs raise local residents' income levels, raise their standards of living, and make them independent by giving them jobs. MSMEs also contribute significantly to the nation's manufacturing and export sectors. The MSMEs contribute 40% of the nation's manufacturing output and 45% of its total exports, which helps it compete with both developed and emerging nations while maintaining rapid economic growth. MSMEs also assist the country's Financial Soundness Indicator (FSI) rise. MSMEs have improved the nation's overall economy as well as the wages of employees and business owners.

(xi) SDG 11: Sustainable Communities and Cities

Making the environment safe, robust, and sustainable is the aim. In both cities and villages, the majority of MSMEs are situated close to residential areas. They have an impact on nearby human life in one way or another. When businesses are established in remote locations, they themselves guarantee a well-developed infrastructure. As a result, they construct roads, guarantee the supply of water, and transport energy to remote locations. As a result, those areas benefit from good water, energy, and transportation infrastructure. They aid in urbanization while advancing industry. MSMEs can contribute to the inclusiveness of cities by working on initiatives like urban redevelopment projects (Kamal-Chaoui, 2017). MSMEs, with their wide range of operations, contribute in a variety of ways to the creation of sustainable cities and communities. MSMEs involved in solid waste management and waste recycling are some good examples in this regard.

(xii) SDG 12: Responsible Production and Consumption

Human demands are limitless, but resources are scarce. The rate of population growth necessitates careful management of the available resources. The key to sustainable consumption and production is the use of resources to meet human needs while minimizing waste. It suggests the creation of effective technologies, increased reliance on renewable resources, less emission of harmful residues, etc. MSMEs can embrace sustainable consumption and production practices because they are more adaptable than large corporations. To cut down on waste, the various production processes must be made more effective. MSMEs could apply Lean manufacturing techniques while adhering to different pollution control rules.

(xiii) SDG 15: Life on Land

Protecting, restoring, and promoting the sustainable use of terrestrial ecosystems, managing forests sustainably, preventing and reversing desertification, and preserving biodiversity are all related to it. The majority of the forest resources are exploited by MSMEs situated in rural areas close to forests. It becomes their duty to safeguard the fauna inside of them. They can make the handmade product with the help of first products and this handicraft sector is the best tool for attaining the aim of SDGs They have a responsibility to preserve the forest ecology. They must promote tree plantation initiatives in exchange for the resource they consume. Additionally, they must make sure that their operations do not degrade the land or harm the plants. The MSMEs must support sustainable agriculture business practices if they engage in agro-business.

(xiv) SDG 17: Collaboration for the Goals

Institutions in the nation must work together to implement all the policies that promote sustainable development. To attain self-sufficiency, governmental and private sector cooperation is required. Additionally, many informal businesses collaborate with official businesses. Given that they jointly improve the value chain; this partnership is crucial. Governments might be able to give MSMEs in our nation more exposure to MSMEs in other nations so that they can share cutting-edge operational techniques.

Improve Sustainability of MSMEs

MSMEs suffer a number of difficulties that reduce their operational effectiveness, particularly in emerging nations like India. If MSMEs can't manage things themselves, we cannot expect them to contribute to sustainable development through their own operations. It is crucial to solve MSMEs' difficulties and encourage them to increase their sustainability in order to get the most out of them in terms of sustainable development.

Significant Problems on MSMEs Confront and Solutions to Secure Their Sustainability

Financial Access

MSMEs' growth is hampered at various stages by a lack of financial facilities. Through a variety of programs and legislative changes, the government should guarantee the availability of funding.

Absence of Helpful Government Policies

The entrepreneurial spirit of young entrepreneurs with talent is also hampered by the absence of supportive government policies and regulatory frameworks. A cogent and encouraging atmosphere must be created by the government for the start-up ecosystem and fresh company concepts.

Lack of skilled labor: The majority of micro and small units lack skilled labor, which has a detrimental effect on their performance. The government should endeavor to provide small business owners and employees who work in their enterprises with skill-development and capacity-building programs through vocational training programs.

Basic infrastructure: The majority of MSMEs in India lack the necessary equipment, structures, technology assistance, etc. to efficiently operate their businesses. It is important to adequately handle this infrastructure issue.

Promoting entrepreneurship: Young people in rural areas and female entrepreneurs could be a powerful weapon in the fight against poverty, gender injustice, and income disparity globally. Unfortunately, rural communities discourage young people and women from starting their own businesses and forbid them from doing so. This way of thinking needs to change. It would be beneficial to promote entrepreneurship in this little area of the economy.

Bringing informal MSMEs into the formal ecosystem: It is important to make efforts to integrate informal MSMEs into the formal ecosystem in order to better monitor their development, contributions, and requirements. Important governmental measures in this area include the MUDRA Bank and the Priority Sector Lending (PSL) regulations.

Some core roles of MSME in sustainable development like as it maintains environmental sustainability, adopting digital technology in the industry for measuring SDG, creating political attraction toward sustainable development, maintaining the traditional craft, upgrading heritage skills alive, and developing the economic situation of the population with cultural connectivity. It has been also represented as a diagram for better understating in the digital era.

Findings

In our study, we discovered that MSMEs have enormous potential for achieving all forms of sustainable development. Especially we have focused on SDGs 8 and 9 because these are directly related to MSMEs and have the capacity to resolve problems of population, empowerment of women, and inclusive growth. MSMEs have a significant impact, either directly or impact on how well the sustainable development goals are progressing in developing countries including India and Bangladesh. Additionally, it has been discovered that due to a variety of issues, micro-businesses themselves struggle to function to their full potential in these economies, because when governments do not promote the MSME and do not care about SDG suggested by UNDP then there is raised issue, how we may achieve the goals. MSMEs must be promoted and supported effectively if they are to participate and contribute as much as possible to the accomplishment of sustainable development goals. It has been found that if microsomal and medium enterprises envelopment is larger in every economy then it will help in the reduction of poverty, women empowerment, improving the lives of people, increasing GDP of countries, and nearly achieve the agenda of SDGs till 2030.

Recommendations

We have several recommendations based on the research paper that will help make MSMEs more sustainable, which would help accomplish the SDGs.

1. The Ministry of MSMEs may adopt a policy strategy to encourage at least small and medium-sized businesses to work toward sustainable development objectives. In this regard, field assistance and official visits would be beneficial. The government should give incentives to MSMEs and direct them to organize awareness programs regarding various topics, such as entrepreneurial development, women's empowerment, financial literacy, etc. Additionally, programs should be created to educate MSME business owners on pollution prevention strategies, safety standards, and other regulations.

2. In fields with significant potential for MSMEs, such as solid waste management, recycling, lean manufacturing, etc., MSMEs entrepreneurs must be inspired to develop novel concepts. The MSMEs' greening should be given priority. Green MSMEs aid in reducing pollution emissions and maintaining a clean environment.
3. It is necessary to improve the institutional credit support system. Along with lending options, a support network must be established so that MSMEs may talk about their issues and seek guidance from authorities. One such support scheme is Mudra Bank's credit plus strategy in India and overseas Gramin Bank in Bangladesh and another similar program in south-east Asian countries. This will facilitate the seamless operation of MSMEs and keep people employed all year long. MSMEs must receive appropriate consulting services.
4. Many aspiring business owners become stuck and fail without the correct advice and direction. Technology developments in business operations and product marketing must be prioritized. Additionally, consulting services, technical recommendations, and other services must be offered to ensure efficient resource usage and minimal waste.
5. The governments should support MSMEs with incentives and advantages to ensure sustainable development. It would be beneficial in this regard if the organizational structure of the ministry's institutions were coordinated. Support for the organization at the federal, state, and local levels should come from

The 2030 Agenda and SDGs Have Opened Up New Prospects for MSMEs in Terms of Business Models and Solutions

Supply chain programs and policies present significant opportunities for businesses to scale up their sustainability efforts and so promote the SDGs. In order to verify the validity of sustainability claims in the fields of human rights, labor (including health and safety), the environment, and anti-corruption, traceability is the capacity to recognize, distribute, trace the history and location, parts, and materials and application of products. Adopting traceability initiatives and schemes can expand a company's market reach and appeal, often enabling MSMEs to recoup the costs of doing so by charging higher rates for superior-quality goods and services.

The SDGs will be accomplished with the support of new business strategies, possibilities, and models for MSMEs. The development of global value chains is related to an integrated global economy as envisioned by this objective (GVCs). GVCs and digital transformation present MSMEs with fresh chances for global economic integration. MSMEs may have a competitive edge in international markets because of their increased flexibility, ability to modify products, and capacity to differentiate products due to their good potential to respond fast to transforming market situations and investigation of products in a shorter time. MSMEs cover some natural root for utilization of the market and connect to the people with the global market, and the creation of micro and small businesses are a very important shareholder or cooperative unit for developing a new product. For instance, in Germany, small and midsize businesses account for the majority of the country's net foreign trade surplus and hold between 70 and 90% of the global market share in some specialized manufacturing categories. Access to information, expertise, technology, money, trade facilitation, and connection are some of the challenges faced by MSMEs while trying to internationalize. Evidence suggests that MSMEs are embracing digital technology more slowly than other industries. While the gap is generally small for basic connectivity and web presence, it widens when taking into account involvement in e-commerce and, particularly, more advanced apps. In many nations, there is also a significant adoption gap for cloud computing, which, for example, allows smaller businesses to use powerful computers by renting them from an outside supplier. While removing some of the obstacles brought on by the high fixed costs of ICT investment. Compared to larger companies, MSMEs are also disproportionately impacted by company taxes, compliance requirements in export destinations, and trade facilitation/administrative costs. The difficulties MSMEs have integrating into GVCs are further exacerbated by the limited lending opportunities they have. MSMEs will assist in achieving the larger vision envisioned in international trade and investment cooperation under this goal by addressing these problems. The role of digital technology in MSME can help to achieve the SDGs in developing countries through digital banking, a single window system to open an account for workers, digital skill training and a capacity-building program for the workers involved in small industries.

Conclusion

The SDGs are lofty objectives designed to improve both the earth and humankind's future. It will be difficult to accomplish them in fifteen years, though. Through various pledges, several countries, including India, are working toward these objectives. There are numerous aspects of sustainable development that can be accomplished through the efforts of groups and individuals in various forms. One of the sectors with significant potential to influence how most of these goals are progressing is the MSME sector. MSMEs are key participants in this effort due to their operational flexibility and diversity. If MSMEs succeed in the coming years, they will undoubtedly contribute to increasing employment, fostering equality, resolving regional imbalances, reducing poverty, etc., and making the world a better place. In this highly competitive environment, it is essential that MSME challenges are systematically discussed through sustainable policy implementation and initiatives by the governments and that entrepreneurial abilities are well-nourished.

References

Bhattacharya, A., & Bijapurkar, A. (2017). *India: Growth and jobs in the new globalization*. The Boston Consultancy Group. https://media-publications.bcg.com/BCG-CII-India-Future-of-Jobs-Mar-2017.pdf

Blue Orchard Academy. (2017). *SMEs and SDGs: Supporting small and medium enterprises to achieve sustainable development goals—Insights from a Blue Orchard Survey*. www.blueorchard.com. https://www.blueorchard.com/wp-content/uploads/Blue-Orchard-pages-11-12.pdf

https://www.devalt.org/images/L2_ProjectPdfs/MGNREGS_SDGs_DA.pdf

Division for Sustainable Development Goals (DSDG). (n.d.). *Sustainable Development Goal 13*. https://sustainabledevelopment.un.org/sdg13

Freimann, A., Ham, M., & Mijoc, J. (2014). Measuring objective well-being and sustainable development. *Journal of Knowledge Management, Economics and Information Technology, IV*(2). http://www.scientificpapers.org/wp-content/files/1443_FRAJMAN_IVKOVIHAMMIJOMeasuring_Objective_Well-Being_and_Sustainable_Development_Management.pdf

Ion Oncioiu, F., Raluka, B., & Petrescu, A. G. (2017). The potential contribution of creative Indian SMEs to long-term growth. *Euro Economia, 2*(1). http://journals.univ-Danubius.ro/index.php/euroeconomic/article/view/4177/4040

Kamal-Chaoui, L. (2017). *Unlocking the potential of SMEs for the SDGs*. OECD Development Matters. https://oecd-development-matters.org/2017/04/03/unlocking-the-potential-of-SMEs-for-the-SDGs/#fn2

Liu, K. C. (n.d.). *Policy brief: The role of micro-small and medium enterprises in achieving SDGs*. https://sustainabledevelopment.un.org/content/documents/18349policy_brief_27_04final.pdf

Ministry of MSME. (2017). *Annual Report 2016*. Government of Bangladesh.

Ministry of MSME. (2018). *Annual Report 2017*. Government of India. https://msme.gov.in/sites/default/files/MSME-AR-2017-18-Eng.pdf

Organization for Sustainable Development in the Netherlands (Corporate SNV) and the World Business Council for Sustainable Development. (2007). *Fostering SME development for sustainable development*. www.wbcsd.org/Programs/People/Social/Impact/Resources

PIB. (2018, December). *The MSME sector has an established export promotion council*. http://pib.nic.in/PressReleaseIframePage.aspx?PRID=1557901was retrieved as the source.

Rena, R. (2010). The impact of global economic crisis on the namibian economy: Issues and challenges. *Review of Business and Technology Research, 3*(1), 475–488.

Shah, A. (2009, November 18). *Introduction to Sustainable Development*. http://www.globalissues.org/article/408/sustainable-development-introduction

Sharma, R. (2009). Where are we on the path to sustainable development? *Indian Journal of Community Medicine, 34*(4), 276–278. https://doi.org/10.4103/0970-0218.58381

Tambunan, T. T. H. (2018). The impact of the economic crisis on micro, small, and medium enterprises and their crisis mitigation measures in Southeast Asia with reference to Indonesia. *Asia & the Pacific Policy Studies, 6*(1). https://ssrn.com/abstract=3339683

World Business Council for Sustainable Development (WBCSD) and Corporate SNV—Netherland Development Organization. (2007). *Promoting small and medium enterprises for sustainable development*. Ordering Publications. https://www.wbcsd.org/Programs/People/Social-Impact/Resources/Promoting-Small-and-Medium-Enterprises-for-Sustainable-Development

Yadav, U. S., Tripathi, R., & Tripathi, M. A. (2020). Strategies for development of handicraft sector small industries in India. *Small Enterprises Development Management and Extension Journal, 49*(2). https://doi.org/10.1177/09708464221096983

Yadav, U. S., Tripathi, R., & Tripathi, M. A. (2022). Digital analysis of the transformation of institutions in the knowledge and innovation system of the handmade carpet industry. *Small Enterprises Development Management and Extension, 47*(1). https://journals.sagepub.com/doi/full/10.1177/09708464221096903

CHAPTER 10

The Role of MSMEs in Achieving the SDGs, Especially on Poverty Reduction in 5 ASEAN Countries

Ernestus Florian Lamaile and Nuning Trihadmini

Introduction

The Sustainable Development Goals (SDGs) are a global action plan agreed upon by world leaders to end poverty, reduce inequality, and protect the environment. The SDGs contain 17 Goals and 169 Targets that are expected to be achieved by 2030. The SDGs cover a wide range of social and economic development issues. These include poverty, hunger, health, education, climate change, water, sanitation, energy, environment, and social justice. Of the 17 goals, the first goal of the SDGs is to end poverty in all its forms. The World Bank defines absolute poverty as living on an income below $1/day, and medium poverty as an income of $2 a day. Absolute poverty refers to a consistent standard, not affected by time

E.F. Lamaile · N. Trihadmini (✉)
Faculty of Economics and Business, Atma Jaya Catholic University of Indonesia, Jakarta, Indonesia
e-mail: nuning.trihadmini@atmajaya.ac.id

H. Dasaraju and T.T.H. Tambunan (eds.), *Role of Micro, Small and Medium Enterprises in Achieving SDGs*, Sustainable Development Goals Series, https://doi.org/10.1007/978-981-99-4829-1_10

and place/country. A person is included in absolute poverty if his income is below the poverty line, it is not enough to meet the minimum needs. Meanwhile, relative poverty is a condition of society because development policies have not reached all levels of society, causing inequality in income distribution. Several efforts have been made to alleviate poverty, including through the development of small and medium enterprises.

Promising economic growth in Southeast Asian countries (ASEAN) over the last 10 years can be achieved partly because of the support for the development of micro, small, and medium enterprises (MSMEs), through their role in structural reforms. This makes it important to maintain sustainable dynamic growth in the ASEAN region. One of the fundamental problems in economic development, especially in developing countries is poverty alleviation, and this is a major factor in other social problems. According to the World Bank, poverty is defined as a lack of welfare, low levels of education and health, and difficulty in getting access to clean water and sanitation. These conditions are part of the many dimensions of welfare. The Central Bureau of Statistics (BPS) uses the concept of the ability to meet basic needs (basic needs approach) in defining poverty. With this approach, poverty is seen as an economic inability to meet basic food and non-food needs as measured from the expenditure side. So, the poor are people who have an average monthly per capita expenditure below the poverty line.

To overcome the problem of poverty and help the poor move above the Poverty Line, the governments of each ASEAN country have made various programs and policies that directly touch the small communities. One of the government's focuses is to support MSMEs. Various problems of MSMEs that still need to be assisted are the availability of capital, self-development, and capacity building through entrepreneurship training, as well as access to markets, especially the export market. The availability of capital is often a fundamental problem for everyone in starting a business. Various efforts have been made by the government through various capital loan programs. However, it turns out that capital in terms of money alone does not guarantee the continuity of a business. The next problem is the ability of MSME workers to manage their businesses. Some MSME actors, especially on a micro scale, do not have certain abilities or soft skills when running their business, so many do not last long after receiving capital assistance. Therefore, it is very important to provide various pieces of training such as financial management, raw material management, business promotion, or creative thinking. Based on the background, this study aims to analyze the effect of MSME growth on poverty alleviation in 5 ASEAN countries, using the MSME growth variable as the main indicator, the Trade Openness Index, per capita income growth, and the Human Development Index as a control variable. Researchers used 5 ASEAN countries, namely Indonesia, Myanmar, Philippines, Thailand, and Vietnam as samples.

Manzoor et al. (2019) conducted a study on the role of MSMEs in poverty alleviation in the South Asian Association of Regional Cooperation (SAARC) region using panel data analysis. The main objective of the study was to evaluate the progress of poverty alleviation in SAARC countries, through the development of MSMEs, as it is considered a shortcut to poverty reduction. As a measure of poverty in the SAARC region, the researcher uses the indicator of the poorest 20% of incomes as a percentage of GDP (Income share held by the lowest 20%). The development of MSMEs is measured by the share of MSMEs in the GDP in each country. The period of data analysis was from 1990–2015, using the Fixed Effect Model (FEM) and Multiple Regression Models. The results of this study indicate a significant positive effect on the growth of MSMEs in poverty alleviation. Three main factors play an important role in reducing poverty in the SAARC region, namely: MSME growth, trade openness, and social sector development. These three factors are very important and should be the government's priority to reduce poverty in the region.

Nursini (2020) conducted research on the influence of MSMEs on poverty alleviation in Indonesia, either directly or indirectly through employment. The analysis period was carried out from 1997 to 2018. As a measure of poverty, this study uses the Head Count Index, Poverty Gap Index, and Poverty Severity Index. For MSMEs, this study classifies based on the scale, namely small and medium enterprises (SMEs) and micro and small enterprises (MSEs). Economic growth, economic openness, government spending, and investment are the control variables. This study develops an analytical model that is different from previous research by dividing it into two categories, namely the direct influence of MSMEs on poverty alleviation and the indirect effect through employment.

The results show that based on the predetermined MSME scale, there are different implications for poverty alleviation. MSMEs have a greater influence because they not only reduce the number of poor people but also affect the Poverty Severity Index and Inequality. Then from the four control variables, only economic growth has a significant effect on poverty reduction, either directly or indirectly. The conclusion of this study states that the direct influences of SMEs and MSEs are significant on all indicators of poverty alleviation. This shows that the development of MSMEs can reduce the percentage of the poor, the deviation between the average expenditure of the poor and the poverty line, and the disparity in the average expenditure of the poor. Based on these findings, there are several policy recommendations to alleviate poverty, namely: (i) the need for marketing support for MSME products to support the production cycle through digitalization; (ii) MSME products must continue to be promoted to foreign countries so that economic openness can reduce poverty, and (iii) investment should also be made available in the MSME sector and not only in large-scale industries. UNDESA (2020) states that the role of MSMEs in achieving SDGs and reducing poverty is as follows:

i. through job creation. Most of the poor in developing countries are not working or do not earn enough to lift themselves out of poverty. Job creation in the private sector has proven to be a key driver in the fight against poverty. Over the past 30 years, the private sector has contributed to a sharp decline in the share of the population in developing countries living below the poverty line;
ii. MSMEs are generally informal companies and function as a source of livelihood, involving the poor as business actors. The informal sector has flexibility that provides opportunities for the poor including women and youth to earn money. According to the International Labor Organization (ILO), most of those working in the informal economy lack social protection, rights at work, and decent working conditions. Informal enterprises offer avenues of intervention to formalize MSMEs and provide access to social protection for those who are employed. Given the informality of employment, especially in developing countries, support in formalizing MSMEs will be a step to achieve; and
iii. individual MSMEs have the potential to adopt business practices in their business. They can also recruit, train, and employ local community members, including those living in poverty, and integrate them into the MSME value chain. In addition, the adoption of a new business model will be beneficial to increasing its business capacity. Although the purchasing power of the poor is limited, the cumulative effect is large.

Methodology

In this study, the authors use secondary data which is annual data, obtained from several website sources such as Our World in Data, Asian Development Bank, Macrotrend Data, Asian Regional Integration Center, and Human Development Report. The data that has been collected and compiled into a panel dataset consists of 5 ASEAN countries, namely Indonesia, Myanmar, the Philippines, Thailand, and Vietnam from 2014–2019. The specifications of the model used in this study are as follows:

$$YP_{it} = \alpha_{0it} + \alpha_1 GSME_{it} + \alpha_2 TOI_{it} + \alpha_3 GIPC_{it} + \alpha_4 HDI_{it} + \varepsilon_{it}$$

where:

YP:	The proportion of poverty below the international poverty line
I:	Cross-section
T:	Period
GSME:	Growth of small (including micro) and medium enterprises
TOI:	Trade openness index
GIPC:	Growth of income per capita
HDI:	Human development index
ε:	Error term.

To measure poverty across countries consistently, the World Bank applies a common standard that is measured by the average size of consumption expenditure per capita in the world's poorest countries. The extreme poverty line is currently set at $1.90 per day, which represents the average consumption expenditure per capita in the 15 poorest countries. The growth of SMEs is calculated using a simple formula: (final value − initial value)/initial value × 100%.

Where the initial value is the number of MSMEs in the initial year and the final value is the number of MSMEs at the beginning of the following year. For example, to find out the growth of MSMEs in 2015, the calculation is as follows: (number of MSMEs in 2015 − number of MSMEs in 2014)/number of MSMEs in 2014 × 100%. The Trade Openness Index is one indicator to determine the economic openness of a country by calculating the proportion of trade to GDP. The trade-to-GDP ratio is often used to measure the importance of international transactions relative to domestic transactions. This indicator is calculated in each country as a simple average, i.e., the average of total trade (amount of exports and imports of goods and services) relative to GDP. BPS calls this the degree of openness. The greater the degree of openness indicates the more open the economy of a country. The formula for calculating the degree of openness is:

$$\frac{(\text{Export} + \text{Import})}{\text{GDP}} \times 100$$

To become a developed country, per capita income is a benchmark and also an illustration of the economic condition of a country. Per capita income is a measure of the amount of money each person earns in a country or geographic area. Per capita income is calculated so that the government can provide solutions in the form of policies that are appropriate to the economic conditions and needs of the population. Per capita income is calculated from the National Income divided by the total population of a country or a certain geographic area.

Data Analysis

The data analysis that will be presented in this section begins with descriptive statistical analysis and then analyzes the panel data output using the Fixed Effect Model. Descriptive statistics were conducted to provide a detailed description of the data used in the study. The results of descriptive data analysis can be seen in Table 10.1. The variable proportion of consumption expenditure below the International Poverty Line from 5 ASEAN countries has the lowest value of 0.024, namely Thailand, and the highest value of 13.30 from Myanmar. The proportion below the international poverty line in the Philippines has the highest average of 8.37, while Thailand has the lowest average of 0.049. This indicates that the poverty rate as measured by the proportion below the international poverty line is very high in the Philippines, while in Thailand it is very low.

The variable growth of MSMEs from 5 ASEAN countries has the lowest value of −9.2%, namely Thailand which shows a decline in the number of MSMEs, and the highest value of 52.4% comes from Myanmar. The growth of MSMEs in Vietnam has the highest average of 11.11% while Thailand has the lowest average of 0.96%. Based on

Table 10.1 Descriptive statistics of the proportion of consumption expenditure below the International Poverty Line

	POV_IN	*POV_MY*	*POV_PH*	*POV_TH*	*POV_VI*
Mean	6.9189	5.9115	8.3710	0.0490	2.4265
Median	5.9979	5.318	8.4355	0.0357	2.3860
Maximum	13.287	13.305	12.160	0.1328	4.0277
Minimum	2.878	0.8669	3.8139	0.0240	1.3522
Std. Dev	3.3512	4.511	3.1231	0.033	0.8129
Observations	10	10	10	10	10

Source Output Eviews 10 (reprocessed)

the maximum value of growth, Myanmar experienced the largest MSME growth of 52.4% and Indonesia had the smallest maximum growth of 4% (Table 10.2). This means that the increase in the growth of MSMEs in Indonesia is relatively the least among the 5 ASEAN countries.

The openness of trade shows the ratio of the sum of the value of exports plus imports then divided by GDP. A value of more than 100 indicates a country's total trade is greater than that country's GDP. The trade openness variable from 5 ASEAN countries has the lowest value of 29.01, namely Myanmar, and the highest value of 158.63 coming from Vietnam. Trade Openness in Vietnam has the highest average, which is 134.75. Meanwhile, Indonesia has the lowest average of 36.41. Vietnam in this period is experiencing a significant change in economic growth and trade openness, while Indonesia has the lowest average in trade openness (Table 10.3).

Table 10.2 Descriptive statistics of MSME growth

	GSME_IND	*GSME_MY*	*GSME_PH*	*GSME_TH*	*GSME_VIet*
Mean	2.44	7.38	2.64	0.96	11.11
Median	2.4	0.7	0.8	1.05	9.65
Maximum	4	52.4	15.2	8.6	18.2
Minimum	2	−2.8	−4.9	−9.2	6.9
Std. Dev	0.592	16.61	5.673	4.307	3.793
Observations	10	10	10	10	10

Source Output Eviews 10 (reprocessed)

Table 10.3 Trade openness statistics

	TOI_IND	*TOI_MY*	*TOI_PH*	*TOI_TH*	*TOI_VI*
Mean	36.41465	43.02807	47.431	106.6273	134.753
Median	36.99428	44.72956	46.89993	106.9203	133.8969
Maximum	42.67813	54.09692	53.57066	121.1416	158.6347
Minimum	29.22795	29.01043	42.96968	88.99961	108.5839
Std. Dev	4.833202	9.386644	3.555183	10.29007	18.81479
Observations	10	10	10	10	10

Source Output Eviews 10 (reprocessed)

Per capita income is defined as GDP at constant prices divided by the population. The per capita income growth variable from 5 ASEAN countries has the lowest and highest values of −4.58% and 38.08%, both of which are grown in income per capita from Indonesia. On average, Vietnam's per capita income growth was the highest at 8.41%, while Myanmar had the lowest average at 6.03%. The above results show that Vietnam's per capita income has increased along with the increase in trade openness in the country. Meanwhile, the opposite result was experienced in Myanmar (Table 10.4).

Human development and improving the social welfare of a population require a set of abilities. The human development index (HDI) is another measure of the well-being of a country's population, focusing on three basic dimensions of human development: the ability to live a long and healthy life as measured by life expectancy at birth; the ability to acquire knowledge as measured by the average length of schooling and

Table 10.4 Descriptive statistics of per capita income growth

	GIPC_IND	*GIPC_MY*	*GIPC_PH*	*GIPC_TH*	*GIPC_VI*
Mean	6.835	6.036	6.318	6.568	8.414
Median	3.835	2.995	5.35	6.92	8.09
Maximum	38.08	21.2	16.35	20.49	15.72
Minimum	−4.58	−2.46	1.4	−3.51	2.7
Std. Dev	12.6898	7.45365	4.81997	6.7963	3.8484
Observations	10	10	10	10	10

Source Output Eviews 10 (reprocessed)

the expected length of schooling; and the ability to achieve a decent standard of living as measured by gross national income per capita. As can be seen in Table 10.5, the HDI from 5 ASEAN countries has the lowest value of 0.51 namely Myanmar and the highest value of 0.77 coming from Thailand. The Human Development Index in Thailand has the highest average, which is 0.74. Meanwhile, Myanmar has the lowest average of 0.55. The above results indicate that Thailand has succeeded in improving the quality of life of its population and Myanmar is still trying to improve the quality of its population.

The next analysis is data estimation using the panel fixed effect method. As shown in Table 10.6, the results of testing the effect of MSME growth, HDI, per capita income growth, and trade openness on poverty alleviation in 5 ASEAN countries. All variables significantly affect poverty reduction, except for the growth of MSMEs, which is significant at 10% alpha. The results of the F-test also show that all variables have a significant effect. The results of this study are in line with the research conducted by Manzoor et al. (2019) where they found a significant effect between the growth of MSMEs on poverty alleviation in the SAARC region or South Asia. The growth of MSMEs has the smallest effect compared to other variables, with a negative relationship direction. This means that the more MSMEs are involved, the less poverty will be.

The relatively small influence of MSMEs can be explained that there possibility that poverty alleviation is more influenced by other variables that have a stronger or greater influence than the growth of MSMEs. The growth of MSMEs does not have a major influence on poverty when compared to the Trade Openness Index, growth in per capita income, and also the HDI. On average, the condition of poverty in these 5 ASEAN

Table 10.5 Descriptive statistics of the Human Development Index

	HDI_IN	*HDI_MY*	*HDI_PH*	*HDI_TH*	*HDI_VI*
Mean	0.6931	0.5521	0.696	0.7489	0.6853
Median	0.6925	0.5535	0.6985	0.7455	0.6855
Maximum	0.718	0.583	0.718	0.777	0.704
Minimum	0.665	0.515	0.671	0.724	0.661
Std. Dev	0.0171881	0.0229174	0.0154056	0.0180027	0.0135486
Observations	10	10	10	10	10

Source Output Eviews 10 (reprocessed)

Table 10.6 Estimation results of fixed effect model

Dependent Variable: POV
Method: Panel EGLS (Cross-section SUR)
Sample: 2010 2019
Periods included: 10
Cross-sections included: 5
Total panel (balanced) observations: 50
Linear estimation after one-step weighting matrix

Variable	*Coefficient*	*Std. Error*	t-*Statistic*	*Prob*
C	89.72018	2.946056	30.45433	0.0000
GSME	−0.013423	0.007776	−1.726337	0.0918
TOI	−0.028591	0.004581	−6.241528	0.0000
GIPC	0.046444	0.007589	6.120274	0.0000
HDI	−123.1417	4.663542	−26.40519	0.0000
Weighted Statistics				
R-squared	0.98824	Mean dependent var		12.62233
Adjusted *R*-squared	0.985945	S.D. dependent var		19.37578
S.E. of regression	1.063686	Sum squared resid		46.38853
F-statistic	430.6641	Durbin-Watson stat		1.664794
Prob(*F*-statistic)	0			
Unweighted Statistics				
R-squared	0.884229	Mean dependent var		4.735423
Sum squared resid	97.78363	Durbin-Watson stat		0.248465

Source Output Eviews 10 (reprocessed)

countries has decreased every year. So, it can be said that people who are currently below the poverty line are those who need assistance with basic needs, such as access to health, education, and the economy which are included in the HDI. Therefore, it is not the current growth of SMEs that has a significant influence on poverty reduction, but also assistance to obtain fulfillment of basic needs.

The economic analysis of each variable will be explained as follows. The Trade Openness Index has a significant effect on poverty alleviation. This is in line with the research results of Manzoor et al. (2019) where trade openness influences the lowest/poorest 20% of the income community. According to the study, increasing trade openness will also lead to strengthening economic activity. To meet the increasing demand for goods, both exports and imports, more labor is needed and it directly affects the increase in employment. If employment increases, it will certainly affect the poverty rate.

The second variable that affects poverty reduction is income per capita. In this study, there is a positive relationship between per capita income and poverty reduction. The direction of the relationship between these two variables should be negative, but in this study it is positive. It can be explained that in measuring the poverty line, the indicator used is the amount of consumption expenditure used to meet basic needs. When there is an increase in per capita income, the increase in income may not be used to increase consumption expenditure but be used for other expenditures so that it does not affect the poverty line.

The HDI has a significant influence on poverty alleviation. People who are currently below the poverty line are those whose basic needs are not met, namely education, health, and the economy. The results of previous research conducted by Manzoor et al. (2019) showed the same results as this study. They show that improvements in social and economic indicators, such as literacy rates, life expectancy, and GDP per capita have a significant positive role in reducing poverty. Human development has an important role in alleviating poverty and increasing the level of community welfare. The government's political will to promote human development is one of the keys to success in poverty reduction. Improved human development performance can be implemented through greater attention to human development, especially by increasing the budget and planning expenditures for the sector and creating adequate employment opportunities, especially for the lower middle class.

Conclusion

Research on the role of MSME growth in achieving the SDGs goal of poverty reduction, using panel data analysis in 5 ASEAN countries in the 2014–2019 period shows that there is a significant relationship between MSME growth and poverty alleviation. Likewise with the Trade Openness Index, per capita income growth, and the HDI. The three dimensions contained in the HDI illustrate that poverty alleviation must begin with the fulfillment of basic human needs. The Trade Openness Index will increase economic activity through increasing job opportunities, thereby and indirectly contributing to increasing people's per capita income so that the poverty rate can be reduced.

Overall, the results of this research may give a strong message that the sustainable production growth of MSMEs is one or even the main prerequisite for the ability of these enterprises to play their role in achieving the SDGs.

References

ASEAN Secretariat. (2019). *ASEAN Key Figures 2019*. ASEAN Secretariat.

Manzoor, F., Wei, L., Nurunnabi, M., & Subhan, Q. A. (2019). Role of SME in poverty alleviation in SAARC region via panel data analysis. *MDPI, 11*, 1–14. https://pdfs.semanticscholar.org/ecd1/5a252f20cd8ff01ff3712804ccbfae27d70c.pdf

Nursini, N. (2020). Micro, small, and medium enterprises (MSMEs) and poverty reduction: Empirical evidence from Indonesia. *Development Studies Research, 7*(1), 153–166. https://doi.org/10.1080/21665095.2020.1823238

UNDESA. (2020). *Micro-Small, & Medium-sized Enterprises (MSMEs) and their Role in Achieving the Sustainable Development Goals*. UN.

Others

https://knoema.com/atlas/Indonesia/Poverty-rate-at-national-poverty-line. Indonesia—Poverty headcount ratio at the national poverty line.

https://www.adb.org/publications/series/basic-statistics. Growth of Income Percapita.

https://aric.adb.org/database/integration. Trade of Openness Index.

https://www.worldbank.org/in/news/press-release/2016/10/02/tackling-inequality-vital-to-end-extreme-poverty-by-2030. The World Bank.

https://ipm.bps.go.id/page/ipm

CHAPTER 11

Contribution of MSMEs to SDGs Through Corporate Social Responsibility

Shweta Bapat

Introduction

Over the past few decades, the MSME sector has developed as a highly promising and dynamic sector in almost all developing economies . It contributes meaningfully to the whole eco-system of business as well a significant contribution can be noticed in the economic as well as social development of the country. According to the World Bank, there are about 365–445 million MSMEs in emerging markets: 25–30 million are formal SMEs, 55–70 million are formal micro, and 285–345 million are informal enterprises. Start-ups and young firms, which are generally small or micro firms, are the primary source of net job creation in many countries and are the driving force of innovation and sustainability in the private sector. With very low capital investment, by generating large employment and nurturing entrepreneurship opportunities MSMEs have created their own place in the economy of each country. As large

S. Bapat (✉)
Business Ethics Foundation, Pune, India
e-mail: shwetambapat@yahoo.co.in

H. Dasaraju and T.T.H. Tambunan (eds.), *Role of Micro, Small and Medium Enterprises in Achieving SDGs*, Sustainable Development Goals Series, https://doi.org/10.1007/978-981-99-4829-1_11

industries play an important role in the growth of the country, MSMEs also have their own role to play by complementing large industries . As ancillary units, this sector contributes expressively in the industrial development of the country.

A country that expects inclusive growth can look forward to a healthy MSME sector as it creates jobs across regions and sectors, including for skilled, semi-skilled, and unskilled labor. In fact, through its variety of business processes MSMEs provide skills development opportunities to society and their own employees. A recent study says MSMEs are widening their domain across sectors of the economy, producing a diverse range of products and services to meet the demands of domestic as well as global markets.

The most talked about the recent concept of 'sustainability' is not only applicable to large businesses but also is more relevant to MSMEs as they face different challenges at all stages of their lifecycle.

SDGs

The SDGs are ambitious and require revolutionary changes in business activities at the national level. This transformation may demand doing business differently like adapting new business models, bringing in new innovation , and adopting new technology. This opens up new business opportunities for the business world in general and MSMEs in particular.

It is also expected that businesses should operate more sustainably and ethically. Seventeen SDGs can be roughly divided into 3 categories, Environment—SDG-6, 13,14, 15, Society SDG-1, 2, 3, 4, 5, 7, 11, 16, and Economy—SDG-8, 9, 10, 12, SDG 17th Can be a part of all three categories. Corporate social responsibility (CSR) activities can address most of these SDGs. CSR is the way through which businesses contribute toward societal development. Every country has aligned its CSR guidelines or laws with SDGs to achieve these global goals at the national level. CSR activities of all MSMEs together can contribute largely to SDG achievement for the Nation. The goals set out in the SDGs cannot be achieved without the contribution of the types of businesses therefore MSMEs also play an important role along with large businesses. MSMEs due to their large number can contribute substantially in achieving SDGs. This chapter is aimed at focusing the role of MSMEs toward SDGs mainly through CSR.

Concept of CSR

As early as 1916 in the Journal of Political Economy J. M. Clerk mentioned that "if men are responsible for the known results of their actions, business responsibilities must include, the known results of business dealings, whether these have been recognized by law or not."

It is believed the concept of CSR emerged in the 1930s to 1940s, Prof. Theodore Kreps introduced the subject of business and social welfare to Stanford in 1931 and used the term 'social Audit' for the first time in relation to companies reporting on their social responsibilities. The term CSR became formalized in 1953 with the publication of the book named 'Social Responsibilities of the businessman' the term CSR became popular in the 1990s.

Corporations while operating their business get involved in many things which give an impact on the internal as well as external environment of the organization. Any action in the business is generally initiated after a thorough analysis of its impact on the internal and external environment of the business. As mentioned by David Crowther and Gulier Aras in their book on CSR "the external environment includes both the business environment in which the firm is operating and the societal environment in which the organization is located." the impact can be positive as well as negative.

The concept of CSR was initially interpreted as an effort any organization takes to minimize the negative impact of its actions on the societal environment. Although the roots of CSR lie in philanthropic activities, the concept of CSR nowadays includes all related concepts such as the triple bottom line approach, corporate citizenship, corporate sustainability, business responsibility, etc.

Though it is believed that the concept of CSR became formalized in 1953, it is not at all new to India. The concept of CSR in India can be traced back to the ancient times of Vedas which were composed approximately between about 1500 and 1000 BCE. Corporate Philanthropy in India was one of the earliest activities and it was voluntary. Being voluntary it was totally dependent on the willingness of the businessmen to donate for society. This research paper is an attempt to take a review of the journey of CSR in India from the concept of 'Dharma' as voluntary CSR of the ancient times of Vedas to the mandatory CSR as per the Companies Act 2013.

Though there is no universally accepted definition of CSR, numerous sincere attempts were made since 1953 to define CSR Formally. Following are some of the definitions of CSR: "*CSR is a concept whereby companies integrate social and environmental concerns in their business operations and their interaction with their stakeholders on a voluntary basis*"—(Commission Of The European Communities, 2002).

Carroll's CSR Pyramid

It is suggested in this model that four kinds of social responsibilities constitute total CSR, economic, legal, ethical, and philanthropic. Furthermore, these four components of CSR might be portrayed as a pyramid (Fig. 11.1).

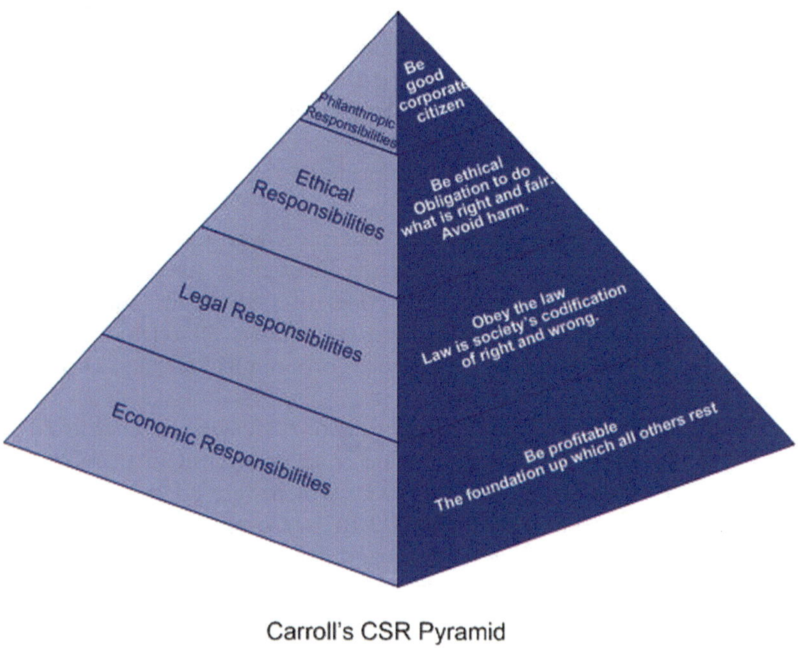

Fig. 11.1 Four components of CSR

- Economic Responsibilities: historically the principal role of business organizations was to produce goods and services that consumers needed and wanted and to make an acceptable profit in the process. Before it was anything else, the business organization was the basic economic unit in our society. The remaining business responsibilities are grounded upon the economic responsibility of the firm.
- Legal Responsibilities: it is part of the contract between business and society organizations are expected to chase their economic achievements within the framework of the law. Legal responsibilities can be termed codified ethics; they symbolize basic notions of fair operations.
- Ethical Responsibilities: it includes those activities and practices that are expected by society even though they are not codified into law. Ethical responsibilities can be termed as standard norms of fair treatment to all stakeholders of the organization.
- Philanthropic Responsibilities: to prove businesses be good corporate citizens philanthropy does not only include financial help to the community but those corporate actions or programs which promote inclusive growth.

Approaches to CSR

According to "Altered Images: the 2001 State of Corporate Responsibility in India Poll," a survey conducted by Tata Energy Research Institute (TERI), the evolution of CSR in India has followed a chronological evolution of 4 thinking approaches:

- Ethical Model (1930–1950): one significant aspect of this model is the promotion of "trusteeship" that was revived and reinterpreted by Gandhiji. Under this notion the businesses were motivated to manage their business entity as a trust held in the interest of the community. The idea prompted many family-run businesses to contribute to socio-economic development. The efforts of Tata group directed toward the well-being of society are also worth mentioning in this model.
- Statist Model (1950–the 1970s): under the aegis of Jawahar Lal Nehru, this model came into being in the post-independence era. The era was driven by a mixed and socialist kind of economy. The

important feature of this model was that the state ownership and legal requirements decided the corporate responsibilities.

- Liberal Model (1970s–1990s): the model was encapsulated by Milton Friedman. As per this model, corporate responsibility is confined to its economic bottom line. This implies that it is sufficient for businesses to obey the law and generate wealth, which through taxation and private charitable choices can be directed to social ends.
- Stakeholder Model (1990s–present): The model came into existence during 1990s as a consequence of the realization that with growing economic profits, businesses also have certain societal roles to fulfill. The model expects companies to perform according to the "triple bottom line" approach. The businesses are also focusing on accountability and transparency through several mechanisms.

Case of India

Contribution of MSMEs

In India , the new MSME classification has come into effect on 1st July 2020. The earlier criteria of classification of MSMEs under the MSMED Act, 2006 were based on investment in plant and machinery/equipment. It was different for manufacturing and service units. Indian economy recently has undergone significant changes. A revision in MSME criteria of classification was announced under Aatma Nirbhar Bharat package in May 2020.

This change in the criteria has been done in order to be realistic with time and establish an objective system of classification and to provide ease of doing business. The following image can be studied to understand the situation in India, where the number of SMEs in India is more than 36 million units, and 120 million persons are involved to produce more than 6000 products (Fig. 11.2).

Mandatory CSR Spending

In India Companies' bill 2011 was considered and approved by the Rajya Sabha on 8 August 2013. It received the President's assent on 29 August 2013 and has now become the Companies Act, 2013. The Companies Act 2013 is legislation that officially embarked on one of the world's largest experiments of introducing the concept of CSR as a mandatory provision.

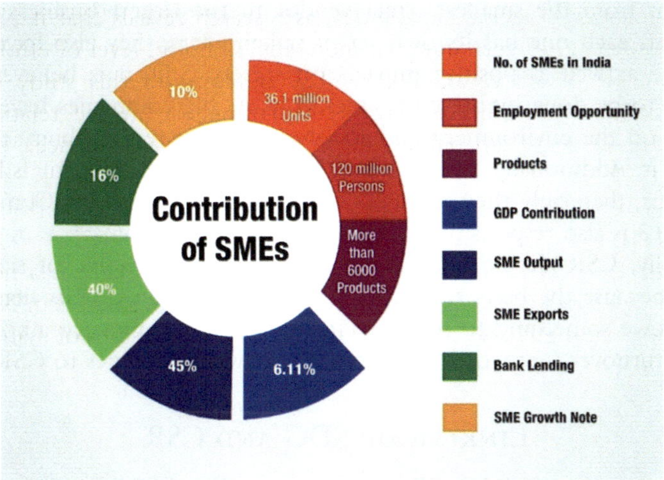

Fig. 11.2 Contribution of MSMEs in India (*Source* https://sktuligroup.com/2018/08/19/importance-of-small-medium-business-in-india/.html)

As per Section 135 of the act, the companies fitting into the specified criteria shall ensure that they spend every financial year at least 2% of the average net profits made during the three immediately preceding financial years of the company.

Section 135 of the Act provides for the applicability of the CSR provisions to corporate. Sub-Section (1) of section lays down that every company having a net worth of Rs. 500 Crore or more; or a turnover of Rs. 1000 Crore or more; or a net profit of Rs. 5 Crore during any financial year shall be compulsory constitute a CSR Committee of the Board consisting of three or more directors, out of which at least one director shall be an independent director.

The Companies Act 2013 introduced a revolutionary move of mandatory spending on CSR. India became the first country to make CSR spending mandatory.

In this background, many times the question is asked, Are MSMEs 'expected' to get involved in to CSR? But rather than asking this question can we think about searching for the answer of our MSMEs need' to get involved in CSR

Right from the smallest street vendor to the largest business establishment, each one has its own set of stakeholders, they also have both negative as well as positive impacts on society. Still, it is believed that CSR is to be done by big companies only. As big companies have more impact on the environment and society, it is their responsibility to take care of it. Additionally, there is also a belief that if the company is having a surplus, then only they are supposed to do CSR. The new Companies Act 2013 is also responsible to spread this belief in India.

Ideally, CSR should apply to organizations irrespective of size and sector because the basic reason for getting involved in CSR is that businesses owe something to society. The size of the business or amount of profit/turnover/net worth must not matter when it comes to CSR.

Linking of SDG and CSR

As mentioned by KPMG, SDGs and the Indian CSR regulation both were implemented/formulated around the same time and seemingly have a tremendous potential to develop a cohesive sustainable growth model. Moreover, the SDGs and CSR thematic development areas share a lot of overlap in terms of activities needed to achieve either.

The CSR regulation in India in the form of Schedule VII of Companies Act 2013 has set a broad framework and gives direction for a better sustainable future and the SDGs set tangible well defined targets to measure the outcome of activities.

The SDGs bring together players from all sectors for pursuing a common vision of sustainable development. It is expected that it will address the socio-economic and environmental challenges.

In the initial stage of the new CSR law, there was confusion regarding activities to be considered as CSR, the guidance in the form of Schedule VII of the Act was established. This also brings in the opportunity of collaboration for the achievement of SDG. Schedule VII explicitly gives overall direction to businesses.

In a document published by KPMG, at the time of the first SDG summit held in 2017, leveraging CSR to achieve SDGS, Schedule VII items are mapped with SDGs. The following examples can prove how Indian CSR can leverage achieving SDGs (Fig. 11.3).

Point No 1 of Schedule VII talks about "*Eradicating hunger, Poverty and malnutrition, promoting preventive healthcare and sanitation including to Swach Bharat Kosh set up by central government for the*

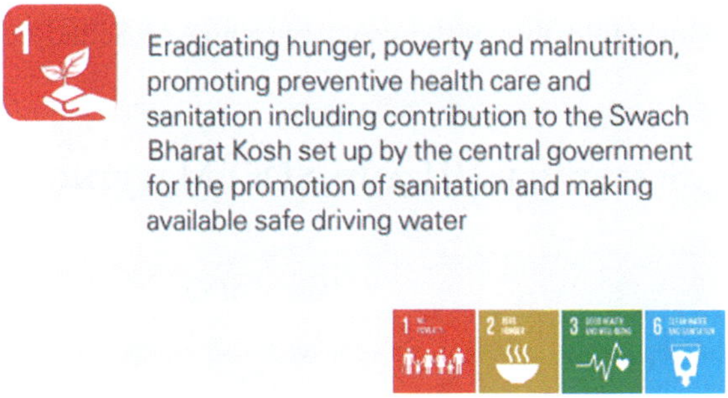

Fig. 11.3 Point 1 Schedule VII mapped with SDGs

promotion of sanitation and making available safe drinking water." It can be seen in the image that this point actually tries to address four different SDGs i.e. SDG-1: No Poverty, SDG-2: Zero Hunger, SDG-3: Good Health and Wellbeing and SDG-6: Clean Water and Sanitation (Fig. 11.4).

The second example of point no 4 can be taken into consideration to witness again how CSR can help in achieving SDGs. As shown in the above image point no 4 addresses 7 different SDGs which include SDG-6, 7, 9,11,13,14 & 15.

The above examples show that India has chosen CSR as one of the ways to achieve SDGs effectively and has also implemented it in a systematic manner.

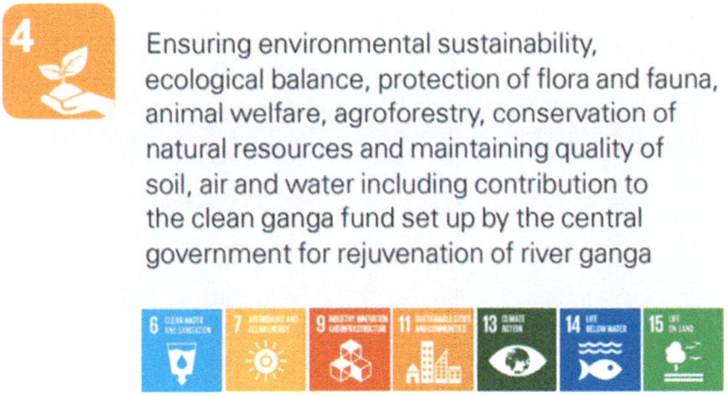

Fig. 11.4 Point 4 Schedule VII mapped with SDGs

The following table shows the exact mapping of Schedule VII points to SDGs.

Schedule VII and SDG Mapping

Schedule VII	*Mapped SDGs*
Point 1—Eradicating hunger, poverty and malnutrition, promoting preventive health care and sanitation including contribution to the Swach Bharat Kosh set up by the central government for the promotion of sanitation and making available safe driving water	SDG-1 End poverty in all its forms everywhere SDG-2 End hunger, achieve food security and improved nutrition, and promote sustainable agriculture SDG-3 Ensure healthy lives and promote well-being for all at all ages SDG-6 Ensure availability and sustainable management of water and sanitation for all
Point 2—Promoting education, including special education and employment enhancing vocation skills especially among children, women, elderly and the differently abled, and livelihood enhancement projects	SDG-1 End poverty in all its forms everywhere SDG-2 End hunger, achieve food security and improved nutrition, and promote sustainable agriculture SDG-4 Ensure inclusive and equitable quality education and promote lifelong learning opportunities for all SDG-8 Promote sustained, inclusive, and sustainable economic growth, full and productive employment and decent work for all
Point 3—Promoting gender equality, empowering women, setting up homes and hostels for women and orphans; setting up old age homes, day care centers and such other facilities for senior citizens and measures for reducing inequalities faced by socially and economically backward groups	SDG-1 End poverty in all its forms everywhere SDG-5 Achieve gender equality and empower all women and girls SDG-10 Reduce inequality within and among countries

(continued)

(continued)

Schedule VII	*Mapped SDGs*
Point 4—Ensuring environmental sustainability, ecological balance, protection of flora and fauna, animal welfare, agroforestry, conservation of natural resources, and maintaining the quality of soil, air, and water including contribution to the clean ganga fund set up by the central government for rejuvenation of river ganga	SDG-6 Ensure availability and sustainable management of water and sanitation for all SDG-7 Ensure access to affordable, reliable, sustainable, and modern energy for all SDG-9 Build resilient infrastructure, promote inclusive and sustainable industrialization, and foster innovation SDG-11 Make cities and human settlements inclusive, safe, resilient, and sustainable SDG-13 Take urgent action to combat climate change and its impacts SDG-14 Conserve and sustainably use the oceans, seas, and marine resources for sustainable development SDG-15 Protect, restore, and promote sustainable use of terrestrial ecosystems, sustainably manage forests, combat desertification, and halt and reverse land degradation and halt biodiversity loss
Point 5—Protection of national heritage, art, and culture including restoration of buildings and sites of historical importance and works of art; setting of public libraries, promotion and development of traditional arts and handicrafts	SDG-9 Build resilient infrastructure, promote inclusive and sustainable industrialization and foster innovation SDG-11 Make cities and human settlements inclusive, safe, resilient, and sustainable
Measures for the benefit of armed forces veterans, war widows, and their dependents	SDG-3 Ensure healthy lives and promote well-being for all at all ages SDG-4 Ensure inclusive and equitable quality education and promote lifelong learning opportunities for all SDG-8 Promote sustained, inclusive, and sustainable economic growth, full and productive employment, and decent work for all
Point 6—Training to promote rural sports, nationally recognized sports, Paralympic sports, and Olympic sports	SDG-8 Promote sustained, inclusive, and sustainable economic growth, full and productive employment, and decent work for all SDG-10 Reduce inequality within and among countries

(continued)

(continued)

Schedule VII	*Mapped SDGs*
Point 7—Contribution to the prime ministers national relief fund or any other fund set up by the central government for socio-economic development and relief and welfare of the scheduled castes, the scheduled tribes, and other backward classes, minorities, and women	SDG-1 End poverty in all its forms everywhere SDG-2 End hunger, achieve food security and improved nutrition, and promote sustainable agriculture SDG-3 Ensure healthy lives and promote well-being for all at all ages SDG-4 Ensure inclusive and equitable quality education and promote lifelong learning opportunities for all SDG-6 Ensure availability and sustainable management of water and sanitation for all SDG-9 Build resilient infrastructure, promote inclusive and sustainable industrialisation, and foster innovation SDG-15 Protect, restore, and promote sustainable use of terrestrial ecosystems, sustainably manage forests, combat desertification, and halt and reverse land degradation and halt biodiversity loss
Point 8—Contributions or funds provided to technology incubators located within academic institutions which are approved by the central government	SDG-9 Build resilient infrastructure, promote inclusive and sustainable industrialisation, and foster innovation SDG-12 Ensure sustainable consumption and production patterns SDG-13 Take urgent action to combat climate change and its impacts
Point 9—Rural development projects	SDG-1 End poverty in all its forms everywhere SDG-2 End hunger, achieve food security and improved nutrition, and promote sustainable agriculture SDG-3 Ensure healthy lives and promote well-being for all at all ages SDG-4 Ensure inclusive and equitable quality education and promote lifelong learning opportunities for all SDG-9 Build resilient infrastructure, promote inclusive and sustainable industrialisation, and foster innovation

(continued)

(continued)

Schedule VII	*Mapped SDGs*
Point 10—Slum area development	SDG-1 End poverty in all its forms everywhere SDG-2 End hunger, achieve food security and improved nutrition, and promote sustainable agriculture SDG-3 Ensure healthy lives and promote well-being for all at all ages SDG-4 Ensure inclusive and equitable quality education and promote lifelong learning opportunities for all SDG-6 Ensure availability and sustainable management of water and sanitation for all SDG-7 Ensure access to affordable, reliable, sustainable, and modern energy for all SDG-9 Build resilient infrastructure, promote inclusive and sustainable industrialization, and foster innovation

Barriers to CSR

Thus, from the literature review, it can be concluded that the SMEs face following barriers/challenges in engaging with CSR activities:

- Lower awareness of CSR benefits: CSR provides lots of benefits to any size and type of organization. An organization getting involved in CSR creates a different impression in the mind of all stakeholders. Awareness regarding CSR benefits is very meager among MSMEs and therefore they do not go ahead with engaging themselves in CSR voluntarily.
- Lack of time: MSMEs need to devote significantly high time to make their business economically sustainable. Therefore, it is observed that taking out time for any non-profit-making activity is seen as a burden and loss of productive time by MSMEs.
- Lack of resources: limited resources challenge faced by MSMEs becomes a hurdle in their day-to-day business activities, similarly the type of resources required for implementing CSR activities might also be scarcely available.

- Conservative mindset: the mindset of MSME owners is also a big barrier to CSR. Many organizations think that only if the company is making a profit beyond a certain limit then only they should get involved in CSR. As the majority of MSMEs earn limited profits they do not even think of CSR.
- Top management commitment and values: CSR needs to become an inevitable part of the culture of the company. In the case of MSMEs, Top management pays more attention to business goals and the main motive is to earn profit for the sustainability and growth of a business, in turn, management lacks in inculcating values in the culture. This becomes a barrier to CSR

Why Should MSMEs Undertake CSR in Spite of Barriers?

It is definitely acceptable that MSMEs have to prioritize their business needs to be sustainable and profitable. MSMEs to survive have to pay attention to their economic needs. But a lot of research done in the field of CSR shows that CSR can be used as a tool to achieve organizational benefits. The way large organizations enjoy the advantages of being socially responsible, even MSMEs can enjoy, if they get involved in CSR meaningfully. CSR is proven beneficial in achieving financial gains, employee engagement and retention, attracting investors, customer loyalty and satisfaction, enhancing corporate image, cost reduction, increased trust in stakeholders', and improved production along with social, cultural, and environmental betterment. Therefore, even when it is voluntary to get involved in CSR, MSMEs can get benefitted and can also contribute to achieving SDGs .

- Positive business reputation: companies' reputation in the market is a very crucial and delicate concern for any type of firm. The company's involvement in meaningful corporate social responsibility helps them to enhance the image and reputation of the company. Increased reputation can solve many basic problems faced by the company for regularizing and growing their business activity.
- Increased trust: socially responsible companies always gain trust of all types of stakeholders, internal as well as external. When a company is able to gain the trust of stakeholders like investors employees,

customers, suppliers, and distributors, business problems can be solved easily.

- Increased sales and customer loyalty: it is seen that customers believe in organizations that are socially responsible, and they are also ready to pay a sufficient price for the product of such companies. Meaningful CSR can help organizations in increasing sales and customer loyalty and ultimately profit.
- Better financial performance: Because of the responsible image of the company investors trust more on the company. They show their readiness to invest. Suppliers do extend support by giving credit to the company. Customers show readiness to pay more for the product of such a responsible company. In all this helps the company to increase its financial performance.
- Greater ability to attract talent and retain staff: as per previous research it is seen that employees prefer to join the socially responsible organization even at a comparatively low salary. Companies having a strong CSR base have observed a low employee turnover ratio as a result of CSR.
- Organizational growth: company's Involvement in CSR attracts many advantages. Due to this the main problem of business sustainability and required profit earning can be solved to a certain extent. Once the main hurdle is crossed by these companies, it can become easy to pay attention to organizational growth. The support required for the growth can also be initiated by them as they have already gained trust of various stakeholders.

Concluding Remarks

All businesses including MSMEs are producing goods and offering services to society at large and keeping the economic balance of society, but it is impossible to do business without the help of society. To operate successfully, a business needs various resources that are available in society and the environment. Each and every action of business have either a good or bad impact on the society and environment, therefore business is expected to take responsibility of it. Corporate social responsibility allows businesses to take societal responsibility. CSR can be viewed as an effective way to achieve SDGs .

Small businesses have added advantage of legal freedom of doing the CSR activities that they think fit. They do not have any pressure of having CSR policies or compulsion of spending a specific amount on CSR. As these businesses are having small negative impacts on society small CSR activities are enough for them to be socially responsible. It is not the size of the initiative that matters but it's company's willingness of doing something for socialmatters.

It is being proved from various research studies that CSR is definitely beneficial to any type of business. As large business units enjoy some benefits because of their involvement in CSR small businesses also enjoy similar benefits like customer retention, improving business reputation, attracting, retaining, and engaging responsible employees, creating a brand image and a social image of business, etc. The biggest benefit of CSR to MSMEs can be; that they are likely to get the contracts of large businesses because large businesses hold their entire supply chain accountable for ethical practices. It is clear that MSMEs can enjoy more benefits including economic benefits or business development opportunities because of CSR. Therefore, MSMEs can take up the small challenge of getting involved in CSR to increase the chances of getting big business opportunities.

Although for large businesses the CSR resources available are huge and it becomes comparatively easy for them to take big CSR initiatives like adopting a village, small businesses with their limited resources can also make a noticeable difference. Many MSMEs have the willingness to get involved in CSR but because of limited manpower and funds, it becomes a challenge for them. Collaborative CSR activities can be a better idea for small businesses to get involved in CSR to make a big social impact. A group of MSMEs can collaborate with each other and take up CSR activities as per their combined agenda. This method will help all the entities to contribute toward the community work with their limited resources and funds. In this way, MSMEs will contribute toward SDG 17.

It is observed that MSMEs face problems related to the resources and therefore they are found reluctant to get involved in CSR. Availability of resources is definitely a challenge in front of MSMEs but CSR doesn't need separate and ample resources, in fact, businesses are expected to utilize existing resources to make a difference in society. There can be different ways to do CSR; if the business is small then they can carry on small CSR activities as per their capacity to make difference. For example, painting a single classroom of a nearby school, teaching life skills to the kids of poor people, or facilitating teaching in government schools can also be counted as CSR. If the impact of CSR activities is considered, then achieving SDGs is not far. Provided along with large-scale organizations MSMEs also adopt CSR practices. We all and especially MSMEs should look at CSR as a tool to achieve sustainable development goals.

CEK

References

Admin. (2018). *Importance of Small & Medium Business in India*. https://sktuligroup.com/2018/08/19/importance-of-small-medium-business-in-india/

Angel, G. (2018). *SMEs are key for more inclusive growth*. OECD. https://www.oecd-ilibrary.org/docserver/25aa3d56-en.pdf?expires=1663738795&id=id&accname=guest&checksum=8309AEDBD4C06EB80784A0D0BD7A6A1F

Bapat, S. (2015). The review of corporate social responsibility of selected large scale establishments in India post companies act 2013. *International Journal of Management and Social Sciences Research, 4*(7), 26–34. http://www.irjcjournals.org/ijmssr/July2015/4.pdf

Bapat, S. (2017). A study of employee participation in CSR and its impact on employee engagement. *International Journal of Management and Social Sciences Research, 6*(8), 40–46. http://www.irjcjournals.org/ijmssr/Aug2017/8.pdf

Bapat, S., & Upadhyay, P. (2021). Implications of CSR initiatives on employee engagement. *Social Responsibility Journal, 17*(2), 149–163. https://doi.org/10.1108/SRJ-05-2018-012

Clark, J. (1916). The changing basis of economics responsibility. *Journal of Political Economy, 24*, 218–224.

Commission Of The European Communities. (2002). *Communication from the Commission concerning corporate social responsibility: A business contribution to sustainable development*. https://eur-lex.europa.eu/LexUriServ/LexUriServ.do?uri=COM:2002:0347:FIN:en:PDF#:~:text=The%20Green%20paper%20defined%20CSR,leads%20to%20sustainable%20business%20success

Crowther, D., & Aras, G. (2008). *Defining corporate social responsibility, Bookboon learning's*. E-book, Bookboon Learnings. https://bookboon.com/en/defining-corporate-social-responsibility-ebook

Dudovskiy, J. (2012). *Carroll's CSR Pyramid and its applications to small and medium-sized businesses*. Business Research methodology. www.research-methodology.net: https://research-methodology.net/carrolls-csr-pyramid-and-its-applications-to-small-and-medium-sized-businesses/

Fareeha, Z., & Farooq, M. (2014). *Corporate Social Responsibility in Small and Medium Enterprises to Achieve Organizational Benefits*. https://citeseerx.ist.psu.edu/viewdoc/download?doi=10.1.1.428.4694&rep=rep1&type=pdf
Hopkins, M. (2011). *Definition of corporate social responsibility*. http://mhcinternational.com: http://mhcinternational.com/articles/definition-of-csr
ICSI. (2015). *Corporate social responsibility, an engine for inclusive growth*. Indian Institute of Company Secretaries of India.
KPMG. (2008). *Corporate social responsibility—Towards a sustainable future* (White Paper).
KPMG. (2017). *Sustainable development goals: Leveraging CSR to achieve SDGs*. https://assets.kpmg/content/dam/kpmg/in/pdf/2017/12/SDG_New_Final_Web.pdf
Kumar, R., Murphy, D. F., & Balsari, V. (2001). *Altered images: The 2001 state of corporate responsibility in India poll. Understanding and encouraging corporate responsibility in South Asia: Update one*. Tata Energy Research Institute. https://insight.cumbria.ac.uk/id/eprint/3459/
Ministry of Corporate Affairs. (2011). *National Voluntary Guidelines*. http://www.mca.gov.in/Ministry/latestnews/National_Voluntary_Guidelines_2011_12jul2011.pdf
Ministry of Micro, Small and Medium Enterprises. (2022). *Annual Report 2021–2*. https://msme.gov.in/sites/default/files/MSMEENGLISHANNUALREPORT2021-22.pdf
Mirvis, P. (2012). Employee engagement and CSR: Transactional, relational, and developmental approaches. *California Management Review, 54*, 93–117. http://www.gncc.org/sites/gncc.org/files/Mirvis_Employee%20Engagement%20and%20CSR_2012
Sushil, K. D. (2020, August). *Barriers to corporate social responsibility: An Indian SME perspective*. https://www.lbsim.ac.in/uploads/image/741imguf_12.pdf
Tata Energy Research Institute. (2001). *Altered images, the 2001 state of corporate responsibility in India poll*. https://www.scribd.com/document/7019127/Altered-Images-The-2001-State-of-Corp-Resp-in-India-Poll
Towers Perrin. (2007). *Global workforce survey*. Towers Perrin Company. https://employeeengagement.com/towers-perrin-employee-engagement/
Tyagi, R. (2012). *Impact of corporate social responsibility on financial performance and competitiveness of business: A study of Indian firms* (PhD thesis). IIT Roorkee.
World Bank. (n.d.). *Small and medium enterprises (SMEs) finance*. https://www.worldbank.org/en/topic/smefinance

PART III

The Operational Management or Development of MSMEs

CHAPTER 12

Increasing the Role of Small Businesses in Achieving the SDGs Through Strengthening Managerial Capabilities: The Case of Nigeria

Timilehin Olasoji Olubiyi

Introduction

Small businesses can be considered as the central part of every nation's economic activity, and their role is one of the most important aspects for every nation that wants to thrive and flourish economically. Small enterprises are vital, and central to the efforts to achieve sustainable development goals and they contribute to broader social economic objectives, including employment creation making this sector a significant priority area for achieving the SDGs. Globally, MSMEs, often known as small businesses, may be regarded as the growth engine of many countries.

T.O. Olubiyi (✉)
Department of Business Administration and Marketing, School of Management Sciences, Babcock University, Ilishan-Remo, Ogun State, Nigeria
e-mail: olubiyi0070@student.babcock.gov.edu

H. Dasaraju and T.T.H. Tambunan (eds.), *Role of Micro, Small and Medium Enterprises in Achieving SDGs*, Sustainable Development Goals Series, https://doi.org/10.1007/978-981-99-4829-1_12

The performance of these small businesses has continued to gain significant momentum in recent years (Akbar et al., 2021a, 2021b; Khurana et al., 2021; Qureshi et al., 2021), particularly with the novel coronavirus (COVID-19) pandemic and its attendant consequences. The persistent aggravation of Nigeria's fundamental problems (such as a high poverty rate and poor educational quality, among others) by poor leadership and weak governance systems is quite alarming. Due to this, given the COVID-19 outbreak, fragile progress has been made in SDGs. Therefore, the 2030 Agenda may be challenging to achieve, particularly the SDGs in Nigeria.

The exceptional situation created by COVID-19 at the beginning of 2020 is influencing the country's commitment to SDGs and undermining the general approach of achieving the 17 SDGs. It has also continued to alter the course of development and performance of MSMEs in the country. Hence, the 2030 Agenda may be difficult to reach, especially, the attainment of SDGs in Nigeria. Small businesses are exposed to more sustainability-related risks than large organizations recent times the coronavirus (COVID-19), has exposed this inadequacy in small businesses or MSMEs and a number of challenges have continued to threaten their continuity and in them increasing their role in achieving the SDGs in Nigeria. Significantly, COVID-19 has brought about changes in the global business world and has caused businesses to experience uncertainty. SMEs are disproportionately impacted by market failures, impediments, and inefficiencies in the business environment and policy realm due to internal constraints. The contributions of MSMEs are also contingent upon their access to strategic resources such as skills, capabilities, and financing, as well as public investments in areas such as education and training, innovation, and infrastructure. The capabilities and abilities of managers have been seen as a key factor in how businesses make strategic changes and adapt to changes in the environment. Despite the fact that some small businesses in Nigeria have grown to become large firms and have achieved success, by integrating sustainability into their strategies. However, in a large number of cases, others have seen a fall and/or stagnation.

According to a review of the available literature, there are more small enterprises dissolving in Nigeria than growing. This would be a huge decline in SDGs particularly goal 8 and goal 9 social economic objectives (SDG 8- promote inclusive and sustainable economic growth, employment, and decent work), and SDG 9—improve sustainable industrialization and foster innovation). The failure is due to many reasons. But management problems are the most important (Cao et al., 2020; Hernández-Linares et al., 2021; Ihenetu & Kpurugbara, 2021). According to Mohammed and Obeleagu-Nzelibe (2014) and Ihenetu and Kpurugbara (2021), Nigerian small business entrepreneurs lack sufficient and/or formidable management capabilities and skills. This issue hinders the development and performance of small businesses, and due to the many problems, that these small businesses face in Nigeria, it has been difficult for businesses to bring value to the economy or sustain businesses with the consequences of COVID-19. In a newspaper publication in Nigeria on April 3rd, 2022, the changing narrative of small-scale business was highlighted, with significant issues such as management incapability, inefficiency in general company management, poor record keeping, competition, and lack of market knowledge being the bane of the majority of Nigerian small businesses.

From context observations, many small businesses in Nigeria have experienced weak performance or failure with the COVID-19 implications. It is important to continue to find ways to improve their performance. According to the literature, managerial capabilities have a beneficial impact on performance (Hussain et al., 2022; Kumbure et al., 2018). However, what motivates small business owners to use capabilities as a strategy is not entirely understood since there has not been enough research undertaken to explain this link. Academics generally believe that entrepreneurial self-efficacy is significant in explaining management capabilities and business performance. Many distinct mediating and moderating factors have been examined to understand the strategy by which management capabilities increase the performance of small businesses or the entity of MSMEs (Ihenetu & Kpurugbara, 2021). The fact that there are not many studies on internal organizational mediators that explain the link between managerial capabilities and business performance (Wang et al., 2021; Wales et al., 2013) is a gap in this field of research.

Shahab et al. (2018) found entrepreneurial self-efficacy to be an important mediator in determining both the strength of entrepreneurial intentions and the likelihood that those intentions would result in entrepreneurial actions and business performance. Rachmawan et al. (2015) believed that in starting a new business, entrepreneurial self-efficacy is a crucial factor in increasing the likelihood of business start-up activity. Hence, following this line of discussion, this paper hypothesized that entrepreneurial self-efficacy does not significantly mediate the relationship between managerial capabilities and the performance of small businesses in Lagos State, Nigeria. Therefore, this paper seeks to investigate the impact of entrepreneurial self-efficacy as a mediator between managerial capabilities and the performance of small businesses in Lagos State, Nigeria. It includes an analysis of four independent sub-variables (profitability, market share, sales growth, and competitive advantage) and four dependent sub-variables (managerial capabilities).

Literature Review

Sustainable Development and Linkage to MSMEs

Various studies imply that the implementation of corporate social responsibility and systematic sustainability in the business sector is a crucial aspect of attaining sustainable development. In both developed and developing nations, the vast majority of private sector businesses and economic activity are conducted by MSMEs. The International Trade Centre report "SME Competitiveness Outlook" (2019) indicates that MSMEs can contribute to the achievement of the SDGs through four channels: employee impacts (Goals 1, 2, 3, 8), business practice impacts (Goals 5, 8, 9, 10, 12, 13, 14, 15, 16), sectoral impacts (Goals 2, 3, 4, 5, 7, 9, 11), and national economy impacts (Goals 2, 3, 4, 5, 7, 9, 11). (Goals 1, 8, 9, 10, 17). This confirms the experts' observations that MSMEs play a key role in achieving the economic SDGs, as well as fostering innovation and sustainable industrialization, in addition to their contribution to the promotion of human development through promoting social progress and addressing income inequalities by providing good-quality jobs and working conditions. In other terms, it suffixes to state that MSMEs constitute a "major engine" of economic growth and societal development. Consequently, their potential contribution to the attainment and implementation of the SDGs is essential.

Managerial Capabilities

Managerial capabilities describe a set of skills, related knowledge, and attributes that allow an individual, such as a manager/owner of a small-scale business, to perform a task or an activity within a specific function or a job (Kato & Tsoka, 2020), are considered a critical factor that contributes to the performance and survival of small-scale businesses. The precise definition of managerial capabilities is heavily contested. The concept of managerial capabilities is incomplete without an understanding of managerial competencies and skills. For Mishra (2019), competencies refer to "observable and applied knowledge, skills, and behaviors that create a competitive advantage for an organization or business." As such, managerial competencies describe those demonstrable knowledge, skills, abilities, and behavioral traits that an owner or manager of a small business should possess to effectively accomplish their entrepreneurial and managerial duties to generate financial value for the business. In the context of small businesses, managerial capabilities are defined as sets of knowledge, skills, behaviors, and attitudes that can contribute to the personal effectiveness of managers/owners of small businesses (de Souza & Forte, 2021).

Managerial capabilities are a set of qualities and attributes in the personality of the manager that enable them to effectively manage the work of the organization (Hernandez-Perlines, 2018). Also, management capabilities are linked to symbolic factors like communication skills, the ability to adapt to new situations, openness to the outside world, teamwork, strong technical skills, stress management, the ability to work well with others, social intelligence, and an appreciation for cultural diversity (Beta Badri, 2020; Isaga, 2018).

Small Business Performance

Performance has been defined in terms of how well an organization is managed (Khurana et al., 2021), and the value the organization delivers to customers and other stakeholders. Business performance can be defined and measured in terms of profitability, growth, market value, the total return on the shareholder, economic value added, and customer satisfaction, based on the stakeholders' expectations (Mihaela, 2017). Small business performance entails the provision of values to its stakeholders, which includes the owners, customers, government, and society (Pangarso et al., 2022). According to Owolabi and Makinde (2012), the performance of an organization can be measured by improved occupational health and safety performance, including increased customer satisfaction. According to Pangarso et al. (2022), Small business performance is considered to be the outcome of adapting useful management processes. Businesses, as well as entrepreneurs, value performance. According to

Khurana et al. (2021), the performance of a small business is defined by its ability to contribute to job and wealth creation through enterprise start-up, survival, and growth. Extant literature has been prolific in generating the difference in the measurements of small business growth and highlighting their merits and demerits. A business organization could measure its performance using financial and non-financial measures. The financial measures include profit before tax and turnover (Bonsu & Kuofie, 2019; Maldonado-Guzman, 2017; Milfelner et al., 2008), returns on sales (Keats et al., 2018), and returns on equity (Haber & Reichel, 2005; Miller et al., 1988), while the non-financial measures focus on issues pertaining to market share, customer satisfaction, and customer referral rates, delivery time, sales growth, waiting time, and employee turnover (Batra & Dhir, 2019; Roschk & Hosseinpour, 2020). Based on the review of relevant literature, this study has adopted a four-dimension non-financial performance measure. These are a competitive advantage, sales growth, market share, and profitability.

SDGs, Small Business and Economic Performance

Every country's small business sector plays a significant and critical role in economic and industrial growth (Karagiorgos et al., 2020; Khurana, et al., 2021; Mohamed et al., 2020; Muriithi, 2017). Global phenomena that are a central focus of the SDGs have been drastically altered, bringing mindfulness to the new realities and forms of existence we could not have imagined before particularly with MSMEs. Mobility and migration were severely hindered by lockdown measures, disruption to the supply chain, and restrictions that incurred huge human and monetary costs. The consequences are still very visible, chiefly the high inflation and weakening of the SDG 12 (consumption and production) including SDG 1 and 2 putting a large number of people in hunger and poverty among other.

According to Muriithi, (2017) and Karagiorgos et al. (2020), small businesses have a higher chance of strengthening the domestic economy by producing products and services that can accelerate Nigeria's economy. According to Owolabi and Makinde (2016), small businesses contribute to over 90% of commerce, contributing to job creation, employment, tax revenue, and GDP according to Owolabi and Makinde (2016). Similarly, Owolabi and Makinde (2016) stated that small businesses are an important source of poverty reduction, job creation, and economic growth in Nigeria. Supportably, Muhammad, and Isah (2018) posit those small businesses significantly improve living conditions as well as considerable local capital generation and high levels of productivity. Small enterprises in Nigeria pave the way for wealth creation for entrepreneurs. This might be one of the reasons Nigerians pay attention to small business operations for their livelihood. The MSME sector is a "major engine" of economic growth and social progress. Therefore, their potential contribution to the realization and implementation of the SDGs is vital.

Coronavirus, SDGs, and Performance of Small Businesses in Nigeria

The development of the coronavirus pandemic and its devastating effects on the global economy seems to have significantly slowed global economic activity (Gerald et al., 2020; Rahman et al., 2022). The virus was discovered in Wuhan, Hubei province. The virus spread over Europe, America, Africa, and most parts of the globe since its discovery in Wuhan, China. Businesses have adjusted and adopted tactics to limit the pandemic's effects. Amidst the COVID-19 pandemic consequences, small enterprises in Lagos State, Nigeria have continued to face exacerbating challenges such as policy inconsistency, managerial incompetence, poor infrastructure, lack of government incentives, a poor regulatory environment, and low institutional quality (Rahman et al., 2022). MSMEs play a crucial role in the attainment of the SDGs and are largely involved in the implementation, but recent events have diminished their intensity and impact. All the recent realities and challenges have continued to threaten the performance of small businesses and their survival in Nigeria.

Entrepreneurial Self-Efficacy

In the current research, the lens of the entrepreneurial self-efficacy theory of Bandura (1977, 1993, 2001, 2006) was adopted, based on the notion that socioemotional elements such as attitudes, identity, and interest are likely to be more important for long-term business outcomes (Fortus & Touitou, 2021; Maltese & Tai, 2011; Nguyen & Riegle-Crumb, 2021; Tytler & Osborne, 2012). Entrepreneurial self-efficacy is the level of confidence shown by an individual toward the undertaking of a task (Bandura, 2000). Entrepreneurial self-efficacy (ESE), refers to an individual's belief in effectively completing the many duties and responsibilities of entrepreneurship aimed at a positive entrepreneurial outcome (Newman et al., 2019). It is widely agreed that it is a major factor in whether or not people go into business for themselves and act in an entrepreneurial way.

Why is entrepreneurial self-efficacy so important in entrepreneurship and why do we need a robust definition? The entrepreneurial self-efficacy construct is good for studying entrepreneurship because it is task-specific and measures a person's confident beliefs about internal (personality) and external (environment) constraints and opportunities. It is also close to the action and action intentionality (Newman et al., 2019). Self-efficacy is a useful concept for explaining human behavior as research reveals that it plays an influential role in determining an individual's choice, level of effort, and perseverance (Chen & Zhou, 2017). According to Mcgee et al. (2009), the factors affecting entrepreneurial self-efficacy are (1) experience; (2) modeling; (3) social persuasion; and (4) psychological factors. The most important factor for deciding a person's entrepreneurial self-efficacy is mastery experience. This is the experience of attaining a belief in one's own capabilities through past mastery (Bandura, 2009). Success raises entrepreneurial self-efficacy, while failure lowers entrepreneurial self-efficacy.

Empirical Review

The COVID-19 pandemic has had implications for all economic sectors and organizations, especially small businesses (Hasanat et al., 2020; Rahman et al., 2022). Many countries in sub-Saharan Africa (SSA) including Nigeria have been confronted with vast social, economic, and political challenges (including the economic crisis, political instability and conflicts, corruption, bad leadership, and food shortages). According to studies and observations, small businesses are faced with these challenges and a variety of difficulties due to post-COVID-19 consequences, most importantly their performances. Because of this, many small businesses were not able to handle the situation (Oyewale et al., 2020; Ozili, 2020).

Several studies have focused on the role of entrepreneurial self-efficacy in various facets of entrepreneurship during the past two decades, spanning from entrepreneurial goals to business performance (Baum & Locke, 2004; Chen & Zhou, 2017; Khalil et al., 2021; Khedhaouria et al., 2014; Kickul et al., 2009). Several studies have been conducted on company performance (Baum & Locke, 2004; Hmieleski & Baron, 2008; Khedhaouria et al., 2014; Kickul et al., 2009; Torres & Watson, 2013).

Robin et al. (1993) found that entrepreneurial self-efficacy mediated the effect of interpersonal communication outcomes on the performance of firms. Interpersonal communication competence directly affected the rewarding, satisfying communication of managers in engendering improved performance, in Robin et al. (1993). In the study carried out by Buenaventura-Ver (2017), the entrepreneurial self-efficacy of the leaders positively mediated the innovative workplace behavior, an indicator of entrepreneurial orientation and performance relationship. Furthermore, the entrepreneurial self-efficacy of the leaders mediated positively on communication measured as team-member exchange and on performance (Buenaventura-Ver, 2017; Egwakhe Osabuohien, 2010). According to previous research, performance and entrepreneurial self-efficacy are linked (Cumberland et al., 2015; Torres & Watson, 2013), as well to Zubair and Khan, (2021) managerial capabilities and performance seem to be linked in the majority of studies, and various research has shown that with a mediator or moderator performance may have both negative and positive consequences depending on the situation or circumstances. As a result, the primary goal of this research is to look into the mediating impact of entrepreneurial self-efficacy on managerial capabilities and small business performance connection. It is pertinent to mention that limited investigation has been done on testing this relationship in literature.

Theoretical Review

Penrose proposed the Resource-Based View (RBV) in 1959. Wernerfelt turned it into a coherent theory in 1984, and Barney spread the word about it in 1991. According to the theory, the scarce, valued, non-replaceable, and imperfectly imitable business resources and capabilities are the foundation of a business's sustainable competitive advantage. The RBV contends that the company's present resources and competencies affect its growth and performance (Muthee & Ngugi, 2014). RBV proposes that a business may maintain a competitive advantage by fostering the development of firm-specific capabilities, building complex social interactions, rooted in the firm's history and culture, and generating tacit organizational knowledge and capabilities (Lee, 2016). Human resources' potential position as a strategic asset in the corporation has long been supported by the resource-based perspective of the enterprise (Wright & McMahan, 1992).

The basic argument of the RBV, as explained by (Wright et al., 1994) is that the firm's human resources confer enduring competitive advantage to a firm to the extent that they remain scarce or hard to duplicate, have no direct substitutes, and enable companies to pursue opportunities. This view of the firm infers that firms create competitive advantages by implementing unique combinations of resources and business practices that are difficult for competitors to imitate. A resource-based view of a firm explains its ability to deliver sustainable competitive advantage when resources are managed such that their outcomes cannot be imitated by competitors, which ultimately creates a competitive barrier (Mahoney & Pandian [1992] cited by Hooley & Greenley, 2005). The RBV explains that the internal resources, assets, capabilities, and knowledge of a firm are the key determinants of its competitive position (Barney, 1991). Where a firm's resources are rare and valuable, it is able to create a competitive advantage and earn above-average returns. Since resources constitute both tangible and intangible assets, then both strategic entrepreneurship and business profitability may also be considered valuable resources or capabilities (Bakar & Ahmad, 2010). The RBV considers the role of the firm's resources and capabilities in deciding the scope of its operations and laying the groundwork for its long-term strategy.

Conceptual Model

Operationalization of Variables

In order to conduct research on the constructs of this chapter, it is necessary to operationalize the variables in the study. Small business performance (SBP) is the dependent variable, and managerial capabilities (MC) is the independent variable. The mediating variable is entrepreneurial self-efficacy (ESE). The mathematical relationships among the variables based on the research hypotheses are stated in this section.

Independent Variables

$$X = \text{managerial capabilities (MC)}$$

$$Y = \text{small business performance (SBP)}$$

$$X = (x_1, x_2, x_3, x_4)$$

where:

$$x_1 = \text{communication (CO)}$$

$$x_2 = \text{marketing (MA)}$$

$$x_3 = \text{negotiation (NE)}$$

$$x_4 = \text{people management (PM)}$$

$$Y = (y_1, y_2, y_3, y_4)$$

where:

$$y_1 = \text{profitability (PR)}$$

$$y_2 = \text{sales growth (SG)}$$

$$y_3 = \text{market share (MS)}$$

$$y_4 = \text{competitive advantage (CA)}$$

and

$$M = \text{entrepreneurial self} - \text{efficacy (ESE)}$$

Functional Relationships

$$y_1 = f(x_1, x_2, x_3, x_4)$$

$$y_2 = f(x_1, x_2, x_3, x_4)$$

$$y_3 = f(x_1, x_2, x_3, x_4)$$

$$Y = f(X, M)$$

The equations of the study based on the research hypotheses are as follows:

Hypothesis

$$Y = f(X, M)$$

$$Y = \beta_0 + \beta X + \beta M + e_i \tag{12.1}$$

$$\beta_0 = \text{Constant term}$$

Figure 12.1 shows the workings of the variables as they relate and are stipulated in the hypotheses of the study:

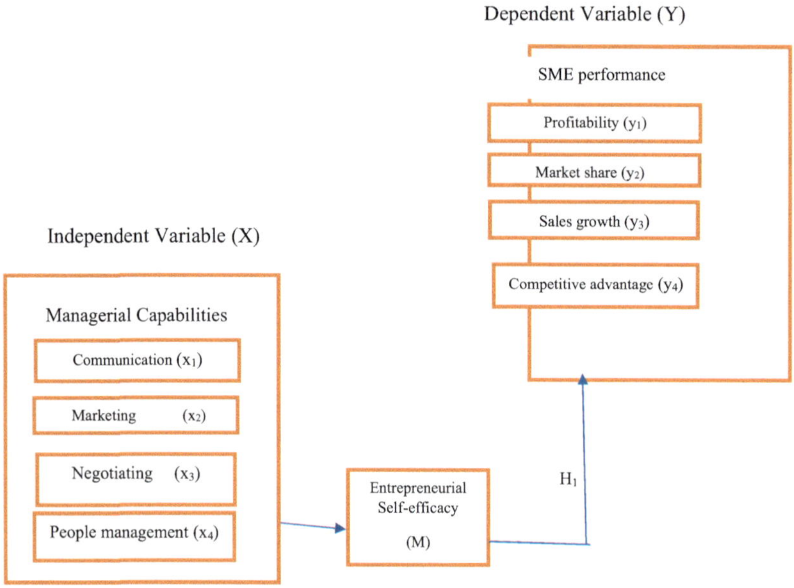

Fig. 12.1 Research conceptual framework

Methodology

Research Design

This study used a survey research approach to investigate the relationship between management qualities and small business performance in Lagos State, Nigeria. In this paper, the survey research design was cross-sectional, and the technique was quantitative. The adoption of this design was influenced by the research problem and its corresponding research questions. The use of this design was predicated on the work of Asikhia (2010) and Makinde (2015), and was also consistent with the studies of Adisa et al. (2016), Allen (2011), Egwakhe (2014), Egwakhe and Kabasha (2016), and Olubiyi et al. (2019), which generate detailed and reliable explanations of the relationship between variables.

The population of this study comprised the four thousand, five hundred and thirty-five (4,535) owners/managers of small businesses, which are the formally constituted and registered MSMEs operating in Lagos State and registered with the small and medium enterprise development agency of Nigeria (SMEDAN). The justification for the focus of the study on small businesses is that these companies generally suffer from resources and managerial capabilities constraints and, as a result, there is a need to closely study their internal functions to assist their performance (Buli, 2017; Olubiyi et al., 2019). Lagos State was chosen because of its unique position in Nigeria as the commercial nerve (Apulu et al., 2011; Olubiyi et al., 2019). Also, the highest number of small businesses operating in Nigeria exists in Lagos State (Olubiyi et al., 2019; Wang & Maduako, 2018). Table 12.1 presents the list of small and medium enterprises by five divisions in Lagos State.

The unit of analysis for this study was the owners/managers of formal and registered small businesses in Lagos State but distributed over the

Table 12.1 List of small businesses by divisions in Lagos States Nigeria

Small business sector	*Ikorodu*	*Badagry*	*Ikeja*	*Lagos Island*	*Epe*	*Total*
Manufacturing	102	24	492	363	31	1012
Real Estate	120	110	48	27	200	505
Agriculture	281	161	–	–	175	617
Service	424	69	1356	388	164	2401
Total	927	364	1896	778	570	4535

five geo-political zones/divisions of the State. The study's sample size, as determined by the population, was 354 using the Cochran formula approach to arrive at a representative sample from the general population. According to Israel (2013), there is a need to give an allowance of a 30% non-response rate to enhance the sample size. This is to ensure incomplete questionnaires, un-returned questionnaires, and any form of uncertainty are covered. Therefore, the non-response rate allowance of 30% amounts to 106, which was added to the statistically calculated sample size of 354, to arrive at 460. The researcher adopted the primary source of data collection in carrying out this study. The primary data were chosen because they made it easier to find out what the respondents thought and felt (Belaya & Hanf, 2013; Olubiyi et al., 2019) about all the things that were used to measure managerial capabilities, entrepreneurial self-efficacy, and small business performance.

Data Analysis and Results

The researcher distributed 460 copies of the questionnaire to the respondents. 457 copies of the distributed questionnaire were duly filled and returned and were used for the analysis. This represents a response rate of about ninety-nine percent (99%), which was considered adequate in this paper. Table 12.2 presents the results of the response rate.

Next, Tables 12.3a, 12.3b, 12.3c, 12.3d, 12.3e, 12.3f and 12.3g show the summary of stepwise regression analysis for mediating effect of entrepreneurial self-efficacy Between managerial capabilities and performance of small businesses in Lagos State.

The hypothesis sets out to investigate the mediating effect of entrepreneurial self-efficacy on the relationship between managerial capabilities and the performance of small businesses in Lagos State, Nigeria. To test the hypothesis, stepwise process analysis was employed.

Table 12.2 Response rate

Category	*Frequency*	*Percentage %*
Completed usable copies of questionnaire	457	99
Unreturned/Incomplete copies of questionnaire	3	1
Total received	457	100

Table 12.3a Summary of the mediating effect of outcome variable-entrepreneurial self-efficacy

R	*R-sq*	*MSE*	F	*df1*	*df2*	*P*
0.826	0.682	0.190	974.303	1	455	0.000

Outcome variable: Entrepreneurial self-efficacy

Table 12.3b Summary of regression between managerial capabilities and entrepreneurial self-efficacy

	Beta	*Standardized coefficients Coeff*	*Se*	*T*	*P*	*LLCI*	*ULCI*
constant	−1.005		0.162	−6.223	0.000	−1.322	−0.687
Managerial capabilities	1.266	0.826	0.041	31.214	0.000	1.187	1.346

Outcome variable: Entrepreneurial self-efficacy

Table 12.3c Model summary of regression analysis between managerial capabilities, entrepreneurial self-efficacy, and small business performance

R	*R-sq*	*MSE*	*F*	*df1*	*df2*	*P*
0.380	0.144	0.396	38.309	2	454	0.000

Outcome variable: Performance

Table 12.3d Model of regression analysis between managerial capabilities, entrepreneurial self-efficacy, and small business performance

	Beta	*Standardized coefficients Coeff*	*Se*	*T*	*P*	*LLCI*	*ULCI*
constant	2.004		0.243	8.253	0.000	1.527	2.4814
Managerial capabilities	0.081	0.060	0.104	0.784	0.433	−0.123	0.2855
Entrepreneurial self-efficacy	0.289	0.329	0.068	4.271	0.000	0.156	0.4221

Outcome variable: Performance

Table 12.3e Model summary of regression analysis between managerial capabilities and small business performance

Model Summary	*R*	*R-sq*	*MSE*	*F*	*df1*	*df2*	*P*
	0.332	0.110	0.411	56.243	1	455	0.000
Outcome variable: Performance							

Table 12.3f Model of regression analysis between managerial capabilities and small business performance

	Beta	*Se*	*T*	*P*	*LLCI*	*ULCI*
constant	1.7136	0.2375	7.2152	0.0000	1.2469	2.1804
Managerial capabilities	0.4475	0.0597	7.4995	0.0000	0.3303	0.5648
Outcome variable: Performance						

Table 12.3g Indirect effect(s) of *X* on *Y*

	Effect	*BootSE*	*BootLLCI*	*BootULCI*
Entrepreneurial self-efficacy	0.3661	0.0808	0.2304	0.5509

Table 12.3b revealed a statistically significant relationship between managerial capabilities and entrepreneurial self-efficacy (= 1.2663, t = 31.2138, p 0.05). The adjusted R square value of 0.682 shows that entrepreneurial self-efficacy explains 68.2% of the changes in managerial capabilities.

Table 12.3d presents the results of the role of entrepreneurial self-efficacy between managerial capabilities and small business performance. The results show that entrepreneurial self-efficacy (= 0.2891, t = 4.2712, p 0.05) has a significant relationship with performance, while managerial capabilities (= 0.0814, t = 0.7844, p 0.05) have a positive but not significant relationship with small business performance in Lagos State, Nigeria. Table 12.3f presents the result of the relationship between managerial capabilities and performance. The result shows that managerial skills (= 0.448, t = 7.4995, p = 0.05) have a strong effect on how well small businesses in Lagos State, Nigeria do. The result of the r-square Table 12.3e of 0.110 indicates that managerial capabilities only explain 11% of the changes observed in performance at the selected small businesses in Lagos State. Table 12.3g presents the result of the process analysis conducted on the mediating effect of entrepreneurial self-efficacy on the relationship between managerial capabilities and the performance of small businesses in Lagos State, Nigeria.

For mediation, the mediating effect is the product of the coefficients of the interaction between managerial capabilities and entrepreneurial self-efficacy (= 1.2663, p 0.05) and the interaction between small business performance and entrepreneurial self-efficacy (= 0.2891, p 0.05). The mediation effect is thus 0.3661. The results of the process analysis conducted revealed that the indirect effect of X on Y with a value of 0.3661 indicates that entrepreneurial self-efficacy has a strong mediating effect between managerial capabilities and small business performance. Furthermore, the results revealed that there is no value of 0 between the confidence interval values, which further firms that entrepreneurial self-efficacy has a strong mediating effect between managerial capabilities and small business performance in Lagos State, Nigeria.

The results indicate that if entrepreneurial self-efficacy is poor, managerial capabilities will negatively affect small business performance. This implies that small businesses should consider characteristics such as entrepreneurial capabilities, personality traits, need for achievements, building an innovative environment, and creation, delivery, and capture of value as these mediate the relationship between managerial capabilities and performance (Fig. 12.2).

Discussion of Findings

The hypothesis set out to investigate the mediating effect of entrepreneurial self-efficacy on the relationship between managerial capabilities and the performance of small businesses in Lagos State, Nigeria. To test the hypothesis, process analysis software was employed. The result of the analysis revealed the indirect effect of managerial capabilities on performance to be 0.3661. The value of 0.3661 indicated that entrepreneurial self-efficacy has a strong mediating effect between managerial capabilities and small business performance in Lagos State, Nigeria.

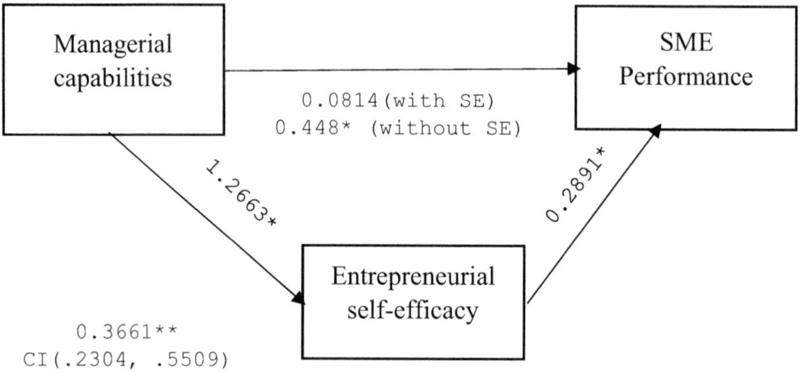

Fig. 12.2 Results of the relationship between managerial capabilities and performance

The findings of this study are in consonance with the view of Bandura (2012), who insists that entrepreneurial self-efficacy is the level of confidence shown by an individual toward the undertaking of a task or position which favorably influences small business performance. The findings of this study are consistent with those of Ngek (2015), Dessyana and Riyanti (2017), Adolfina and Lumintang (2018), Budiman and Pangestu (2018), Cumberland et al. (2015), Yusuff et al. (2019), Oyeku et al. (2020), Islam et al. (2020), Kale, (2020), Khalil et al. (2021), McGee and Peterson, (2019), Torres and Watson, (2013) which demonstrate that entrepreneurial self-efficacy positively and significantly affects business performance. Self-efficacy is a useful concept for explaining human behavior and managerial capabilities, as research reveals that it plays an influential role in determining an individual's choice, level of effort, and perseverance (Khalil et al., 2021). The outcome of this study is that entrepreneurial self-efficacy plays a mediating role in the relationship between managerial capabilities and small business performance.

Conclusion and Recommendation

With the COVID-19 crisis, everything changed. So, there is a need to reinforce managerial capabilities for sustainability in businesses in Nigeria. Therefore, the study also concludes that all four dimensions of managerial capabilities- effective communication skills, marketing skills, negotiation skills, and people management skills are key and required by owners/ managers of small businesses to develop and achieve reasonable performance. They have to review and change operational models and strategic plans, in order to stay competitive in the industry as well as gain a competitive advantage and impact SDGs positively. So, those dimensions must be taken headlong for positive and significant change in the MSME sector. Current development shows that SDGs and MSMEs have synergistic potential. MSMEs play a significant role in the implementation of these sustainable development goals because it stimulates the search for and implementation of creative solutions to social and ecological challenges from the standpoint of managerial capabilities. Therefore, self-efficacy and the managerial capabilities of the owner-managers and MSME operators can provide opportunities for innovative corporate development and helps solve existing difficulties. The pandemic's problems highlight the worries about the attainment of the 2030 Agenda for Sustainable Development and therefore SMEs have a significant role to play.

As such, it is important that strong partnerships between the government and industry be developed to improve the managerial capabilities of MSME operators. This to a large extent will reduce skill gaps and impact positively on the demands of SDGs. Finally, the results of the process analysis conducted revealed that entrepreneurial self-efficacy has a strong mediating effect between managerial capabilities and small business performance. Based on this finding, the study recommends that owners/ managers of small businesses must possess entrepreneurial capabilities, and personality traits, driven by a need for achievement, build an innovative environment, and create, deliver, and capture value as these have been found to mediate the relationship between managerial capabilities and performance of small businesses. This research makes many theoretical and practical contributions. To begin with, the outcomes of this research add to the body of knowledge on entrepreneurial self-efficacy. Entrepreneurial self-efficacy is critical because it mediates the relationship between managerial capabilities and small business performance. The importance of SMEs in achieving the SDGs has always been emphasized.

Limitations and Future Directions

It recognized that the findings and implications of this paper are situated within Lagos, Nigeria, and primarily small businesses as the focus of its attention. The research can be expanded with similar studies conducted across large businesses or listed firms in Nigeria or other climes. The sample size can be raised by considering more respondents with innovative data-gathering techniques, to achieve more generalizability and trustworthiness. In this research, the absence of data in the majority of small businesses in Lagos was the greatest challenge to obtaining objective performance measurements, for this reason, non-financial performance indicators were adopted. According to Khalil et al. (2021), subjective measurements may be produced in the absence of objective measures. Nonetheless, the limitations of this research can be recognized as research gaps for future studies, and further research is advised to survey the influence of business owners'/managers' characteristics on entrepreneurial behaviors instead of intention.

References

Akbar, A., Jiang, X., Qureshi, M. A., & Akbar, M. (2021a). Does corporate environmental investment impede financial performance of Chinese enterprises? The moderating role of financial constraints. *Environmental Science and Pollution Research, 28*, 58007–58017.

Akbar, M., Akbar, A., & Draz, M. U. (2021b). Global Financial crisis, working capital management, and firm performance: Evidence from an Islamic Market Index. *SAGE Open* 11 (21582440211015705).

Bandura, A. (1977). Self-efficacy: Toward a unifying theory of behavioral change. *Psychological Review, 84*(2), 191–215. https://doi.org/10.1037/0033-295X.84.2.191

Bandura, A. (1993). Perceived self-efficacy in cognitive development and functioning. *Educational Psychologist, 28*(2), 117–148. https://doi.org/10.1207/s15326985ep2802_3

Bandura, A. (2000). Cultivate self-efficacy for personal and organizational effectiveness. *Handbook of Principles of Organization Behavior, 2*, 11–21.

Bandura, A. (2001). Social cognitive theory: An agentic perspective. *Annual Review of Psychology, 52*(1), 1–26. https://doi.org/10.1146/annurev.psych.52.1.1

Bandura, A. (2006). Guide for constructing self-efficacy scales. Entrepreneurial self-efficacy beliefs of adolescents. *Information age publishing*. https://doi.org/10.1017/CBO9781107415324.004

Batra, I., & Dhir, S. (2019). The effects of psychic distance and inter-partner ft on the performance of international joint ventures. *Asian Academy of Management Journal, 24*(1), 151.
Beta, E. A., & Badri, M. S. (2020). Managerial cognitive capabilities, organizational capacity for change, and performance: The moderating effect of social capital. *Cogent Business & Management, 7*(1), 2–23, 184. https://doi.org/10.1080/23311975.2020.1843310
Bonsu, S., & Kuofie, M. (2019). Small business survival. *Journal of Marketing & Management, 10*(1), 51–63.
Cao, X., Ouyang, T., Balozian, P., & Zhang, S. (2020). The role of managerial cognitive capability in developing a sustainable innovation ecosystem, a case study of Xiaomi. *Sustainability, 12*(17), 7176. https://doi.org/10.3390/su12177176
Chen, Y., & Zhou, X. (2017). Entrepreneurial self-efficacy and firms' innovation behavior: The negative mediating role of social capital. *Social Behavior and Personality: An International Journal, 45*(9), 1553–1562.
Cumberland, D. M., Meek, W. R., & Germain, R. (2015). Entrepreneurial self-efficacy and firm performance in challenging environments: Evidence from the franchise context. *Journal of Developmental Entrepreneurship, 20*(01), 1–19.
Fortus, D., & Touitou, I. (2021). Changes to students' motivation to learn science. *Disciplinary and Interdisciplinary Science Education Research, 3*(1), 1. https://doi.org/10.1186/s43031-020-00029-0
Gerald, E., Obianuju, A., & Chukwunonso, N. (2020). Strategic agility and performance of small and medium enterprises in the phase of Covid-19 pandemic. *International Journal of Financial, Accounting, and Management, 2*(1), 41–50.
Hasanat, M. W., Hoque, A., Shikha, F. A., Anwar, M., Hamid, A. B. A., & Tat, H. H. (2020). The impact of coronavirus (Covid-19) on E-business in Malaysia. *Asian Journal of Multidisciplinary Studies, 3*(1), 85–90.
Hernandez-Perlines, F. (2018). Moderating effect of absorptive capacity on the entrepreneurial orientation of international performance of family businesses. *Journal of Family Business Management, 8*(1), 58–74. https://doi.org/10.1108/JFBM-10-2017-0035
Hernández-Linares, R., Kellermanns, F. W., & López-Fernández, M. C. (2021). Dynamic capabilities and SME performance: The moderating effect of market orientation. *Journal of Small Business Management, 59*(1), 162–195. https://doi.org/10.1111/jsbm.12474
Hussain, A., Minhas A., Arfan S., Petra P., Ahsan A, & Rohail H. (2022). E-commerce and SME performance: The moderating influence of entrepreneurial competencies. *Administrative Sciences, 12*, 2–16. https://doi.org/10.3390/admsci12010013

Ihenetu, H., & Kpurugbara, N. (2021). Entrepreneurs and real sector development in Nigeria. *African Journal of Business and Economic Development, 1*(9), 13–24.

International Trade Centre. (2019). *SME competitiveness outlook 2019: Big money for small business—Financing the sustainable development goals*. ITC.

Isaga, N. (2018). The relationship of personality to cognitive characteristics and SME performance in Tanzania. *Journal of Small Business and Enterprise Development, 25*(4), 667–686. https://doi.org/10.1108/JSBED-02-2017-0067

Khalil, M., Khan, M., Zubair, S. S., Saleem, H., & Tahir, S. (2021). Entrepreneurial self-efficacy and small business performance in Pakistan. *Management Science Letters, 11*, 1715–1724. https://doi.org/10.5267/j.msl.2021.2.011

Kato, A. I., & Tsoka, G. E. (2020). Impact of venture capital financing on small- and medium-sized enterprises' performance in Uganda. *The Southern African Journal of Entrepreneurship and Small Business Manaement, 12*(1), a320.

Keats, E. C., Macharia, W., Singh, N. S., Akseer, N., Ravishankar, N., Ngugi, A.K., Arjumand, R., Khaemba, E. N., Tole, J., & Bhutta, Z. A. (2018). Accelerating Kenya's progress to 2030: understanding the determinants of under-five mortality from 1990 to 2015. *BMJ Global Health, 3*(3), 727–742.

Khurana, S., Haleem, A., Luthra, S., & Mannan, B. (2021). Evaluating critical factors to implement sustainable oriented innovation practices: An analysis of micro, small, and medium manufacturing enterprises. *Journal of Cleaner Production, 285*, 125377. https://doi.org/10.1016/j.jclepro.2020.125377

Karagiorgos, A., Gaitanaki, A., Ignatiou, O., & Terzidou, A. (2020). Role and contribution of administrative accounting to small and very small businesses. *Journal of Accounting and Taxation, 12*(2), 75–84. https://doi.org/10.5897/JAT2020.0399

Maldonado-Guzman, G., Martínez-Serna, M.C., & PinzónCastro S. Y. (2017). Entrepreneurial orientation and growth in Mexican small business. *Journal of Management and Sustainability, 7*, 94–102.

Maltese, A. V., & Tai, R. H. (2011). Pipeline persistence: Examining the association of educational experiences with earned degrees in STEM among U.S. students. *Science Education, 95*(5), 877–907. https://doi.org/10.1002/sce.20441

McGee, J. E., & Peterson, M. (2019). The long-term impact of entrepreneurial self-efficacy and entrepreneurial orientation on venture performance. *Journal of Small Business Management, 57*(3), 720–737.

Mihaela, H. (2017). Drivers of firm performance: Exploring quantitative and qualitative approaches. *Studies in Business and Economics, 1*, 79–84.

Milfelner, B., Gabrijan, V., & Snoj, B. (2008). Can marketing resources contribute to company performance? *Organizacija, 41*(2), 3–13.

Miller, A., Wilson, B., & Adams, M. (1988). Financial performance pattern of new corporate ventures. *Journal of Business Venturing, 3*, 287–300.

Mishra, C. S. (2019). Does managerial ability drive firm innovativeness? *IEEE Transactions on Engineering Management, 68*, 1139–1154.

Mohamed, W., Yasseen, Y., & Nkhi, N. (2020). International financial reporting standards for small and medium-sized entities: A survey showing insights of South African accounting practitioners. *Journal of Economic and Financial Sciences, 13*(1), a489. https://doi.org/10.4102/jef.v13i1.489

Muriithi, S. M. (2017). African small and medium (SMEs) contributions, challenges, and solutions. *European Journal of Research and Reflections in Management Sciences, 5*(1), 36–48.

Muthee, M., & Ngugi, K. (2014). Influence of entrepreneurial marketing on the growth of SMEs in Kiambu Town-CBD. *Kenya. Journal of Business Management, 1*(11), 361–377.

Newman, A., Obschonka, M., Schwarz, S., Cohen, M., & Nielsen, I. (2019). Entrepreneurial self-efficacy: A systematic review of the literature on its theoretical foundations, measurement, antecedents, and outcomes, and an agenda for future research. *Journal of Vocational Behaviour, 110*, 403–419.

Olubiyi, T. O., Egwakhe, A. J., Amos, B., & Ajayi, A. A. (2019). Entrepreneurial orientation and firm profitability: Evidence from Lagos State Nigeria. *IOSR Journal of Business and Management, 21*(6), 42–54.

Owolabi, S. A., & Makinde, O. G. (2012). The effects of strategic planning on corporate performance in university education: A Study of Babcock University. *Arabian Journal of Business and Management Review, 2*(4), 28–44.

Oyewale, A., Adebayo, O., & Kehinde, O. (2020). Estimating the impact of COVID-19 on small and medium scale enterprise: Evidence from Nigeria. *Open Journal of Business and Management, 9*(5), 1–19.

Ozili, P. (2020). COVID-19 in Africa: Socio-economic impact, policy response and opportunities. *International Journal of Sociology and Social Policy.* https://doi.org/10.1108/IJSSP-05-2020-0171

Pangarso, A., Sisilia, K., Setyorini, R., Peranginangin, Y., & Awirya, A. A. (2022). The long path to achieving green economy performance for micro small medium enterprise. *Journal of Innovation and Entrepreneurship, 11*, 16. https://doi.org/10.1186/s13731-022-00209-4

Qureshi, M. A., Minhas, A., Ahsan, A., & Petra, P. (2021). Do ESG endeavors assist firms in achieving superior financial performance? A case of 100 best corporate citizens. *SAGE Open, 11*, 21582440211021600.

Rachmawan, A., Lizar, A. A., & Mangundjaya, W. L. (2015). The role of parent's influence and self-efficacy on entrepreneurial intention. *The Journal of Developing Areas, 49*(3), 417–430.

Rahman, M. S., AbdelFattah, F. A., Bag, S., & Gani, M. O. (2022). Survival strategies of SMEs amidst the COVID-19 pandemic: Application of SEM and fsQCA. *Journal of Business &Industrial Marketing*. https://doi.org/10.1108/JBIM-12-2020-0564

Roschk, H., & Hosseinpour, M. (2020). Pleasant ambient scents: A meta-analysis of customer responses and situational contingencies. *Journal of Marketing, 84*(1), 125–145.

Shahab, Y., Chengang, Y., Arbizu, A. D., & Haider, M. J. (2018). Entrepreneurial self-efficacy and intention: Do entrepreneurial creativity and education matter? *International Journal of Entrepreneurial Behavior & Research, 12*, 499–522.

Torres, J. L. N., & Watson, W. (2013). An examination of the relationship between manager self-efficacy and entrepreneurial intentions and performance in Mexican small businesses. *Contaduría y Administración, 58*(3), 65–87.

United Nations (UN). https://www.un.org/sg/en/content/sg/statement/2013-09-05/secretary-generals-remarks-g20-working-dinner-sustainable

Wales, W. J., Gupta, V. K., & Mousa, F. T. (2013). Empirical research on entrepreneurial orientation: An assessment and suggestions for future research. *International Small Business Journal, 31*(4), 357–383.

Wang, W., Lu, W., Kweh, Q., & Lee, J. (2021). Management characteristics and corporate performance of Chinese chemical companies: The moderating effect of managerial ability. *International Transactions in Operational Research, 28*(2), 976–996.

Zubair, S. S., & Khan, M. (2021). Entrepreneurial self-efficacy and small business performance in Pakistan. *Management Science Letters, 11*(6), 1715–1724.

CHAPTER 13

Configuring Ecosystems for MSMEs to Achieve SDGs: A Co-Design Study in Botswana and Malaysia

Badziili Nthubu, Sivakumari Supramaniam, and Susela Devi K. Suppiah

Introduction

MSMEs generally face several issues, challenges, and impediments that hinder their growth. Furthermore, MSMEs experience problems ensuring their sustainability and remaining competitive in the market and changing environment. Globally, the presence and importance of a large MSME

B. Nthubu
Department of Mechanical, Energy & Industrial Engineering, Botswana International University of Science & Technology, Palapye, Botswana
e-mail: nthubub@biust.ac.bw

S. Supramaniam
Department of Marketing Strategy & Innovation, University Sunway University, Bandar Sunway, Malaysia
e-mail: sivakumaris@sunway.edu.my

H. Dasaraju and T.T.H. Tambunan (eds.), *Role of Micro, Small and Medium Enterprises in Achieving SDGs*, Sustainable Development Goals Series, https://doi.org/10.1007/978-981-99-4829-1_13

sector comprising the majority of enterprises in all economies are observed. In developing economies, MSMEs, in particular, are at different stages of evolution in their respective economies. Therefore, their relative roles and contributions also differ. Thus, the constraints they face and the corresponding policies aimed at strengthening their competitive performance may vary as well. Nevertheless, there are also basic similarities in the broad challenges faced by MSMEs in these economies, for example in terms of access to finance, technology, human resources, market information, and, above all, in adjusting to both the opportunities and threats of globalization (United Nations ESCAP, 2021).

This chapter argues that MSMEs have an important role in driving the sustainable development goals (SDGs) initiative because the SDGs cannot be achieved without MSMEs. MSMEs' potential strength lies in their strong ties with stakeholders, which enables MSMEs to be agile, adaptive, and pursue socially responsible behavior with stakeholders (Smith et al., 2022). This chapter emphasizes the value of networking with stakeholders (communities, governments, and other stakeholders), facilitating MSMEs' pursuit of sustainable practices, which otherwise form a barrier to implementing independently (Vrontis et al., 2020). There is a need for capacity-building MSMEs to engage with their stakeholders and to design a conducive ecosystem in emerging economies. Given the limited studies on designing ecosystems, this chapter discusses a potential co-design approach to help MSMEs configure their ecosystems to enhance entrepreneurship, creativity, and innovation.

The remaining discussion in this chapter is organized as follows: Sect. 13.2 discusses the background emphasizing the relevance of SDGs to MSMEs and the importance of MSMEs to the global economy. It also compares the economic setting of both Botswana and Malaysia. Section 13.3 discusses the extant prior studies on MSMEs and SDGs and identifies the research gap and research objective. Section 13.4 examines the theoretical underpinnings of the research. Section 13.5 discusses the research methodology and approach. Section 13.6 discusses the findings. Section 13.7 concludes.

S. D. K. Suppiah (✉)
Department of Accounting, Sunway University, Bandar Sunway, Malaysia
e-mail: suselas@sunway.edu.my

Background

Importance of SDGs

On 25 September 2015, the General Assembly of the United Nations adopted the resolution "Transforming our World: The 2030 Agenda for Sustainable Development". The Agenda provides a plan of action for people, the planet, and prosperity, seeking to strengthen universal peace. It recognizes that the greatest global challenge is eradicating poverty in all its forms and dimensions, including extreme poverty which is critical for sustainable development. Achieving this mammoth task requires the collaboration of all countries and all stakeholders, acting in partnership to free the human race from the tyranny of poverty and to heal and secure our planet.

The agenda calls for bold and transformative steps urgently needed to shift the world onto a path of sustainability and resilience. This collective journey is the responsibility of all global citizens who endeavor to pledge that no one will be left behind. The Agenda announced 17 SDGs (Table 13.1) and 169 targets to build on the Millennium Development Goals and complete what they did not achieve. SDGs seek to realize the human rights of all and to achieve gender equality and the empowerment of all women and girls. They are integrated and indivisible and balance the three dimensions of sustainable development: the economic, social, and environmental. The UN Agenda has stimulated action in areas of critical importance for humanity and the planet.

To ensure the SDGs are achieved, the United Nations issues progress reports on the achievement of regional groups. Unfortunately, the UN's Economic and Social Commission for Asia and the Pacific (ESCAP) (2021), in its fifth progress report on the achievement of SDGs, highlighted that the Asia–Pacific region is not on track to achieve any of the 17 SDGs by 2030. Based on its current trajectory, it is estimated that this region may achieve less than 10% of the SDG targets. Hence, there is an urgency to ensure that responses to the pandemic in the region and at the national level accelerate progress toward the 2030 Agenda.

Further, it is reported the pandemic has impacted the region adversely and hindered the progress to achieve SDGs. The most notable is the impact on jobs, SMEs, and informal sector workers. Unemployment increased by 15 million in the region in 2020. Compared to 2019, workers in the region lost 7.1% of their labor income in 2020, more than $1.0 trillion. In April 2020, lockdown measures impacted some 829

Table 13.1 MSMEs in Botswana and Malaysia: A comparison

Category/ Criteria	*Botswana*		*Malaysia*		
	Manufacturing	*Services & Other Sectors*	*Manufacturing*	*Services & Other Sectors RM (USD)*	*No of firms (%)*
Micro		26,256			964,495 (78.6%)
Sales turnover	< BWP 300,000 (<USD23,624)	< BWP 300,000 (<USD23,624)	< RM300,000 (< USD 69,767)	< RM300,000 (< USD 69,767)	
No of employees	< 6	< 6	< 5	< 5	
Small		Missing data			242,540 (19.8%)
Sales turnover	BWP 300,000 < BWP 1.5mil (<USD23,624 < USD117,512)	BWP 300,000 < BWP 1.5mil (<USD23,624 < USD117,512)	RM 300,000 < 15 mil (USD 69,767 < 3.5mil)	RM 300,000 < 3 mil (USD 69,767 < 0.7mil)	
No of employees	6 < 25	6 < 25	5 < 75	5 < 30	
Medium		Missing data			19,459 (1.6%)
Sales turnover	BWP 1.5mil < BWP5mil (<USD 117,512 < USD 391,708)	BWP 1.5mil < BWP5mil (<USD 117,512 < USD 391,708)	RM 15mil < 50mil (USD 3.5mil < 11.6 mil)	RM 3 mil < 20mil (USD 0.7mil < 4.7mil)	
No of employees	< 100	< 100	75 < 200	30 < 75	

Source SME Corp (),Local Enterprise Authority (2022)

million informal workers in the Asia–Pacific region (UN ESCAP, 2021). Hence, there is a need to ensure responses to the pandemic accelerate progress toward SDGs. A focus on the SME sector is, hence. very timely to ensure no one is left behind. Details on SDGs are found in Chapter 1.

Importance of MSMEs

The importance of MSMEs in the economy is widely recognized. MSMEs account for more than 90% of business establishments, around 70% of jobs, and 55% of GDP in advanced economies (World Trade Organization, 2016). This sector also contributes up to 40% of GDP in developing economies and provides 70% of occupational opportunities (Razak, 2018). MSMEs also make up more than 99% of all enterprises in the EU-28 and contribute more than 66% of total jobs (Islam et al., 2022). As envisaged in the Twelfth Malaysia Plan, it is essential for this MSME sector to remain competitive in the current market and achieve long-term survival. However, it is observed that the failure rate is significant. Many observe that nearly half of SMEs fail within the first five years of operation, with nearly two-thirds failing within the first ten years (Yusoff et al., 2018). Furthermore, a significant increase in the failure rate of SMEs under the COVID-19 pandemic of almost 9 percentage points is reported (Kalemli-Ozcan et al., 2020).

The 2030 agenda and goals pose a major challenge in the MSME context. Many entities still grapple with actualizing sustainability in practice for core organizational activities, e.g. strategy, learning and change, production, human resource, and marketing (Haugh & Talwar, 2010; Siebenhüner & Arnold, 2007).

Furthermore, digital transformation impacts sustainable development (Bai et al., 2021; Oliveira et al., 2021). The pandemic witnessed how digitization facilitated a smooth transfer to work remotely from home in critical situations. This affects the number of people commuting, and thus reduces carbon dioxide emissions. In the business environment, digitization minimizes the use of paper and chemicals related to document printing. It has resulted in an awareness of environmental sustainability (Isensee et al., 2020).

While, many studies examine the opportunities for MSMEs to embark on digital transformation and embed sustainability practices (Bai et al., 2021; Marcysiak & Pleskacz, 2021), there is limited conceptualization and theorization to guide the MSME sector to pursue the journey.

Botswana and Malaysia: The Setting

1) **Botswana**

The World Bank classifies Botswana as an upper-middle-income country in sub-Saharan Africa (World Bank, 2021), positioning the country as one of the richest in sub-Saharan Africa. That notwithstanding, Botswana's economy relies on the diamond mining industry to provide revenue for socio-economic development (Government of Botswana, 2016). This reliance on diamonds is highly unsustainable and risky. Botswana uses mining proceeds mainly to provide services such as free education, healthcare, and social welfare to most people in rural areas (Government of Botswana, 2016). However, during the 2008 to 2010 economic meltdown and the recent Covid-19 pandemic, diamond prices plummeted, thus forcing Botswana to divert most of its revenues toward social and business support services.

The uncertainties around diamond prices and markets require Botswana to diversify its economy. Consequently, MSMEs are identified as potential drivers of this envisioned diversification drive to create employment, address the growing inequalities and drive the socio-economic development agenda of the country (Mutoko & Kapunda, 2017). According to the recent World Bank report (World Bank, 2021), Botswana's unemployment stands at 24.7% of the labor force, the highest in the last 20 years. Since MSMEs are central to the economy of Botswana and represent roughly 90% of businesses in the country (International Trade Centre, 2019), Botswana recognizes how much they could play in growing the economy through the vision 2036 agenda, which is aligned with the UN sustainable development goals (The Vision 2036 Presidential Task Team, 2016). Therefore, Botswana is committed to supporting entrepreneurship, creativity, and innovation and encouraging the formalization and growth of MSMEs through government agencies such as the Local Enterprise Authority (LEA) and Citizen Entrepreneurial Development Agency (CEDA).

Although Botswana has massive investments in entrepreneurship and sound policies for growing MSME competitiveness through LEA and CEDA, there are still challenges to MSMEs growth and sustainability (International Trade Centre, 2019). Through a recent study on enhancing manufacturing innovation ecosystems in Botswana, Nthubu (2021b) highlights the need to grow the local ecosystem where MSMEs are embedded to strengthen their linkages with other firms for sustainability. Local ecosystems are a primary source of knowledge, skills, finance, business opportunities, and networks for MSMEs (OECD, 2017). For an economy heavily reliant on non-renewable resources, configuring local ecosystems for MSMEs to achieve UN sustainable development goals is now urgent in Botswana.

2) **Malaysia**

Similar to Botswana, the World Bank has classified Malaysia as an upper-middle-income country in East Asia and Pacific region (World Bank, 2021). Malaysia is a multi-ethnic country with the predominant ethnic groups in Peninsular Malaysia being Malay, Chinese and Indian. In Sabah and Sarawak, the indigenous people represent the majority, which includes Kadazan-Dusun, Bajau, and Murut in Sabah as well as Iban, Bidayuh, and Melanau in Sarawak.

Its population is 32.7 million. Its unemployment rate was highest in 2021 at 4.6–4.8% compared to 3.4% in 2017. Its real domestic product annual growth was 3–4% in 2021 compared to an average of 6.4% for emerging and developing countries (Economic Planning Unit, 2021).

Since gaining independence in 1957, Malaysia has successfully diversified its economy from agriculture and commodity-based to one with significant manufacturing and service sectors. It is a leading exporter of electrical appliances, parts, and components (World Bank, 2022). It is one of the most open economies in the world, with a trade-to-GDP ratio averaging over 130% since 2010. Openness to trade and investment has been instrumental in employment creation and income growth, with about 40% of jobs in Malaysia linked to export activities. Since the Asian financial crisis of 1997–1998, Malaysia's economy has been on an upward trajectory, averaging a growth of 5.4% since 2010. It is expected to achieve its transition from an upper middle-income economy to a high-income economy by 2024.

However, the COVID-19 pandemic has had a major economic impact on Malaysia, particularly vulnerable households. Having revised its national poverty line in July 2020, 5.6% of Malaysian households are currently living in absolute poverty. The Government is focused on addressing the well-being of the poorest 40% of the population ("the bottom 40"). This low-income group remains particularly vulnerable to economic shocks as well as increases in the cost of living and mounting financial obligations. In its Twelfth Malaysian Plan, and its game changers introduced in 2022, initiatives are introduced to accelerate MSMEs development through technology, and digital adoption to embrace new norms in business activities as well as enhance the capacity of MSME entrepreneurs to initiate sustainability practices. There is a need to shift MSMEs from domestic players to global market players. Thus, measures are required to improve the MSME sector competitiveness and business ecosystem to increase this sector's contribution to the economy while advancing sustainability.

This calls for workable solutions for the agencies, such as the Small and Medium Enterprises Corporation, to facilitate the MSMEs to successfully transition to the digital economy while advancing the sustainability agenda. A comparison of the ecosystem in both countries will be insightful in developing appropriate policy initiatives to jumpstart the MSME sector with an appropriate ecosystem that will result in the attainment of the SDGs, specifically, SDG 8: Decent Work and Economic Growth, SDG 12: Responsible Consumption and Production, and SDG 13: Climate Action.

Definition and Population of MSMEs: A Comparison Between Botswana and Malaysia

Table 13.1 compares MSMEs in Botswana and Malaysia. Both countries define MSMEs using similar criteria, except the absolute numbers differ relative to the size of GDP. Malaysia's GDP was USD 366 billion in 2021, and Botswana's GDP was USD 15.78 billion in 2020. Being former colonies of the British empire, both countries inherited similar administrative and legal systems.

Literature Review

SDGs consist of seventeen goals consensually agreed upon by top leading countries in 2015 (UN, 2020). SDGs have a strong foothold to bring a balance in three aspects, in particular to people, profit, and the planet. The commitment and involvement of government, companies, and communities are crucial to driving the SDGs initiatives. While the government spearheads the SDGs, it requires the engagement of the businesses to translate the goals into their business activities (de Villiers et al., 2021).

According to Smith et al. (2022), the involvement of small-size businesses, in particular, MSMEs, is limited even though MSMEs account for 90% of the businesses and contribute 50% of employment globally (UN, 2020). World Bank (2021) forecasted that MSMEs will create 600 million jobs by 2030 and position MSMEs as the backbone for economic growth.

The sustainable development of MSMEs in Asia is focused on two regions, South Eastern and Southern Asia (de Sousa Jabbour et al., 2020). The involvement of MSMEs in progressing toward SDGs is selective in specific areas in developing countries. However, the situation is different in developed countries where MSMEs account for ninety-nine percent of the businesses and contribute significantly to the countries' economy and employment sectors (OECD, 2017). Regardless of the size of MSMEs in developing and developed countries, MSMEs play a significant part in the consumption of world resources. Their responsible actions are important to the people, environment, and organization.

Individually MSMEs face challenges in driving the SDGs initiatives independently mainly to the limited size and resources and lack of awareness and comprehension of the sustainable practices (Lewis et al., 2015). However, MSMEs' strength lies in their ability to work closely with other actors to communicate their needs. Hence, MSMEs as a single unit operation has limited influence on the implementation of the bigger SDGs, but as a collective entity, it brings a higher impact in driving the SDGs initiatives (Crick & Crick, 2021). The role of MSMEs' contribution in driving the SDGs initiative should not be overshadowed by its size but by the strong strength ties, it has with stakeholders enabling MSMEs to be agile, adaptive, and pursue socially responsible behavior with stakeholders (Smith et al., 2022). Vrontis et al. (2020) claim the value of networking with stakeholders (communities, governments, and other stakeholders) assists MSMEs to pursue sustainable practices which otherwise form a barrier to implementing independently. Overall, there is limited discussion of appropriate approaches to designing an ecosystem to capacity-build MSMEs to understand and interact with their stakeholders.

Therefore, this chapter's objective is to explore a co-design approach to help MSMEs configure their ecosystems for entrepreneurship, creativity, and innovation. However, to understand better how the collaboration value with stakeholders helps MSMEs to achieve sustainable goals, we will be discussing the underlying stakeholder theory and social capital in the next section.

Theoretical Perspective

Stakeholder Theory

Stakeholders are important ties connected to the organization that has a legitimate interest in organizational operational activities. Larger organizations recognize their stakeholders based on their corresponding interest in the organization, to name a few suppliers, shareholders, government agencies, community, employees, and others. Each stakeholder has its own intrinsic value in engaging with the organization (Donaldson & Preston, 1995).

The stakeholder theory provides organizations a guide to understanding their stakeholders and how to manage them for the company's betterment. The theory comprises three aspects descriptive, instrumental, and normative (Donaldson & Preston, 1995). The descriptive aspect of the theory helps the organization to recognize its stakeholders and explain the connections they have with them. While the instrumental aspect of the theory offers a more empirical understanding of the linkages to cause and effect in terms of the stakeholder approaches and company performance. The normative aspect of the theory offers to underpin principles or morals that guide organization's actions.

The stakeholder theory is very much applicable to large companies compared to MSMEs due to the complexity of the organizational structures and the nature of external engagement. Large organizations engage with various subsidiaries and engage with a bigger network of stakeholders mainly to benefit from the cooperation through leveraging resources, diversification strategies, the economics of scale, and other responsible behavior (Russo & Perrini, 2010). On the other hand, MSMEs' are considered independent companies with simpler organizational structures that necessitate the owners and employees to adapt to daily challenges and less formal management structures put in place to manage the external stakeholders. Hence, stakeholder theory is more applicable to larger companies that have due responsibilities to manage and work in the best interest of the stakeholders, while MSMEs are not bound to such responsibilities (Russo & Perrini, 2010).

Nevertheless, MSMEs tend to establish a closer bonding or networking with relevant stakeholders (e.g. customers, banks, suppliers, and distributors) for their businesses. Building connections, reputations, and networks help MSMEs establish relational ties that can be leveraged to build their businesses. In this context, networking, and trust, which are essential aspects of social capital, can be fundamental for MSMEs' to manage their relationships with their stakeholders. Thus, Russo and Perrini (2010) claimed both stakeholder theory and social capital can be considered alternative ways to explain companies' responsible behaviour in managing their stakeholders. In the subsequent section, we will explore how social capital and the relational aspects of managing MSMEs' stakeholders help to improve MSMEs' businesses.

Social Capital Theory

Social capital can be viewed from different perspectives. From the perspective of sociology, Bourdieu (1986, p. 20) claimed it as "the aggregate of the actual or potential resources which are linked to possession of a durable network of more or less institutionalized relationships of mutual acquaintance and recognition". Accordingly, social capital has a future value that can be derived from established relationships in the form of gratitude, respect, friendship, and recognition of being known within the network. This aligns with the economic perspective that defines social capital as the "sum of the actual and potential resources embedded within, available through, and derived from the network of relationships possessed by an individual or social unit" (Nahapiet & Ghoshal, 1998).

In other words, social capital has an intrinsic economic value that can be utilized for future benefit deriving from relationships. Putnam (1992) claims that social capital entails key aspects in a relationship, reciprocity, and trust. Reciprocity and trust are very much applicable to MSMEs that leverage on personal relationships they establish in connecting with others (Putnam, 2000). The dependency of MSMEs on the network relationships brings more importance to investing in those relationships that have accrued value for future transactions (Putnam, 1992). Thus, MSMEs work closely with relevant stakeholders and establish relational values (reputation, trust) through established networks (Spence et al., 2003).

The interdependencies of MSMEs working closely with relevant stakeholders help them to establish a collaborative network. According to Abreu and Matos (2010), the collaborative network helps companies to have "survival capabilities" in times of market turbulence. MSMEs which are much more prone to diminish in a hostile environment compared to larger companies with sufficient capital tend to hold on to the collaborative networks for survival purposes. Thus, investing in social capital brings more value for smaller companies to work on common goals, leverage competencies, and obtain new knowledge (Matos and Abreu, 2004).

Methodology

Among the different research approaches, i.e. quantitative, qualitative, and mixed methods, this chapter tackles the ecosystem phenomena using the qualitative approach. The chapter adopts this constructivist worldview to construct and interpret knowledge with social actors in their natural settings (Saunders, 2016). Qualitative methods are reportedly valuable for exploring emerging innovation and entrepreneurship ecosystems, arguably lacking a well-established theory (Roundy et al., 2018). Drawing from the tenets of constructionism, where knowledge creation is a product of a social construction process (Mascolo & Fischer, 2005), the chapter selected two entrepreneurial ecosystem case studies to engage key entrepreneurial actors. As highlighted in the introduction, Botswana and Malaysia entrepreneurial ecosystem case studies were selected as examples of emerging economies characterized by nascent and unknown ecosystem dynamics.

There is a growing need to adopt design methods in businesses to address complex paradoxical and problematic situations involving various sectors such as engineering, IT, business, policy, health, and education (Dorst, 2010, 2011; Razzouk & Shute, 2012). This is partly because design methods are viewed differently, either as processes or tools that can help with innovation or new business ideas (Ramanujam et al., 2021). Fostering a culture where different people are empowered to be creative and innovative is important (Kelly & Young, 2017). In the past, designers thought of themselves as subject matter experts, e.g. furniture designers, architects, and graphic designers. However, this understanding shifted during the last decades from conceiving design based on the artifacts it produces to framing design as a general approach for innovative change (Hargraves, 2018). Dorst (2018) acknowledges current challenges as complex and entangled in the ever-changing networks of various organizations.

In Boland and Collopy (2004), "a design attitude views each project as an opportunity for the invention that includes a questioning of basic assumptions and a resolve to leave the world a better place than we found it" (p. 9). While acknowledging that cultivating design attitudes in public organizations is not always desirable, co-design can help trans-disciplinary teams navigate the complex challenges of working across traditional boundaries of organizations. Specifically, Dorst (2011) delves into how design can create important frames to tackle new challenges and central paradoxes of problem situations in organizations' open and complex systems.

To explore the challenge of configuring ecosystems for entrepreneurship, creativity, and innovation to promote sustainable development, this chapter adopts a co-design approach to reframing the complex and networked structure of entrepreneurial ecosystems across Botswana and Malaysia case studies. These ecosystem cases were selected as different configurations situated in distinct contexts, expected to yield contrasting results but for anticipatable reasons. Co-design is a useful qualitative approach to exploring, interpreting, and understanding ecosystem actors' perspectives and inner workings. This method is also suitable for interpreting the meanings of a complex situation with limited knowledge involving several actors (Saunders, 2016).

What then is co-design or collaborative design? According to Sanders and Stappers (2008), co-design refers to "the creativity of designers and people not trained in design working together in the design development process" (p. 6). Furthermore, Steen (2013) also adds that this kind of design empowers non-professional designers to share ideas and knowledge about the problem and new solutions using creative methods. On the other hand, Islind et al. (2019) discuss co-design as a process that helps generate boundary-spanning objects that act as bridges between intersecting practices. In Coupe et al. (2020), the co-design process was used to capture lived experiences of participants affected by food poverty, thus creating a democratized space for people to share ideas more nuanced and creatively. Recently, Nthubu (2021b) used co-design to analyze manufacturing innovation ecosystems in both the UK and Botswana contexts, where actors were aided in successfully using designer tools to visualize, understand and activate local and micro-ecosystems. Therefore, this chapter demonstrates the value of using co-design to support entrepreneurship, creativity, and innovation through convening various ecosystem stakeholders in one space to engage, share and communicate their ecosystem configurations in new ways.

The Jigsaw Ecosystem Design Framework

Since doing design requires following a framework and using tools to guide designers, this chapter adopts the tried and tested Jigsaw ecosystem design framework (Nthubu, 2021a) as a structure to support SMEs and their stakeholders in the configuration of ecosystems. According to Nthubu (2021a), this ecosystem design framework comprises five recursive pieces of the Jigsaw. In this case, the "Jigsaw" metaphor represents the complexity and the need to find an effective configuration to navigate complex settings (Ibid.). Therefore, the Jigsaw addresses the challenge of the lack of tools and frameworks to help actors configure their ecosystems for entrepreneurship, creativity, and innovation.

The first piece of the Jigsaw is called *Initiate*. This is about convening actors to dialog and develop shared criteria for networks. This piece is used to align actors' varied goals, value interpretations, and different interests. The second piece is called *Design*. This piece visualizes key actors and roles in the network using ecosystem attributes such as weak ties, bridges, main actors, and structural holes. The third piece is called *Review*. This is about analyzing the ecosystem configuration and identifying opportunities such as emerging networks and missing attributes. The fourth piece is called *Activate*. This is about identifying essential resources in actors and promoting collective capabilities toward creating shared values. The last piece is called *Sustain*. This piece is about formulating strategies to sustain new networks and ideas generated from the last four pieces.

As shown in Fig. 13.1, design and visualization tools help actors to characterize the ecosystem in terms of main actors, criteria, roles, ties, positions, clusters, holes, and bridges. These ecosystem attributes are challenging to see without using visualizations and design tools, thus making the design approach more relevant to configure the entrepreneurial ecosystem for entrepreneurship, creativity, and innovation.

The Jigsaw framework is selected as a starting point because it was co-designed in Botswana with manufacturing SMEs and validated with other ecosystems settings (other African and European actors) as a helpful design tool and process to design and reconfigure ecosystems (Nthubu, 2021a).

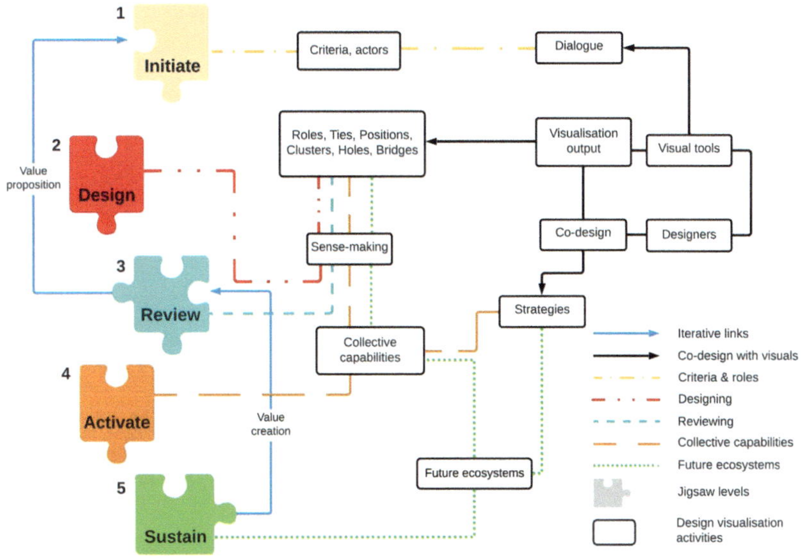

Fig. 13.1 Jigsaw Ecosystem Design Framework (*Source* Nthubu [2021a])

Visualizations as Useful Heuristics

It is imperative here to highlight that design frameworks are inadequate and less useful without tools that can support them. Consequently, design tools are used alongside the Jigsaw framework as shown in Fig. 13.1 to aid the visualization process. This visualization process helps actors to make intangible and tacit knowledge tangible. Visualizations are significant as rigorous heuristics for the exploration of new knowledge about ecosystem configurations and as communication devices or boundary objects across actors' organizational logics. As succinctly put by Padilla et al. (2018), visualizations aid the effective characterization of ecosystem structures for new insights.

Inside the Jigsaw framework, Nthubu (2021b) created a visualization tool to support the infrastructure (mapping and configuration) of ecosystem networks (Fig. 13.2). This visualization tool comprises four central activities that help actors visualize, understand, and activate their ecosystems.

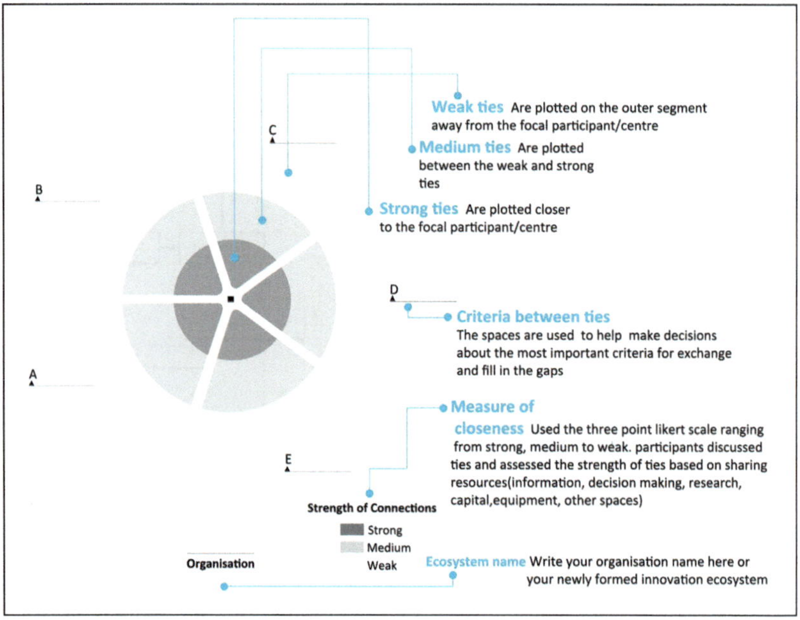

Fig. 13.2 Ecosystem Visualization Tool (*Source* Nthubu et al. [2022])

1. Setting important criteria for networks in groups and choosing a few (at most five) criteria common to everyone in the group.
2. Listing key actors important in achieving the previous criteria for networking.
3. Mapping actors against criteria in groups using distance from the center as an indicator of the strength of that connection (strong versus weak).
4. Evaluating visualization outputs in groups.

As elaborated in Nthubu et al. (2022), the ecosystem visualization tool acts as a translation and communication device to help actors make sense of complex ecosystem structures. For this reason, the design tool is divided into five segments representing the main criteria to be agreed upon by the group (Fig. 13.2). Also, the tool is divided into three dimensions of the strength of ties, with the innermost dimension representing strong ties and the outer dimension representing weak ties (Fig. 13.2).

Participants used the tool spaces to plot actors/stakeholders based on the measure of closeness, thus empowering them to characterize the strength of ties using their perceptions.

Botswana and Malaysian Case Studies: The Co-design Activities

1) **Botswana**

To explore the role of co-design in helping SMEs to configure their entrepreneurial ecosystem in Gaborone, 20 entrepreneurial firms were invited to participate in a co-design workshop at Botswana Innovation Hub on the 7th of February 2020. Botswana Innovation Hub is a government-sponsored incubator for supporting startups with access to finance, subsidized office spaces, and other mentorship programs. Table 13.2 shows the description of the participants. All SMEs listed in Table 13.2 employ six or fewer people and are mostly owned by young people (less than 35 years).

a) **Activity 1: Setting important criteria**

Table 13.2 Description of MSMEs and their groups

Grp name	*SMEs pseudo names*	*Characteristics*
II HUB	SPA, MH, LBN, SL, LH	Data analytics, digital marketing, training and consultancy, Branding, and autonomous solutions
Energy X	CAI, SDS, MOE, OT, GK	Artificial intelligence, Augmented reality and real estate, renewable energy, chemical manufacturing
Innovation links	LAM, SSB, CBH, SBW, SA	Board games, coaching, and training, online stores, renewable energy
Innovation minds	PSHT, MI, TN, KE, RC	Telecommunication, health and wellness, bath products, Web-based solutions, coffee manufacturing

Source Nthubu (2021b)

As the first step of co-design, participants were divided into groups of five to set their requirements for ecosystems (Fig. 13.3). This was a simple creative task aimed at opening actors to discuss and agree on the most important criteria or value of being embedded in local ecosystems. This was an important step to align varied expectations and goals.

b) **Activity 2: Listing key actors**

During the second step, participants listed key actors that can help them achieve their expectations and goals (Fig. 13.4). During this step, participants engaged other actors to find out about their stakeholders.

c) **Activity 3: Mapping actors**

During the third step, participants transferred the main criteria agreed upon in step 1 of the design process into the mapping tool. This was followed by mapping all actors identified in step two on the main mapping spaces shown in Fig. 13.5. This activity allowed actors to map their varied

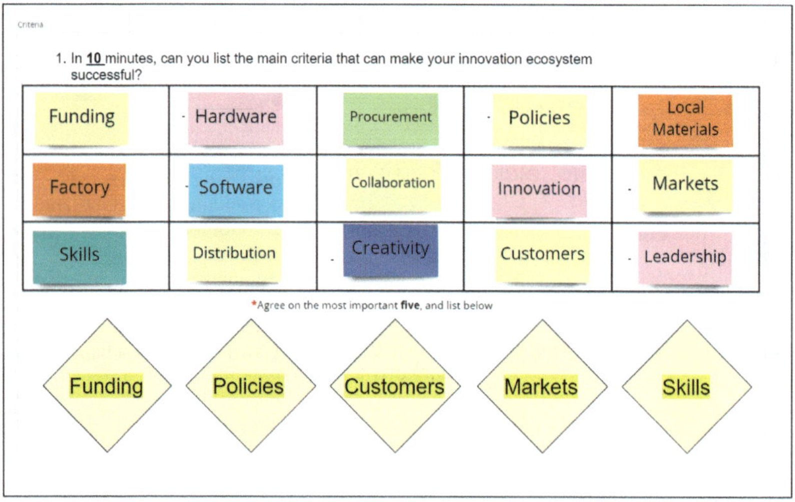

Fig. 13.3 Listing and selecting important criteria for ecosystems

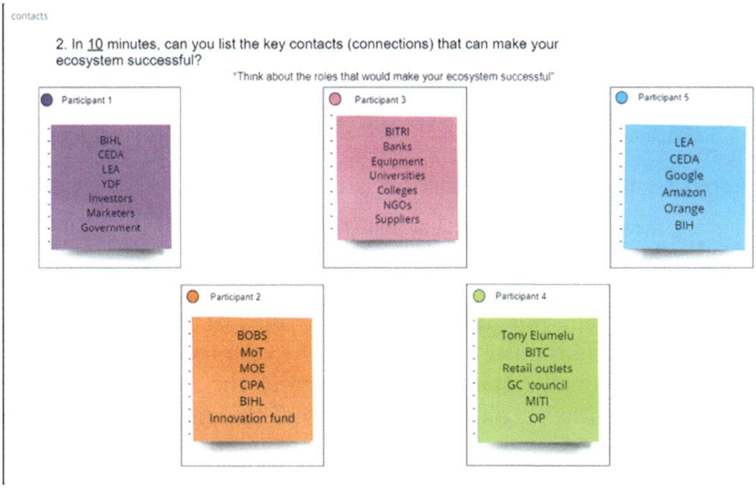

Fig. 13.4 Listing and discussing important stakeholders for ecosystems

stakeholders in the same space, thus prompting discussions and yielding useful visualization outputs.

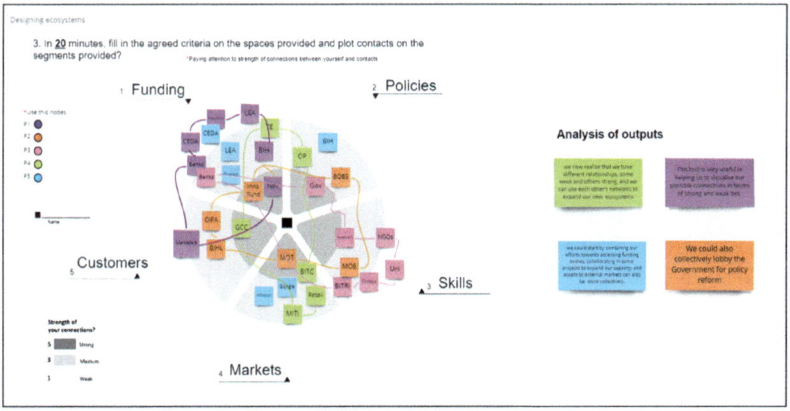

Fig. 13.5 Designing and analyzing the ecosystem networks in groups

d) **Activity 4: Evaluating visual outputs**

The last step was about evaluating visualization output and presenting the findings to the rest of the workshop as shown in Fig. 13.6. Here participants searched for common actors and weak ties to inform their next steps in building more efficient relationships.

2) **Malaysia**

MSMEs have been the main pillars of the Malaysian economy, contributing to a substantial percentage of the country's gross domestic product and national employment. However, the MSMEs business ecosystem has been less explored to understand the diverse actors in the system (i.e. stakeholders) and its interrelationships with the actors. The role of each actor contributes to the functionality of the overall business ecosystem. A workshop was designed and conducted with Malaysian SMEs in collaboration with the Finance Support Agency, Credit Guarantee Corporation (CGC), by employing a co-design framework. The workshop aimed to use the Jigsaw design framework to help MSMEs explore the strength of their ecosystem interrelationships, thus offering

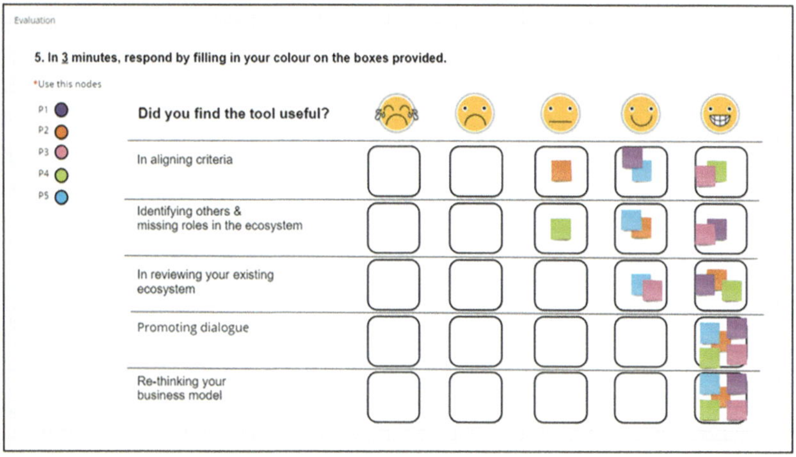

Fig. 13.6 Reflection and evaluation of the co-design process

a deeper insight into the configuration of actors in their business ecosystems.

The workshop was also expected to contribute to identifying, improving, and exploring new ties to provide the MSMEs foundation for actionable insights and solutions. This contribution is important to help MSMEs with deep knowledge of the status quo of stakeholder engagement by exploring the following areas:

- Who are their stakeholders?
- How do they identify them?
- How do they engage with their stakeholders?

In this workshop, participants were approached and recruited by the Credit Guarantee Corporation, 29 MSMEs from various industries (services, retailing, training, manufacturing, food, automobile, and support services) were invited to join the workshop. The workshop consisted of MSME owners and their respective key employees. As the workshop was virtual, it brought a lot of excitement for the participants in terms of engaging on the materials and working closely with the team members.

During the workshop, participants were put into groups to do the task activities to identify and map the important stakeholders for their businesses in a virtual MIRO whiteboard. The participants were expected to pause, think and work independently and with others during the activities to identify important business stakeholders. MSME participants were given time to reflect on their practices and explore the hidden opportunities working with stakeholders.

The workshop was completed within an hour. Most participants gave feedback on how the visualization approach and collaboration or co-creation helped them exchange ways to improve their ties with stakeholders. Furthermore, the workshop opened the door for one of the key stakeholders of Malaysian MSMEs, Credit Guarantee Corporation to explore the benefits of a co-design approach as a tool to engage MSMEs effectively. In CGC's view, these design tools brought fresh ideas and excitement for SMEs to engage better with one another compared to the traditional business training methods.

The co-design activities helped the MSMEs to identify stakeholders and explore the possible opportunities to work closely with them. Additionally, co-design helps MSMEs prioritize the importance of working closely with stakeholders to improve their businesses. For example, the collaborative visualization activities helped MSME actors to better relate with CGC, which is one of the important stakeholders for bridging their financial gaps. The co-design activities were phased out into three stages, as discussed in the subsequent sections.

a) **Activity 1: Setting important criteria**

Working in groups to develop the criteria for networks (15 minutes). In the first activity, participants were asked to write on sticky notes the important criteria for their business connectivity. From the MIRO board, we recognized that as a group, they had identified marketing, sales, and funding as the three most important criteria for the business (Fig. 13.7).

b) **Activity 2: Listing key actors**

Working in groups to identify the important stakeholders (10 minutes). The participants were asked to list the important stakeholders in the

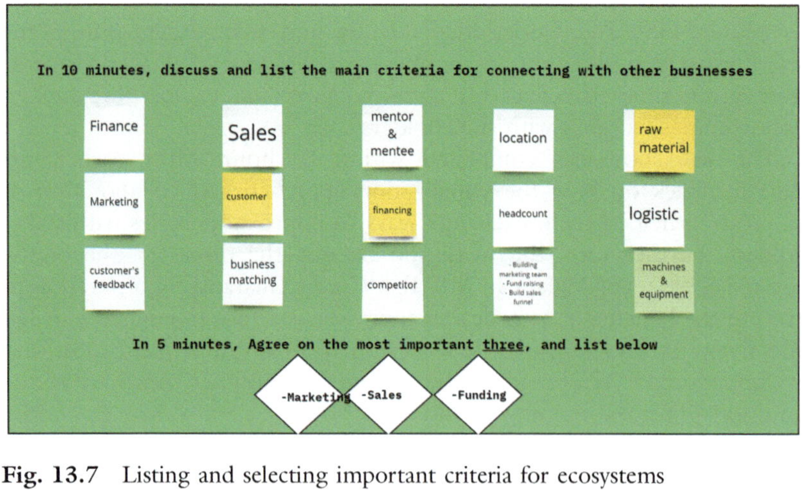

Fig. 13.7 Listing and selecting important criteria for ecosystems

next task. In this exercise, the participants were quicker to list their important connections and start comparing with other members who are their important stakeholders. At b) this juncture, a lot of interaction and reflections took place (Fig. 13.8).

c) **Activity 3: Mapping actors**

Working in groups to design and analyze networks (20 minutes). In the subsequent activities, participants were asked to map the position of the stakeholders based on the three criteria identified in activity 1. The virtual board displayed a diagram indicating stakeholders identified in the inner

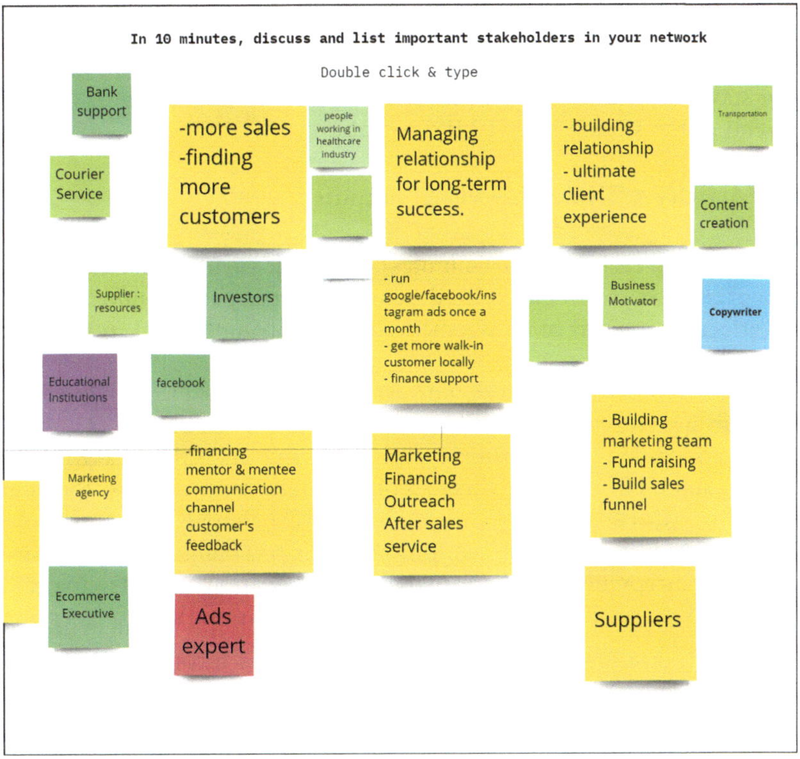

Fig. 13.8 Listing and discussing important stakeholders for ecosystems

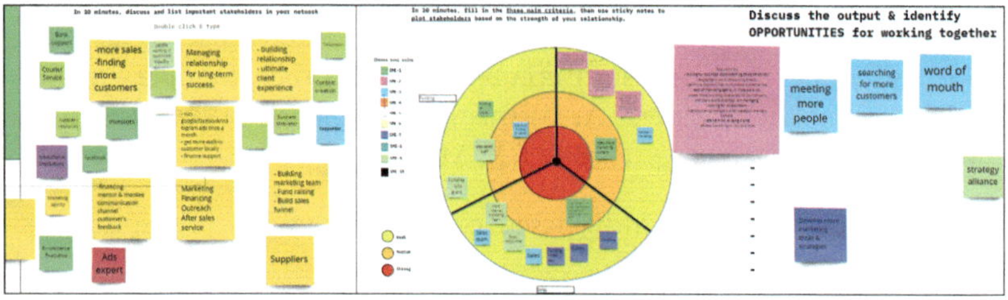

Fig. 13.9 Designing and analyzing the ecosystem networks in groups

circle as stronger relationships and further to the core circle as weaker relationships. Many participants struggle at this point trying to understand the strength of their ties with the identified stakeholders. The co-design activities help the MSMEs to visualize their stakeholders' position and bring out the conversation between and within them on how to improve their stakeholder ties. This reflection is captured in the evaluation stage at the end of the session (Fig. 13.9).

d) **Activity 4: Evaluating visual outputs**

Reflection and evaluation (10 minutes). From the participants' experience in engaging the co-design-related activities, participants found it useful and effective in terms of engagement and identifying the stakeholder ties (Fig. 13.10).

Since the Jigsaw framework aims to help actors visualize, understand, and activate entrepreneurial ecosystems (Nthubu et al., 2022), findings from co-design workshops with Botswana and Malaysia MSMEs can be summed up into three central values that address the aim as follows; (i) visualizing, (ii) recognizing, (iii) simplifying ecosystems. These values were central in assisting MSMEs to configure their ecosystems for entrepreneurship, creativity, and innovation in Botswana and Malaysia. Next section discusses the main findings from the two emerging economies.

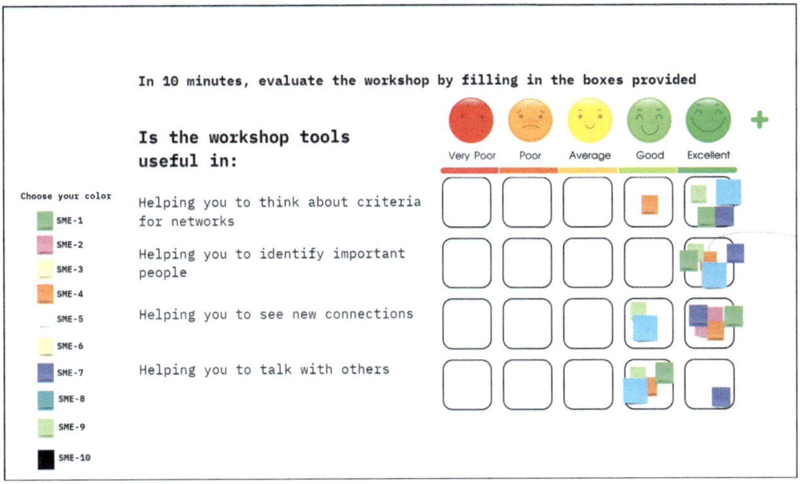

Fig. 13.10 Reflection and evaluation of the co-design process

Findings and Discussions

Botswana

a) **Visualizing**

Based on the feedback from the co-design workshop, the Jigsaw ecosystem design framework tools were helpful to visualize MSMEs' stakeholders and partners in terms of strengths and weaknesses, and secondly, in terms of positions and roles of actors in the ecosystem. Putting all these ecosystem attributes in visual form was important for MSMEs to collaboratively think about which actors are missing in the ecosystem and how to reach out to those actors to improve entrepreneurship, creativity, and innovation.

Visuals were also valuable as boundary objects to start a conversation between MSMEs, which is often challenging when using traditional meetings in transdisciplinary settings.

> This workshop is helping us to visualise our positions and that of our stakeholders, which is important to make us aware of other people's strengths and weaknesses. (Innovation minds)

Although using design tools in this manner was a new experience for Botswana MSMEs, they seemed keen to participate in the visualization process, which was more hands-on and creative. Visualizations helped participants to build new ties with other MSMEs from different sectors.

b) **Recognizing**

Aside from engaging in the visualization process as highlighted above, MSMEs reported that the co-design process enabled them to recognize the significance of critical stakeholders and other institutions like commercial banks in their ecosystem, which they previously took for granted and neglected.

> Using our visual graphs, it is now easy to recognise that Stanbic Bank has a space that we can use for co-creation. This was not clear before this workshop. (Energy X)

Based on MSME participants' feedback, mapping actors in terms of their relative strengths and weaknesses and displaying the visuals for analysis helps recognize how much value everyone has within the ecosystem. MSMEs also realized how important it is to recognize and reconfigure ecosystem structures in groups to take advantage of opportunities embedded within their networks, such as weak ties, structural holes, and bridges. Therefore, seeing network structures in visuals enhances recognition and appreciation of the value/importance of every stakeholder and the characteristics of actors. Thus, recognition is essential to enabling MSMEs to be agile, adaptable, and socially responsible in their decision-making.

c) **Simplifying**

Complexity is inherent in entrepreneurial ecosystems, thus making these structures cumbersome for SMEs to take advantage of. From this co-design workshop, MSMEs reported that the design process helped characterize networks in terms of weak, medium, and strong ties, thus making it plausible to navigate complexity and simplifying the decision-making process regarding which relationships to pursue to improve business performance.

> One thing we don't do well as SMEs is meeting, discussing, and taking decisions on our business interests as a collective, so mapping our network structures in groups makes it easy to engage each other and decide on how to take advantage of opportunities we see displayed through our visuals. (II Hub)

Characterizing ecosystem networks simplifies the structure for configuration, i.e. making decisions on what to do to activate valuable connections. participants reported that without visualizations, it is challenging to see and recognize ties and take meaningful decisions to improve entrepreneurship and innovation. Therefore, simplicity, in this case, was about finding an easier path to use ecosystem attributes for entrepreneurship.

Although other values, such as dialogic thinking, collaborations, knowledge exchange, and trust-building, were realized and reported by participants at the end of the co-design experience, the three central values of visualizing, recognizing, and simplifying ecosystems were paramount to these groups of MSMEs in Botswana. These values can help Botswana MSMEs reconfigure their ecosystems for entrepreneurship, creativity, and innovation. The outcomes of this workshop and other related co-design workshops conducted in Botswana are reported in detail in Nthubu et al. (2022).

Malaysia

In comparison to Botswana in-person workshop, the Malaysia virtual co-design experience also allowed MSMEs to use the Jigsaw design framework to configure their ecosystem. Based on the analysis of the workshop results, similar themes were identified as important, (i) visualizing (who are their stakeholders), (ii) recognizing (how do they identify them) and (iii) simplifying (How do they engage with their stakeholders).

a) **Visualizing**

Participants from Malaysia shared similar experiences that engagement becomes much better using a visualization approach which works better in co-design-related activities. Interaction and working on the visual materials bring more attention to recognizing the important stakeholders who have been least explored in their businesses.

> Analysing the visualization, we can see that all of us places high value in funding as a criterion, so it's an opportunity area we can collaborate on. (MIRO group 1)

participants reported that the co-design activities and visualizations produced thereof exposed them to how other MSME businesses engage with stakeholders and how they can learn from each other.

b) **Recognizing**

Aside from the act of visualization aimed at empowering MSMEs with a vision to see beyond the obvious, MSMEs also highlighted that visual outputs present a framework to recognize hidden clues that provide the basis for collective decision-making. For example, by analyzing their ecosystem design output as a group, they recognized that their ecosystem structure missed a few key stakeholders (main actors) that can help them expand business opportunities. Furthermore, MSMEs also recognized the importance of engaging other MSMEs (weak ties) and how they can combine and benefit from their resources and knowledge. Recognizing their important stakeholders helps MSMEs to understand their current connections and work toward strengthening their stakeholder's connections for betterment.

> Even though we are in the same industry, I have never looked into the social media platforms for my customers. (Beauty Service)

c) **Simplifying**

Like in Botswana' findings, Malaysia MSMEs also acknowledged that visuals make ecosystem network structures accessible and understandable (less complicated). MSMEs primarily belong to smaller networks that pay less emphasis on understanding the business ecosystem they belong to, unlike the larger firms that have more formal connections addressing their stakeholders. The workshop provides an avenue to map the important actors in the network, offering an opportunity to simply reflect, pause and think of the present connections they have with the stakeholders. Although there were challenges in some instances of finding and agreeing on criteria for working together, it was nonetheless easy to engage with strangers on the possibility of collaboration by using visuals as a translation device to point out areas of opportunities.

Many of the participants have gained more networking benefits as well as learning from the co-design activities. The co-design activities helped MSMEs to identify, recognize and reflect on the strength of their stakeholder ties. This initial workshop has brought beneficial outcomes for both SMEs, and CGC paved the path for future follow-up workshops with MSMEs.

> Mapping the important people for my businesses is something I have never done before but definitely it is a good practice to know with whom I engage more. I found this workshop made me think. (Automobile Service)

Discussions

Based on Botswana and Malaysia case studies, it was evident that MSMEs need to tap on their collective capabilities embedded in local ecosystems, particularly networks, to advance sustainable development goals. The chapter found that co-design tools were instrumental in aiding MSMEs to visualize, recognize and simplify ecosystem structures to understand and take advantage of available resources. These findings partly address concerns highlighted by Crick and Crick (2021), that MSMEs as a single unit operation possess the limited capability to achieve SDGs but as a collective entity can achieve a higher impact in driving the SDGs initiatives. In the following subsections, the chapter discusses the three key values of (i) visualize, (ii) recognize and (iii) simplify to demonstrate how MSMEs use co-design to configure ecosystems for achieving SDGs.

a) **Visualize**

Visualizations are often used as representations of data as opposed to tools for exploration. However, this chapter explored how MSMEs can effectively use their network visuals as boundary objects to engage and explore new ideas creatively. Communication and creativity are often challenging when using traditional approaches, i.e. having a speaker or expert do the talking and MSMEs asking questions, particularly in interdisciplinary teams. These workshops also demonstrated that visualizing networks promote active thinking, a new perspective on networks, and collective decision-making. The chapter found that visual outputs are crucial as important frames and heuristics to tackle new interdependent MSME challenges. This finding is in line with Dorst (2011) who emphasizes the importance of creating frames to tackle complex systemic problems.

b) **Recognize**

Recognition in complex systems such as entrepreneurial ecosystems is important to promote informed decision-making. This is because it is often challenging to see and make sense of ecosystem structures without visuals and input from all actors. The Organisation for Economic Corporation and Development (OECD) acknowledge that MSMEs are embedded in local ecosystems where they are exposed to various resources to expand their business. Still, reckoning and accessing those resources is challenging (OECD, 2017). However, through the Jigsaw approach, SMEs can develop capabilities to recognize these local ecosystems to expand their business.

This chapter also emphasizes the importance of bringing all key MSME stakeholders to engage each other in a much more democratized space where power dynamics are almost non-existent. The visual approach allowed MSMEs to recognize each other's structural position, potential, and role in realizing their collective goals, thus determined to unleash the power of collaboration and co-creation in driving creativity.

c) **Simplify**

In this chapter, simplification is not seen as the direct opposite of complexity but as the capability to navigate complexity using simple co-design tools. The co-design tools used during Botswana and Malaysia workshops made abstract ecosystem network concepts such as weak ties, structural holes, and bridges accessible to SMEs. Based on these case studies, using simple co-design tools to characterize network structures help a group of weakly connected MSMEs to find a clear path in their complex web of interrelations, thus effectively framing problems and coming up with new solutions.

This chapter argues that complexity in local MSME ecosystems is a necessary environment for business expansion. Thus, the Jigsaw ecosystem design framework provides an easy-to-use structure to navigate complexity instead of eliminating it. Since the Jigsaw was used as a translation device to help actors move about in understanding and acting upon ecosystem opportunities (Nthubu et al., 2022), it is useful in simplifying the path for MSMEs to configure their ecosystem structures for creativity, innovation, and entrepreneurship.

Practical Implications

This chapter provides a starting point on how MSMEs in emerging economies who are faced with resource scarcity can leverage their local ecosystem resources to achieve sustainable development goals. The chapter provides easy-to-use design tools as a structure to visualize, recognize and simplify ecosystems for new business opportunities. This approach is useful for SMEs in emerging economies facing similar challenges to engage other MSMEs in their local ecosystems. Therefore, aside from further validating the usefulness of the Jigsaw ecosystem design framework, this chapter emphasizes the importance of promoting the three values of visualizing, recognizing, and simplifying ecosystems as the cornerstone for MSMEs to achieve sustainable development goals.

Policymakers and other stakeholders responsible for SMEs in emerging economies may use these findings to enhance their understanding of MSMEs' engagement, particularly on how they can use co-design to foster creativity, innovation, and entrepreneurship activities among MSMEs based on promoting visualizations, recognition, and simplification of ecosystem structures.

Conclusions

Driving sustainable development goals in emerging economies needs MSMEs and their key stakeholders to engage with creative and innovative tools that can help them effectively visualize their network structures, and recognize and simplify hidden opportunities for expanding their business ecosystems. Therefore, this chapter presents an easy-to-use co-design approach based on three central values of visualizing, recognizing, and simplifying ecosystem structures to achieve sustainable development goals.

We suggest that future research be undertaken collaboratively with researchers from developing economies to examine the efficacy of this co-design approach in their settings.

Acknowledgements Funding from SUNWAY UNIVERSITY REWARDING RESEARCH OUTPUT (RRO) 2022. Project Code: GRTIN-RRO-15–2022.

Support and cooperation of Credit Guarantee Corporation. Research assistance by Ms Syazwani.

Funding from Lancaster University through Imagination Lancaster E3 Fund.

References

Abreu, A., & Camarinha-Matos, L. M. (2010). Understanding social capital collaborative networks. *International Conference on Information Technology for Balanced Automation Systems* (pp. 109–118). Springer.

Bai, C., Quayson, M., & Sarkis, J. (2021). COVID-19 pandemic digitization lessons for sustainable development of micro-and small-enterprises. *Sustainable Production and Consumption, 27*, 1989–2001.

Boland, R., & Collopy, F. (2004). *Managing as designing*. Stanford Business Books.

Bourdieu, P. (1986). The forms of capital. In J. G. Richardson (Ed.), *Handbook of theory and research for sociology of education* (pp. 241–258). Greenwood Press.

Camarinha-Matos, L. M., & Abreu, A. (2004, September). A contribution to understand collaboration benefits. *International Conference on Information Technology for Balanced Automation Systems* (pp. 287–298). Springer.

Coupe, G., Whitham, R., Cruickshank, L., Perez, D., & Pearson, B. (2020, August 11–14). Co-designing tools to empower further, independent co-design: Collaborating with diverse individuals with lived experience of food poverty. In S. Boess, M. Cheung, & R. Cain (Eds.), *Synergy - DRS International Conference 2020*. https://doi.org/10.21606/drs.2020.247

Crick, J. M., & Crick, D. (2021). Coopetition and family-owned wine producers. *Journal of Business Research, 135*, 319–336.

de Sousa Jabbour, A. B. L., Ndubisi, N. O., & Seles, B. M. R. P. (2020). Sustainable development in Asian manufacturing SMEs: Progress and directions. *International Journal of Production Economics, 225*, 107567.

de Villiers, C., Kuruppu, S., & Dissanayake, D. (2021). A (new) role for business–Promoting the United Nations' Sustainable Development Goals through the internet-of-things and blockchain technology. *Journal of Business Research, 131*, 598–609.

Dorst, C. (2018). Mixing practices to create transdisciplinary innovation: A design-based approach. *Technology Innovation Management Review, 8*(8), 60–65.

Dorst, K. (2010). The nature of design thinking. Design thinking research symposium, DAB Documents.

Dorst, K. (2011). The core of 'design thinking' and its application. *Design Studies, 32*(6), 521–532.

Donaldson, T., & Preston, L. E. (1995). The stakeholder theory of the corporation: Concepts, evidence, and Implications. *Academy of Management Review, 20*(1), 65–91.

Economic Planning Unit. (2021). Malaysian economy in figures. https://www.epu.gov.my/sites/default/files/2021-12/MEIF%202021.pdf

Government of Botswana. (2016). National Development Plan 11, April 2017–March 2023. in Planning, M. O. F. A. D. (Ed.). Government of Botswana (fao.org).

Hargraves, I. (2018). Care and capacities of human-centered design. *Design Issues, 34*(3), 76–88.

Haugh, H. M., & Talwar, A. (2010). How do corporations embed sustainability across the organization?. *Academy of Management Learning & Education, 9*(3), 384–396.

International Trade Centre. (2019). *Promoting SME Competitiveness in Botswana: A bottom-up approach to economic diversification.* ITC.

Isensee, C., Teuteberg, F., Griese, K. M., & Topi, C. (2020). The relationship between organisational culture, sustainability, and digitalization in SMEs: A systematic review. *Journal of Cleaner Production, 275*, 122944.

Islam, M. A., Mendy, J., Haque, A. A., & Rahman, M. (2022). Green human resource management practices and millennial employees' retention in small and medium enterprises: The moderating impact of creativity climate from a developing country perspective. *Business Strategy & Development*, 5(4), 335–349.

Islind, A. S., Lindroth, T., Lundin, J., & Steineck, G. (2019). Co-designing a digital platform with boundary objects: bringing together heterogeneous users in healthcare. *Health and Technology, 9*(4), 425–438.

Kalemli-Ozcan, S., Gourinchas, P. O., Penciakova, V., & Sander, N. (2020). COVID-19 and SME failures. IMF Working Papers, 2020(207). Retrieved from https://www.elibrary.imf.org/view/journals/001/2020/207/article-A001-en.xml

Kelly, C. J., & Young, A. J. (2017). Promoting innovation in healthcare. *Future Healthcare Journal, 4*(2), 121. https://doi.org/10.7861/futurehosp.4-2-121

Lewis, K. V., Cassells, S., & Roxas, H. (2015). SMEs and the potential for a collaborative path to environmental responsibility. *Business Strategy and the Environment, 24*(8), 750–764.

Local Enterprise Authority. (2022). Small & Medium Enterprise. Small & Medium Enterprises | LEA.

Marcysiak, A., & Pleskacz, Z. (2021). Determinants of digitization in SMEs. *Entrepreneurship and Sustainability Issues, 9*(1), 300–318. https://doi.org/10.9770/jesi.2021.9.1(18)

Mascolo, M. E., & Fischer, K. W. (2005). Constructivist Theories, Theories of Development. *Cambridge Encyclopedia of Child Development*. Cambridge University Press.

Mutoko, W. R., & Kapunda, S. M. (2017). Factors influencing small, medium and micro-sized enterprises' borrowing from banks: The case of the Botswana manufacturing sector. *Acta Commercii, 17*(1), 1–9.

Nahaiet, J., & Ghoshal, S. (1998). Social capital, intellectual capital, and the organizational advantage. *Academy of Management Review, 23*(2), 242–266.

Nthubu, B. (2021a). The value of a co-design visualisation approach: Enhancing the understanding of local entrepreneurial ecosystems. *The Design Journal, 24*(5), 749–760.

Nthubu, B. (2021b). Enhancing the understanding of manufacturing SME innovation ecosystems: A design visualisation approach. PhD dissertation, Lancaster University. https://doi.org/10.17635/lancaster/thesis/1399

Nthubu, B., Perez Ojeda, D., Richards, D., & Cruickshank, L. (2022). Navigating complexity through co-design: Visualising, understanding and activating entrepreneurial ecosystems. *The Design Journal*. https://doi.org/10.1080/14606925.2022.2088096

OECD. (2017). *Enhancing the contributions of SMEs in a global and digitalised economy*. OECD Publishing. https://www.oecd.org/industry/C-MIN-2017-8-EN.pdf

Oliveira, L., Fleury, A., & Fleury, M. T. (2021). Digital power: Value chain upgrading in an age of digitization. *International Business Review*, 101850. https://doi.org/10.1016/j.ibusrev.2021.101850

Padilla, L. M., Creem-Regehr, S. H., Hegarty, M., & Stefanucci, J. K. (2018). Decision making with visualisations: A cognitive framework across disciplines. *Cognitive Research: Principles and implications, 3*(1), 1–25.

Putnam, R. D. (1992). *Making democracy work: Civic traditions in modern Italy*. Princeton University Press.

Putnam, R. D. (2000). *Bowling alone: The collapse and revival of American community*. Simon anchuster.

Ramanujam, H., Ravichandran, B., Nilsson, S. & Ivansen, L. (2021). Barriers and opportunities of implementing design thinking in product development

process of a business to business company. *Proceedings of the Design Society, 1*, 551–560.
Razak, D. A., Abdullah, M. A., & Ersoy, A. (2018). Small medium enterprises (SMEs) in Turkey and Malaysia a comparative discussion on issues and challenges. *International Journal of Business, Economics and Law, 10*(49), 2–591.
Razzouk, R., & Shute, V. (2012). What is design thinking and why is it important? *Review of Educational Research, 82*(3), 330–348.
Roundy, P. T., Bradshaw, M., & Brockman, B. K. (2018). The emergence of entrepreneurial ecosystems: A complex adaptive systems approach. *Journal of Business Research, 86*, 1–10.
Russo, A., & Perrini, F. (2010). Investigating stakeholder theory and social capital: CSR in large firms and SMEs. *Journal of Business Ethics, 91*(2), 207–221.
Sanders, E. B. N., & Stappers, P. J. (2008). Co-creation and the new landscapes of design. *CoDesign, 4*(1), 5–18. https://doi.org/10.1080/15710880701875068
Saunders, M. (2016). *Research Methods for Business Students* [Electronic Resource]. Harlow.
Siebenhüner, B., & Arnold, M. (2007). Organizational learning to manage sustainable development. *Business Strategy and the Environment, 16*(5), 339–353.
Smith, H., Discetti, R., Bellucci, M., & Acuti, D. (2022). SMEs engagement with the Sustainable Development Goals: A power perspective. *Journal of Business Research, 149*, 112–122.
SME Corp Malaysia. (2022a). SME Definition. https://www.smecorp.gov.my/index.php/en/policies/2020-02-11-08-01-24/sme-definition
SME Corp Malaysia. (2022b). MSME Statistics. https://www.smecorp.gov.my/index.phpen/policies/2020-02-11-08-01-24/sme-statistics
Spence, L. J., Schmidpeter, R., & Habisch, A. (2003). Assessing social capital: Small and medium sized enterprises in Germany and the UK. *Journal of Business Ethics, 47*(1), 17–29.
Steen, M. (2013). Co-design as a process of joint inquiry and imagination. *Design Issues, 29*(2), 16–28.
The Vision 2036 Presidential Task Team. (2016). Vision 2036 – Achieving Prosperity for All. In T. V. P. T. Team (Ed.), *Gaborone: Lentswe La Lesedi*. (PTY) LTD.
Trading Economics. (2022). Botswana - New Businesses Registered. Botswana - New Businesses Registered - 2022 Data 2023 Forecast 2004–2020 Historical (tradingeconomics.com).
United Nations. (2015). *Sustainable Development Goals*. https://www.un.org/ga/search/view_doc.asp?symbol=A/RES/70/1&Lang=E

United Nations. (2020). Supporting small businesses through the COVID-19 crisis. https://www.un.org/en/observances/micro-small-medium-businesses-day

United Nations ESCAP. (2021). *SMEs in Asia and the Pacific*. https://www.unescap.org/sites/default/files/7%20%201.%20SMEs%20IN%20ASIA%20AND%20THE%20PACIFIC.pdf

Vrontis, D., Basile, G., Andreano, M. S., Mazzitelli, A., & Papasolomou, I. (2020). The profile of innovation driven Italian SMEs and the relationship between the firms' networking abilities and dynamic capabilities. *Journal of Business Research, 114*, 313–324.

World Bank. (2021). *Small and medium enterprises (SMEs) finance*. https://www.worldbank.org/en/topic/smefinance

World Bank. (2022). World Bank in Malaysia. https://www.worldbank.org/en/country/malaysia/overview

World Trade Organization. (2016). *Levelling the trading field for SMEs*. WTO. https://www.wto.org/english/res_e/publications_e/wtr16_e.htm

Yusoff, T., Wahab, S. A., Latiff, A. S., Osman, S. I., Zawawi, N. F., & Fazal, S. A. (2018). Sustainable growth in SMEs: A review from the Malaysian perspective. *Journal of Management & Sustainability*, *8*, 43.

CHAPTER 14

MSMEs in Achieving SDGs with Reference to Nepalese Education Consultancies Through COVID-19 and Their Status with SDGs for Nepal

Dev Raj Paneru

Background

MSMEs are the leading enterprises that contribute significantly to fulfilling basic needs, financial, economic, educational, technological, developmental, etc. growth or change and likewise, play an important role in achieving SDGs for developing countries. In Nepal, MSMEs include several types of domestic, agro, production, supply, service, construction, mechanical, etc. based enterprises that too supply commodities for the country to earn from exports. As Kharel and Dahal (2020) recorded, MSMEs contribute 22% to GDP of Nepal. Featured as service

D.R. Paneru (✉)
Global College of Management and Global College International, Kathmandu, Nepal
e-mail: devrajpaneru@gci.edu.np

H. Dasaraju and T.T.H. Tambunan (eds.), *Role of Micro, Small and Medium Enterprises in Achieving SDGs*, Sustainable Development Goals Series, https://doi.org/10.1007/978-981-99-4829-1_14

industry in the developing countries such as Nepal, education consultancies have started making their identity as among the leading small and medium enterprises. As such, the arguments are on the fore that Nepalese education consultancies contribute to GDP equally as other MSMEs.

The World Bank (2022) reports that MSMEs represent 90% of total businesses globally. Along with their representation which seems to be large in number and scale, the studies as well assert that the Nepalese MSMEs contribute 50% of the employment in Nepal. From these perspectives i.e. representation in business market and employment majorly, MSMEs are posed to be leading enterprises as concerns their role to achieving economic growth for the developing countries.

Another important dimension is not merely financial but also educational, literacy, empowerment, employment, and so on that the MSMEs are argued to be understood when it comes to assessing their significance in achieving SDGs. In this regard, in line with Ghimire (2021), a number of Nepalese scholars record Nepalese education consultancies being in the category of small and medium enterprises.

Among other benefits, major significance of Nepalese Education Consultancies as integral member of MSMEs is said to be pertinent in their contribution to achieving SDGs for Nepal. Regarding their contribution, some scholarly sources and researchers such as Kandel (2020) report positive rise in the income tax due to a rise in the revenue size. Large part of the rise in review and tax pay has been shared by Nepalese Education Consultancies which were 4000 in number till the year 2019 together with other prospective MSMEs. This is the financial part that the Nepalese education consultancy enterprises are regarded to have been noteworthy for their contribution to GDP which in turn enables Nepal to achieve SDGs.

Another significant contribution of Nepalese Education Consultancies as one leading category of Nepalese MSMEs is accounted of their obvious role on youth education and career empowerment and enhancement of Nepalese youth at national and international level. Nepalese youth in current times are noted to be expert education consultancies for mediation-assistance to get access and enrolment in the institutions and universities of their choice abroad. This allows the Nepalese students to get an access to quality education, transnational and international mobility, and career enhancement opportunities abroad. It too bridges their career to advance toward acquiring an entry to global knowledge communities and thereon developing their profile with traits and skills of global citizenship. Having an opportunity to advance upward global citizenship status, has a significant connection with SDGs achievement for the developing countries. It's particularly on youth empowerment through quality education which is SDG number 4 to be achieved by 2030.

The ongoing COVID-19 pandemic affected the micro, small, and medium enterprises (MSMEs) seriously globally and its impact on these enterprises in the developing countries has been more acute and painful. Scholars record the effects of COVID-Lockdowns on MSMEs quite explicitly that according to Dangol, Chitrakar and Yoo (2020) came far more fiercely to Nepalese MSMEs due to numerous geopolitical and economic reasons and conditions. These situations have played a role to limit the growth and efficiency of MSMEs in Nepal. The result as noted in the World Bank-2020 newsletter, were that Nepal's GDP growth reduced to 2.5 which went down to be negative during the pandemic lockdowns in 2019 and 2020. GDP dropped drastically because the revenue as well as margin of MSMEs along with other income sectors and industries dropped during the pandemic (Dangol et al., 2020) rapidly and contractually. Based on the data as these, MSMEs in general and Nepalese MSMEs in particular are assumed to have fallen among the most hit sectors of business. These could have been the realities, however, without academically validated sources, it would be falsity to conclude the assumptions as given. This argument is relevant as not many researchers have shown their engagement on exploring the problems and their stories of sustainability through COVID lockdowns in Nepal, though just a few studies that in line with Dangol et al. (2020) claim MSMEs to be falling into the hit category. Presumably, among others, as Dangol et al., also point out, encounters with COVID attacks had been fiercer for MSMEs including education consultancies on the same row because of the infrastructure, lack of resources and dependency on foreign materials, lack of alternative resources, halted supply chains, and not the least, lack of innovation shown by their owners in their approach and technique of managing these enterprises.

In Nepal, the list of MSMEs includes education consultancies as very important businesses which have rapidly increased in number to this date. For example, only 254 education consultancies operated in 2013. The number reached 4000 till 2019 when COVID-19 spread globally and the countries imposed pandemic lockdown. The situation as published in the Kathmandu Post (2020, July 29), came as a problem since with the lockdown, many Nepalese education consultancies closed their business. The result was that only 1450 survived to operate amid the COVID-19 lockdown as an obvious loss and problem that the education consultancies together a number of MSMES encountered. Nevertheless, large number of consultancies closing that incurred in loss in number and so triggered financial, and economic difficulties and loss, 1450 education consultancies as well maintained their role even during the lockdowns and beyond.

The financial statements owe Nepalese education consultancies for their significant contribution in the country's revenue every year that in turn tend to have large effect on the economic growth of Nepal. In addition to the economic contribution, the education consultancies are mentioned to be playing a major role in youth empowerment for the country. Their roles are argued to amount in creating environment for youth to get an exposure and opportunity to pursue study at the nationally and internationally renowned institutions of quality education (Kathmandu Post, 2020, July 29). As such, Nepalese education consultancies are said to be contributing substantially in leveraging students, educators, researchers' learning exchanges, and mobility on career prospects as some notable examples of youth empowerment through quality education.

Coupled with MSMEs which are the economic and developmental entities accounted for achieving SDGs, Nepalese education consultancies have evolved largely in recent time that engage multiple stakeholders involved in economic as well as educational endeavors seeing sustainable growth and change. The estimation as this is evident from the experience of the Nepalese education consultancies that were 254 in 2013 in number but increased to 4000 till 2019 when COVID pandemic breakout followed by business lockdowns which in turn brought a decline in the number of operating education consultancies that reduced to 1450 after 2020 and till the end of 2022.

Prolific contribution from the education consultancies in the economic and educational growth is also evident from the reports of Nepal's Ministry of Education. The reports relay 1200 plus No Objection Certificates (NOCs) being issued every day for the students applying abroad on study visa. The flow of outgoing students is the result of consulting assistance primarily contributed by the Nepalese education consultancies. Nevertheless, some kind of drawbacks of consulting services can't be undermined, their share in the economic growth and education for youth empowerment in Nepal self-evident. Their contribution to youth empowerment has turned out to be more substantial in the recent era when aspirations for quality education and economic inclusions have begun to rise on with the new definitions and concepts entered to define quality in education and inclusion in economic development recently.

Given their impactful presence of education consultancies in the economic and educational share, exploring or in order words, identifying some major problems, these enterprises on line with other MSMEs must have faced and the strategies or techniques they could have invented to sustain through and in the post-COVID lockdowns, turns out to be important topic for investigation.

The above-argued problems the survivor education consultancies went through and also invented the solutions, have paved background for the inception of this study. In this regard, explorative approach to identify general if not all types of encounters and the techniques they invented, let alone the importance of any rigorous design on capturing the Nepalese education consultancies in their complexity, was deemed appropriate for analyzing the matter qualitatively. Along with the problems faced, it was felt to be very much required to identify some strategies or techniques the survivor education consultancies could have invented which did work for them to get through the lockdowns and associated economic, communication, and psychological problems.

Concerning the status of Nepalese education consultancies through COVID lockdowns, varyingly contestable thoughts and observations prevail. Common notions are that the education consultancies together with the people involved in their management encountered heavy financial as well as educational losses together with some sort of psychological complications. The opinions allege that Nepal imposed lockdown when COVID was announced to have risen to severity from April 2020 until the later months of 2021. Just as other sectors faced, the lockdowns forced the educational consultancies to be off without business and at the same time had to keep waiting with uncertainties. The prolongation of lockdowns eventually forced a number of consultancies either to go in the dormant state or shut down forever void of any business but with huge financial as well as transactional losses. In short, the results were devastating effects on the students who were waiting to go abroad and also on the investors and employees having their direct role and also dependency on the businesses that the consultancies had acquired.

Nevertheless, general experiences point at the unprecedented types of complexities and decline that few Nepalese education consultancies faced due to varying reasons associated with COVID pandemic and imposed lockdowns, others maintain that those which were in the category of well-established consultancies managed to resume and continue their business and services even amid COVID lockdowns and after 2021. In agreement with the above-mentioned, the Nepalese media too assume virtual modes adopted for transaction by the survived education consultancies in Nepal to have been accountable for sustainability amid the COVID lockdowns.

Given the general perceptions, yet the absence of research-based information about what types of specific problems and what strategic skills and techniques the survivor education consultancies had invented to survive through and resume the services after COVID-19 lockdowns, lent this study a departure.

Focus on the strategic move for sustainability was a key concern toward exploring the approaches, methods, and techniques that the Nepalese educational consultancies could have invented in order to sustain the pandemic till date and also to move on with future prospects.

Challenged with the gaps as indicated above, the study embraced the below-outlined research questions for exploring the phenomenon in academically acceptable design.

1. How are the Nepalese educational consultancies related with the achievement of SDGs for Nepal?
2. What were the problems faced by the Nepalese education consultancies amid COVID-19 lockdowns?
3. What were the major strategic techniques the consultancies invented in order to sustain through COVID lockdown?
4. What have been the implications of their COVID time business status in achieving SDGs for Nepal?

Guided by the outlined research questions, this study was designed to better understand the problems faced by the education consultancies in general and special encounters of economic, educational as well as psychological attacks in particular. To this end, the study executed a review method of analysis and interpretation as described below.

Method

In pursuance to explore and interpret for research understanding of the types of problems faced by and the strategic techniques that the education consultancies of Nepal could have invented, an explorative and interpretive approach to review the secondary and primary sources, was adopted for the study method. For its special focus to identify some major problems and the determinant strategic techniques of interest as invented by the education consultancies for survival through the COVID lockdowns, the explorative design allowed this study to review and interpret the secondary sources-based as well as primary data from the key stakeholders of education consultancies qualitatively.

The results that were supposed to lend sight of strategic techniques which the education consultancies could have invented, were the expected outcomes as regards our purpose to understand how MSMEs such as the education consultancies do sustain the odd situations such as lockdowns. This is important as their maintained status has relation in achieving educational as well as economic growth which in turn has relation with achieving SDGs for a developing country such as Nepal. In this way, the results of this review study are assumed to be significant in understanding what approaches, strategies, techniques, and skills are effective for MSMEs such as education consultancies in achieving sustainability through odds that in turn, contribute to the achievement of SDGs.

For understanding micro, small and medium size enterprises (MSMEs) in their respective features, factors, approaches, techniques, skills, and practices that enable a picture of their category also linking their capabilities to sustain through odd situations, relevant literatures were required which this study reviewed. Also for gathering insights from hand-on experiences of real actors, the owners and experts, involved in the management of education consultancies especially in Kathmandu sector though in small numbers, were reached for open interactions and informant interviews. The major findings of the review and also primary information from the approached informants, have been summarized for referential conclusions regarding the types of problems the consultancies in Nepal were assumed to have faced and the types of strategic techniques they could have invented in order to sustain their services and financial status during and after the COVID-19 pandemic lockdowns in the years from 2020 till 2021.

MSMEs in Developing Countries

How MSMEs are linked with achieving sustainable development goals, is a general question in this study. It broadly guides attention on how MSMEs contribute to achieving the sustainable development goals. It too lends scope to ask what happened to their survival amid the COVID pandemic lockdowns and so, what were the implications as regards their role in achieving sustainable goals during and after the COVID lockdowns for Nepal in particular. In this study, these questions are aimed to be addressed in relation with Nepalese education consultancies as one of the MSMEs in discussion.

In principle, the questions above concern sustainability of MSMEs to keep moving through odd times such as COVID lockdowns. The questions also guide thought on how SDGs were addressed amid COVID lockdown that attacked people who relied on MSMEs here in this study represented by Nepalese education consultancy enterprises.

Before answering the above questions, being familiar with MSMEs to know what category businesses these are, would be required. MSMEs are to be recognized scientifically on the basis of investment, transaction, and profit ranges as major determinants among other indicators.

In this regard, some market-focused perspectives define MSMEs on the basis of the size of the investments and transactions as tabulated (Table 14.1).

Featured with the given investment and annual transaction projections, MSMEs are represented to be startups, retail enterprises, small industries, micro financing, etc. category businesses. These categories survive with small size capital unlike the production industries or large scaled corporates, and national and multinational companies. From the financial perspective, a host of businesses seem to fall in the category of MSMEs. Some notable types of MSMEs include; knitting, weaving, vegetable farming etc., shops, rural cooperatives, groceries, restaurants, hand-skill based domestic occupations, education supplies, services and consultancies, training agencies, and many more… that people manage as basic and easy means of every day subsistence.

MSMEs in Nepalese context are distinguished on the basis of their types of business they do for revenue and transaction. Sources such as Kharel and Dahal (2020) feature a large variety of businesses in the category of MSMEs. A few notable examples as mentioned in Singh (2015), include; knitting occupations, domestic enterprises, retail shops,

Table 14.1 Type of MSMEs

Type	*Investment*	*Transaction or turnover*	*Remarks*
Micro enterprises	Less than 1 crore	5 crore	The turnover includes profit and loss as the result of base money invested, revenue generated, and the expenses incurred
Small enterprises	Less than 10 crore	50 crore	The turnover includes profit and loss as the result of base money invested, revenue generated, and the expenses incurred
Medium enterprises	Less than 20 crore	100 crore	The turnover includes profit and loss as the result of base money invested, revenue generated, and the expenses incurred

Source Data extracted from the Ministry of India (July 1, 2020)

agro businesses specially focused on organic products for export, safety services, education consultancies, small and medium manufacturing enterprises (Kharel & Dahal, 2020), cooperatives, credit agencies, etc. and likewise, restaurants, bars, hotels, and tourism based enterprises, and so on. With the list of hundreds and thousands of small and medium size capital-based businesses which are service and product based, it has been noted that MSMEs contribute 22% to GDP of Nepal (Kharel & Dahal, 2020). European Economic Chamber has included export based and mostly family businesses in the category of small and medium size enterprises. These enterprises, according to EEC-Nepal (2005a, 2005b), have some special features that distinguish them from other types of enterprises and these are as given here:

MSMEs are family-level enterprises set up with family savings and mostly supported by kith and kin.

Credit worthiness of MSMEs is low which impedes their development.

Enterprises rely on entrepreneurial skills and initiatives of the founder/s mainly.

Staff recruitment is often carried out without a proper job analysis or description.

Employees are rarely delegated with authority to decide on their own.

Technology used is usually locally innovated, limited in capacity and efficiency with frequent breakdowns which causes high wastage.

Most MSMEs produce and cater only to the local markets and export-oriented MSMEs are few.

Access to institutional credit for SMEs, despite provisions for the same, is limited due to individual entrepreneur's inability to contribute to equity capital and collateral security.

Another determining component in categorizing MSMEs is the accessibility of an enterprise to bank loans or formal financial sources for capital. For any enterprise small, medium, or large, access to additional reliable and easy sources of capital plays an instrumental role for running transactions by managing contingencies in order to sustain through the odd situations. In this regard, Ministry of India (2020), Kharel and Dahal (2020) and similar studies assert that MSMEs are rarely supported with an access to the bank loans. This means, MSMEs are left to manage their financial viability on their own. This usually involves possible efforts, the MSME owners, which as EEC-Nepal (2005a, 2005b) informs, are the family heads running businesses, make out with their private relations or so.

One of the main reasons of dependency on private sources in Nepal as reiterated in Pant (2016), is due to, "Low financial literacy (p. 1)." Other reasons include, low capital holding people lacking access to the banks or financial institutions(UNESCAP, 2020). The third but very critical reason is related to the existence of cooperatives but that merely aim at catering the owners not common people who come requesting loan assistance to keep their enterprises alive. Shimkhada (2013) has pointed at the tendency of making money for the higher panel or board members of the cooperatives in Nepal's context. This is that cooperatives against their fundamental provision, seek to serve rather the large scale enterprises that might include corporates, large industrial sectors, and so on as their major stakeholders. The basic reason is their money-making orientation that attracts them to connect the large businesses for larger returns.

On the same note, MSMEs, as noted in Singh (2015), function informally and outside the financial regulations without an access to financial support systems but their domestic investments are large. The studies as the one maintain further that MSMEs make significant contribution in GDP of Nepal. There are also few insights MSMEs are operated by women and also working youth whose contribution is not merely financial as the scholars maintain, but also educational and empowerment of youth. From this perspective, as Khatri (2019) asserts, MSMEs, in this study, education consultancies in discussion have played a vital role to help the country address youth empowerment which is one of the SDGs for Nepal. In this way, it results that MSMEs remain deprived from access to capital support and so, are forced to manage from private sources.

As the above discussions shed light, MSMEs in Nepalese context are almost left over to operate on their own in an ad-hoc basis. In the circumstances where the capital is individually and informally managed, its access is not guaranteed nor can it be expected to come to use in the times when needed, it is hence comes suggested that Nepalese MSMEs have usually faced the issue of how to sustain through odd situations. The individually managed investments are usually limited in volume which in turn forces these enterprises to limit their turnover of capital and resources. These as the constraining factors hence, keep MSMEs facing high chances of low sustainability in the odd circumstances such as COVID pandemic lockdowns.

Conceptualizing Enterprise Sustainability

There seem to prevail varying concepts and principles concerning how an enterprise sustains or gets sustainability through odd situations. Among others several knowledge bases seem to capture sustainability linking it with the meaning and concept of Resilience. It is basically an ability to cope with the ups and downs of somebody's private and personal life situation, in our workplace surrounding, in profession, business, and environment, and so on. Craig (2019) asserts, "Resilience is the ability to bounce back." Following this concept, it seems important to realize that as and when unpleasant situations such as the lockdown of COVID come upon in our business process, exploring new ways that are significant to boost the actors' ability to bounce back when different associated challenges are encountered as part of life, is crucial.

Scholarly views linking small and medium businesses, corporate businesses, education, job sectors, bureaucracy, and more... advocate resilience as an indispensable quality to be successful through any circumstances such as COVID-19 and associated catastrophes. The other important feature of resilience is that it is a skill that enables one to adapt to and create changes so as not only to bounce back from the challenges that descend on us but also convert them into opportunities. From this perspective, resilience emanates to be a strategy or strategic skill. Strategy by nature is an intellectual skill that guides individual thoughts to be employed for solving problems of any kind.

The idea of exploring what makes any business sustainable embodies in connection with SDGs that started engaging the world governments, business leaders, development agencies, professionals, experts, corporates, businesses small, medium, and large, and all common citizenry from the arrival of 1990s when the whirlwind of globalization began to rise in general and from 2015 when the world leaders in an assembly of UN formalized it as a central issue in particular (Inc. 2U Brand, 2020). Sustainable development for the global communities is conceptualized in 17 major goals. Among them, quality education for global citizenship and career enhancement, affordable and clean energy, decent work and economic environment (Freedman, 2022), industry innovation and infrastructure, strong health systems, and innovative partnerships (Gavi, 2020), have direct relations when it comes to hew out the long-term sustainability of businesses.

Economy-focused opinions point at the indispensability of developing sustainable models for businesses. It is as Inc. 2U Brand (2020) elucidates, "Sustainable business model may be a successful business model."

Considering the crucial role of sustainability models for the success of businesses in developing economy also linking them in achieving sustainable development goals, some basic yet critical questions engage multiple stakeholders in general. But among others, the question of sustainability designs for the success of businesses, do very much connect the researchers whose goal it is to invent new concepts, content, and methods of sustaining enterprises. Researchers are usually intrigued with a question of how to develop capability in designing sustainable businesses. As such, inventing some models of businesses sustainability, in real sense then comes linked with exploring factors and approaches that shall foster the capability of an enterprise. It too concerns with inventing new technologies, techniques, tools, media, and skills that should contribute to constituting new yet more effective systems in filling the businesses with the caliber required to sustain through the odd situations.

On the stated requirements, just a few fundamental concepts have been referred to defining sustainability that seems applicable in projection of small and medium sized enterprises too. Theoretically, sustainability is defined to be an ability to continue. This concept is further extended in UNESCO (2020) that translates the meaning of sustainability for operational purpose as here:

> Sustainability is a paradigm for thinking about the future in which environmental, societal and economic considerations are balanced in the pursuit of an improved quality of life. For example, a prosperous society relies on a healthy environment to provide food and resources, safe drinking water and clean air for its citizens. (UNESCO, 2020, p. 1)

Following the idea outlined here, it comes into view that any phenomenon, system, task or technique can be called sustainable if that in concept and/or mechanism and action, is adequate to nurture and preserve the resources (natural, manmade and human resources) for future by causing minimal depletion while obtaining maximum out of its use (Freedman, 2022).

Though there are numerous perspectives engaged to answering what, how and why sustainability, the adopted concept is relevant as concerns our purpose and it is to define sustainability as a term. Primarily, the relevance of outlined notion pertains to defining sustainability in terms of the capability that an entity or phenomenon, a system or process should bear to balance, cope and continue to get through the existing circumstances in general and also through the ones that might emerge in course of time in particular.

Maintaining capability as central to capturing sustainability of the business for survival through the circumstances, scholars have proposed numerous ways concerns how any situation or a system such as the economic and/or business systems of today could be modeled to be sustainable. Centering to this basic question of search, several scholars have proposed a variety of competitive approaches and methods that could be deemed relevant for developing the businesses of today into sustainable models as vital for achieving sustainable development goals. Below, a few special concepts reverberated in diverse knowledge bases have been referred in pursuance to the quest what helps our business sustain through the circumstances which is embedded with answering how sustainability takes a form for a business practice.

Inc. 2U Brand (2020) proposes nine under-listed as key measures helping the businesses sustain through situations:

1) Sketching or business plan that connects with the need which are real and long term

2) Repairing social trust and it consists of practicing business within the partnership model that helps connecting society, employees, governments, suppliers, consumers, and communities with some form of ownership
3) Adopting sustainability as a core principle at input, process, and output levels
4) Constantly researching to evaluate existing practices and explore new ideas
5) Innovating that means changing strategies in response to the demand of times and contexts
6) Incorporating diverse leaderships
7) Setting holistic and long-term vision throughout
8) Ensuring operations, service delivery, administrations, and other execution aspects with accountability
9) Embracing competitors as collaborators

From the outlined discussion, it is the primary information that sustainability is a very demanding concept. Among others as the above outlines lend idea, sustainability results from sharp and effective intellectual strategies. It too demands building capabilities of diverse people in an organization or business that may include diverse levels such as people involved for technical handling, managerial work, and of course, on top of all, leadership level whose concern it is to develop the business in model that substantially fits to and is effective to copying with as well as survive through any emergent contexts such as COVID-19 contexts.

Crafting sustainable models for businesses, as Ludeke-Freund et al. (2019) maintain, demands planning as a basic yet that very important action that needs to be done with a bigger picture in mind. Planning triggers construction of new patterns or designs that in turn can result into the development of models which are effective for any particular business to survive through odds. The second equally important element of sustainability is modeling. As suggested in Ahmed and Sundaram (2012), modeling is centrally a job of creating partnership model. It is crafting a design in that the business operates in the cooperative partnership of different players. This design among others can use nearer society, government agencies, suppliers, consumers, investors, and experts, etc. in the collaborations or partnership with ownership of each.

Sustainable strategy according to Burton (2020) results from resilience as a very important skill and strategic power for the success of business. Resilience shapes in the flexibility of intellectual thinking, planning, and devising strategies. These endeavors play instrumental role for constructing not only the ideas against a situation but also actions that should contribute to solving the emergent problems of any kind.

Adding more, as concerns how to create resilience for the businesses in operation, Craig (2019) stresses on the role of practice that is not only consistent but also strategically artful and skilled one. Practice as the researcher suggests, should be relevant and best fit in a given context particularly for tackling with the situations efficiently let alone the supremacy of its scientific design that helps averting risks of loss for effectiveness in achieving the aimed outcome.

Trading experts, such as Dam van (2009), define sustainability in an ability of businesses to sustain a situation that results from the practitioners' ability in crafting high probability trading strategies. Scholars maintain that high probability trading strategies build only if the business practitioners can bring out rigorous trading plans in written form and consist of concrete steps and processes for an outcome.

In brief, the reviewed theoretical bases indicate that businesses are conditioned in general to get through diverse circumstances that come as challenges for survival as a basic problem. Alongside the challenges, it is as well true that new circumstances bring new opportunities associated as a common law of the phenomena or nature. As the emerged circumstances also bring along a number of possibilities, then it becomes necessary for the stakeholders to explore them as the opportunities so that the challenges of any kind could be coped with and their solutions worked out. In this way, it is a suggestion from the review o that businesses need either to follow or if possible create sustainability in their system and process which shall be instrumental to build in their new capability for the businesses to sustain through any odd situations that usually can hit the businesses abruptly.

Education Consultancies in Achieving SDGs for Nepal

Scholarly sources such as Kandel (2020) report positive rise in the income tax payment in Nepal due to the rise in inland and also remittance revenue size. Linking this as an important component of GDP having a role achieving SDGs, scholars on the same note, draw Nepalese Education Consultancies that were 4000 in operation and paying significant amount of income tax till 2019. This is linked with the financial contribution that the Nepalese education consultancies made. Also as highlighted in some manuscripts that consent with 2022 fiscal report of Nepal Central Bank, Nepalese education consultancy enterprises have served considerably to GDP financially. Whatever the size of revenue and growth in GDP, it is in turn connected with achieving SDGs from financial perspective.

Although amid the COVID-hit era, only 1450 plus consultancies survived (Kathmandu Post, 2020, July 29), as recorded in the report of Central Bank of Nepal-2020, their financial participation persisted to be far more important as compared to other businesses for revenue and GDP. Similar reports linking education consultancies' revenue with GDP claim their impactful role in achieving SDGs for Nepal. In this way, in agreement, some scholars also consenting with Singh (2015), acknowledge MSMEs and on the same line, Nepalese education consultancies as being the very influential players which make a large contribution in raising the Inland Revenue that helps building financial status leading to economic growth for the country. As such, it is concluded that the Nepalese education consultancies are impactful in raising the Inland Revenue, and so, these enterprises have started to be one of the most engaging sectors due to their impactful presence in the linearity of economic growth and changes toward achieving SDGs from financial perspective.

Along with the financial contributions, other significant role which the studies report about Nepalese Education Consultancies as one leading category of MSMEs include their obvious roles in the educational and career enhancement that trigger empowerment of youth as another determinant on addressing SDGs for Nepal as a developing country. Even amid COVID pandemic lockdown and after its normalization in Nepal, the expert education consultancies were reported to have remained preoccupied in mediating admissions for the learner communities in the foreign Universities which in case of Nepalese students increased considerably. The media reports such as Kathmandu Post (2020) have relayed that the Nepalese education consultancies that survived also helped the students getting enrolled at a number of universities that were recognized for their promising academic profile as a must condition in lending the scholars access to quality education. This is another significant component as regards achieving SDG from access to quality education and empowerment perspective that the Nepalese education consultancies have been accounted for their major role.

Access to quality education as one major determinant in achieving SDGs, is claimed to have been immensely contributed by Nepalese education consultancies as asserted in the studies that consent with Expert Education (2021). The studies highlight that the Nepalese Education Consultancy enterprises even through COVID pandemic lockdowns managed to keep a large number of Nepalese students intact to continue toward career enhancement with abroad studies. This is argued to have been significant in bridging the career gaps of Nepalese students and thereon, take an advance leap in developing their profile with new intellectual traits and global citizenship skills earned out of their active participation in the academic and career endeavors with global knowledge communities.

Nepalese education consultancies focused to availing Nepalese youth with an opportunity to advance in their status and transform to be global citizens through their access to international education, have significantly helped Nepal in addressing SDGs from empowerment perspective. The studies pinpoint that it's particularly youth empowerment, resulting from an access to quality education achieved by means of active facilitation roles played by Nepalese education consultancies which is a contribution to achieving SDG- No.4 by 2030.

In short, as based on the above discussions, it is summarized here that together with other many types of relations, the enrolments of Nepalese students in large number in the renowned universities as well as placements found abroad as mediated by the Nepalese Education Consultancies have been significant in the career growth of Nepalese youth. As career growth is the result of empowerment it is regarded to be another eminent sub-determinant upon achieving SDGs for Nepal from an empowerment perspective. Thus, it is here finally argued also agreeing with Expert Education (2021) that Nepalese Education Consultancies have been the major enablers for Nepalese youth to gain financial niche with their academic and career profile enhanced at par the international status which has ultimately contributed to the country's revenue and educational enhancement. Thus, the role of Nepalese Education Consultancies has been considerable as concerns achieving SDGs for Nepal from financial as well as empowerment perspectives.

Problems Faced by the Nepalese Education Consultancies

In recent times, education consultancies have come in the Nepalese education market in a mushroom growing fashion. For instance, there are the reports that show 254 consultancies legally operated in 2013 reached 4000 till the breakout of COVID-19 and from 2020 the sources mention 1450 plus consultancies still legally serving in Nepal (The Himalayan Times, 2020, May 21). Together with the notable rise in the number of consultancies, it is as well to be seen that the trend of Nepalese students going abroad on study, has exacerbated in an unprecedented pace. These critical developments clubbed with an abrupt rise in the number of education consultancies that brought associated unwarranted competitions. The volatilities associated with unwarranted increment in number and resulting unhealthy competitions gave birth to different types of complexities for the educational consultancies to cope with. Whereas the odd situations such that came associated with and after the outbreak of COVID-19 pandemic lockdowns, forced the economic activities of the individual states and global levels to stop at large. This eventually shut the economic mobility down to be nearly null for example evidenced from Nepa's GDP that reached negative level of 2.2% (Gaire & Koirala, 2021) in 2020. The downfall as reported, attacked local, regional, national and international businesses yet accounting the education consultancies that turned out to be among the most hit enterprises amid the COVID lockdowns.

The attack of COVID-19 pandemic on the Nepalese small and medium businesses under that featured the education consultancies as one category started facing yet another stingier problem. It was fierce deadlock to their services which invited a problem of how to pay for rentals for their offices let alone the payments to their employees and staff which couldn't be imagined amid the circumstances.

It was visible not only in terms of loss borne by the student candidates who got stranded from going abroad due to border and global locks but also financial losses forced the education consultancies to encounter acute situation that included; loss in the resources (Kathmandu post, 2020, July 29). It eventually led many consultancies to shut down since they were unable to pay even the minimum expenses. That too constrained the education consultancies to decide employee layoff. The sources maintain that 3 out of 5 employees who lost their jobs due to COVID-19 in

Nepal. In addition to the losses of lives due to lack of medical supplies, rise in death toll not only from Corona Attack but also from psychiatric issues (Poudel & Subedi, 2020), hunger, and so on…

The Nepalese education consultancies were reportedly ignored by Nepal as a state from its responsibility to support when they fell victim to financially acute situations. Among several reasons, as Serchan (2019), the expert consultant, pointed out pertained to the problem of how to retain the expert resources on whom the consultancies all depended for their transaction and success to help the students seeking to fly abroad for study. It too related to the question of how to operate the offices and sustain till their "business could come to as usual." The scholar Serchan (2019) claimed, "Meeting the operational costs which are so high in Kathmandu and retaining our highly trained human resources who are our biggest asset, is the challenge (p. 2)."

In an argument concerning how deep and long-term impacts of COVID-19 lockdowns have been more acute on education consultancies than other business, few scholars point at the issue of downfall of the economy globally that majorly affected the students who sought to go abroad with the hope to study on scholarship support and job wages. Since the economic dripping made these opportunities of scholarship and earning narrower, the situation directly affected the students and Nepalese consultancies. It was simply because Nepalese students tend to depend largely on scholarship assistance and work-wages. Serchan (2019) has the following observation:

> With global economy plummeting, and Nepalese students highly dependent on scholarships, part-time jobs, co-ops and internship, the bounce back will entirely depend on how the host countries come out of this crisis and how long would respective economy take to recover. It will also be guided by the new immigration rules and foreign student policy of the respective countries/colleges.

According to a few observers as published in The Himalayan Times (2020, May 21), among many hurdles, the education consultancies faced during the COVID lockdowns rather, centered on the problem of transaction blockages, deadlocks in communication with the students and also outside with the universities due lockdowns and vacuum it created. These problems resulted for example as mentioned in the THT (2020, May 21), from deference in the admissions of the abroad going students in the American and other country universities. As such, the problems as the details in the news coverage suggest, didn't only limit to the blockages in business but also, the problems rose to the level that forced many normal consultancies to close their agency as a whole.

Another problem of COVID-19 as the sources allude, related to the universities that switched the classes online which too affected the waiting students and mediating education consultancies of Nepal. This came to the Nepalese students with a problem on how to continue their classes in the event the internet was not reliable let alone their interest that was to take abroad study with scholarship support and also work in there and get the wages which in fact turned out to be impossible. As such, from admission and continuation of class perspective, the consultancies along with the Nepalese students as the main clientele of education consultancies and universities abroad were forced to face a stop in their business and so, entire transaction.

Ghimire (2021) details a number of problems and strategies invented by the education institutions and also consultancies in Nepal during the COVID havocs. The scholar cites the UNESCO report that maintained over 90% of the world's student population and equally of Nepal who were affected by the pandemic. It is also reported that the online classes were introduced but as the author alleged the government's ignorance and inability to ensure effective IT infrastructure and private ownership of internet, tele-media, and e-learning systems and services, resulted into the ineffective operation of classes and learning engagement. On top of all, it led to the price hike due to unhealthy competition among the media and education suppliers. Evidently, along with educational institutions of the country, the consultancy enterprises were hit badly which too triggered the closing down of a large number of education consultancies that failed to integrate ICT in order to manage classes virtually also their student connections for transactions to move on.

Following the above notes, so many explicit reasons behind the hits suffered by the Nepalese consultancies, also recorded in though just a few research papers, include varying barriers.

One of the barriers faced by the education consultancies together with the universities as Serchan (2019) asserted, was an absence of student candidates who couldn't be brought into contact by a number of education consultancies. The scholar indicated that the blockages of contact with the candidates was suffered by the small size consultancies directly.

It was because the small size consultancies lacked resources and virtual media that were required to keep their clientele intact (Ghimire, 2021).

In addition to the outlined problems, this study captured the general market views. According to general market or stakeholder experiences, the education consultancies of Nepal, as Thiemes (2017) a few years prior to COVID-19 had pointed out, suffered not only the acute fangs of economic crises but also collapsing of their agencies due to inability to create resilience. For instance, Serchan (2019) relayed that education consultancies in the majority of the countries have been closed or operating classes digitally but minimally in the first wave of COVID in 2019. On the same magnitude and even worse has been the situation of education consultancies in Nepalese market contexts as noted in the news sources and market updates.

The above problems indicate that a resilience was required for the consultancies to sustain amid COVID lockdowns which 66% consultancies couldn't manage, leading to closedown of their business. This means, a great loss of revenue was faced by the consultancies and also the nation that therefore stood as a barrier to economic growth as a bad impact of COVID-19. Another equally distressing impact of lockdown, the Nepalese education consultancies suffered was the loss of connection in helping the Nepalese students get an admission in the universities abroad. This triggered blockage on youth empowerment which the Nepalese education consultancies were contributing lavishly until COVID lockdown broke out with their role in helping the students get an access to quality education abroad. It is therefore the problem as this is concluded in line with a study by Dawadi, Giri, and Simkhada (2020) that mentions, "Pandemic has had serious impacts on students' learning and well-being, and that it potentially widens the gaps between advantaged and disadvantaged children in their equitable access to quality education."

Techniques Invented for Survival Through COVID-19

Nevertheless, numerous kinds of problems as outlined above, were traced encountered by a large number of education consultancies among which approximately 66% consultancies closed down, the remaining 34% that survived the tough market competitions and with the situation, at the same time succeeded to invent new techniques that contributed to sustainability amid COVID-19 lockdowns. The reality that few consultancies sustained to function even amid the COVID lockdowns, though without very much promising results and performance, offer a scope to research what operating techniques were invented by them which could be interpreted to have been effective to move on if not prosperous through COVID pandemic locks.

Concerning the maintenance of sustainability made out by Nepalese education consultancies amid COVID lockdowns, different stakeholders together with reports of media studies and researchers as well mention varying techniques that the surviving consultancies invented which finally worked, are as briefly elaborated in the subsequent paragraphs hereafter.

Nair, Aedo, Aryal, Sherpa, and Sharma (2020) recorded Nepalese education sector going speedy on the instruction of e-learning system and media which as the stakeholders agreed also stimulated the education consultancies to operate their training classes such as TOEFL, IELTs, etc. supported by these new tools, though without shortcomings faced due to availability of unbranded and ineffective ICT services, media and systems of e-learning.

Considering the role of universities as a major component, some reports published in the national daily newspapers such as THT (2020, May 21) bring into light that the Nepalese consultancies managed to negotiate with the abroad institutions for deferring student admissions. Under this scheme, admissions from current intakes were shifted to the next. Deferring admission means you can postpone your start date for school without having to apply again. This technique as the sources lend idea, eased the process for the students not to go for round of repeatedly applying for admissions and so remain in loop with the Universities. As this was found to be effective saving the students from repeating hassles, it too came as an easier way for the universities to keep the students who had been offered admission in contact and so, not lose their international students. It was witnessed as the reports maintain, many universities approved differing admissions.

The other technique invented by the education consultancies along with the universities included effective negotiation on easing admission process. It shaped right from the application process that was agreed to be made much easier than its practice previously. As pointed out in THT (2020, May 21), "Many institutions will try to lure in international students by either easing application process or introducing attractive funding choices."

In an effort to manage their services move on, as some studies relay, the education consultancies of Nepal brought into fashion virtual interactions alongside the online classes during the COVID-19 lockdowns. The THT (2020, May 21) recorded, "Education consultancies in Nepal are regularly hosting virtual fairs via social media handle." Albeit special advancements came associated in the economic systems with the invention of virtual technologies, as mentioned in some sources, virtual technology-supported activities helped the consultancies keep connections with their students and universities. As opined the stakeholders, this technique did work for the consultancies to keep their services move on even during the lockdowns.

Let alone the problems that arose due to the Nepalese education consultancies that collapsed, the survived consultancies, not merely technical but also psychological techniques were observed with effect that significantly helped the businesses move on. Regarding psychological techniques, few scholars express in line with the extract from Fernandez (2016, June 27):

> Simply put these attitudes are commitment, control, and challenge. As time gets tough, if you hold these attitudes, you'll believe that it is best to stay involved with the people and events around you (commitment) rather than to pull out, to keep trying to influence the outcomes in which you are involved (control) rather than give up, and to try and discover how you can grow through the stress (challenge) rather than to bemoan your fate.

Based on the review of sources and consultations and their findings, the final part of this section concludes whether and how the Nepalese education consultancies managed their role to educational and economic contribution which in turn helped attaining SDGs for Nepal amid COVID-19 lockdowns.

Concluding the Contributions of Nepalese Education Consultancies to SDGs for Nepal

This study aimed at exploring the problems faced by the Nepalese education consultancies as one category of MSMEs which are classified in terms of capital investment and yearly transactions these business stand within Nepalese market. Alongside, this study explored some important techniques of solving their problems as invented by the Nepalese education consultancies in order to sustain their business through COVID-19 linking their business status amid the pandemic and how it affected their economic and educational contribution for Nepal in addressing SDGs.

In this regard, based on the secondary sources and also information regarding the articulated issues gathered from the informal interactions with the consultants and owners of the education consultancies in Kathmandu, the key insights have been derived that 66% Nepalese education consultancies closed down within the first and second wave of the COVID-19 pandemic lockdown in Nepal. The major reasons behind the decline of the recorded number among others as the sources and also interview-based information allowed were the financial crunch that the majority of education consultancies stared facing amid the phenomenon. The main reason behind the occurrence of financial crunch as this study explored was the inability of a number of those consultancies that have now closed eventually to keep their communications and business relations maintained amid the periods of lockdown which ultimately resulted into the loss of confidence and eventually loss not only in business but closing down of the agencies themselves.

Those consultancies that managed to survive, did so, not without the challenges nor without constraints and pains both financial and also non-financial for example psychological as well as personal and professional. As outlined in the afore sections, 34% of the surviving consultancies suffered negative and also least revenue collection in their service transaction, encountered months of blockades due to lack of virtual media and resources that they needed to keep their students in connection together with the universities. Though limited in many respects, the surviving consultancies as the discussions shed light, managed to revive their connections virtually with the universities of enrollment and also the students who sought to go abroad.

First and foremost insights as regards what sustainable techniques the Nepalese education consultancies learnt amid COVID-19 lockdowns to revive their businesses, links with their success in being able to embark on partnership model of sustainability. It was that the agencies as demanded by the situation of capital crunch leant to create an environment of ownership of multiple players. Many agencies that survive till date, are now in the cooperative model and hence, their transactions are moving on as they have been able to attract capital and resource investment along with the entry of the experts not merely as employees but also as owners (Inc. 2U Brand, 2020).

As the reports lent idea, the Nepalese consultancies that survived through COVID-19 lockdowns, are concluded to have learnt adopting virtual channels on business purpose. The new media or channels helped them connect with the new candidates, revive contacts with the student candidates who had been in contact from the time before the lockdowns were imposed and build connections with new universities abroad that came with the new prospects of flexibility in admission process and maintain the previously connected universities to resume mediations and enrolments amid COVID-19 lockdowns. As such, with the resumption of their service, the consultancies that survived through the lockdowns started making revenue, though limitedly in many respects. In this way, it is to the conclusion that the education consultancies that succeeded to adopt virtual channels for communication, were then ones that manage to pay tax though that as the central bank's report alluded, was very minimum, as little as 10% of the amount the consultancies used to pay before COVID-19 pandemic lockdowns.

In addition to the financial contribution that the surviving Nepalese education consultancies managed for the country, also as alluded in the discussions above, the invented techniques that consisted of keeping the clientele in loop such as by deferring the admissions, by counselling the groups for remaining in contact in order not to lose the opportunities but all managed through virtual communications. Together with, revival and continuation of the consultancy transactions that revived their mediating services virtually, did eventually serve educational contribution. In this regard, though minimally, the consultancies not only resumed their services by virtual means but also started making substantial connections such as operating the training and consulting work online virtually, negotiating with the universities to help facilitating admission process with ease and many more. In this way, as evident from the reports of education ministry in Nepal, 1200 NOCs issued on a daily basis during and in the post-lockdown times, it is hence, concluded that though reduced to 1450 plus from 4000 within the lockdown period, the Nepalese education consulting agencies did manage their considerable role enabling the awaiting and new students to get enrolled in the universities abroad. This from the SDG perspective was an empowerment goal which Nepal as a developing country strives to achieve by lending the youth an access, exposure, and participation in quality education that in this study is captured in terms of Nepalese students' success to get admission to pursue the courses of their choice in the universities abroad having their international repute.

References

Ahmed, M. D., & Sundaram, D. (2012). Sustainability modelling and reporting: From roadmap to implementation. *Decision Support Systems, 53*(3). https://doi.org/10.1016/j.dss.2012.02.004

Burton, L. (2020). Resilience in the workplace: What are the benefits and how can businesses develop it? High Speed Training Institute. https://www.highspeedtraining.co.uk/hub/resilience-in-the-workplace/

Craig, H. (2019). *Resilience in the workplace: How to be more resilient at work.* PositivePsychology.com. https://positivepsychology.com/resilience-in-the-workplace/

Dam van, R. (2009, June 21). Minor sustainable development. *Opening Notes,* Universiteit Leiden.

Dangol, J., Chitrakar, S., & Yoo, K. S. (2020). Impact of COVID-19 on Nepalese Small and Medium Enterprises. *Journal of Business and Social Sciences Research,* 5(2). https://www.nepjol.info/index.php/jbssr/issue/view/2252

Dawadi, S., Giri, R., & Simkhada, P. (2020). *Impact of COVID-19 on the education sector in Nepal—Challenges and coping strategies.* ERIC Institute of Education Sciences. https://doi.org/10.31124/advance.12344336.v1

EEC-Nepal. (2005a). *Overview of Nepalese small and medium enterprises.* European Economic Chamber (EEC). http://www.eec-nepal.org.np/trade.php

EEC-Nepal (2005b). *Export oriented small and medium enterprises.* European Economic Chamber (EEC). http://www.eec-nepal.org.np/trade.php

Freedman, M. (2022, Jun 29). Green business ideas for eco-minded entrepreneurs. *Business News Daily.* https://www.businessnewsdaily.com/5102-green-business-ideas.html

Fernandez, R. (2016). *5 Examples of building personal resilience at work.* Harvard Business Publishing, USA.

Gaire, R., & Koirala, Sh. (2021). *Analysis of education budget of fiscal year-2021/22.* NCE-Nepal. https://ncenepal.org.np/wp-content/uploads/2021/07/Analysis-of-Federal-Education-Budget-FY-2021_22.pdf

Ghimire, K. (2021). Problems and challenges of COVID-19 in Nepal education and its solutions. *Academia.* https://www.academia.edu/43293925/

Global Alliances Vaccines Immunes [GAVI]. (2020). SDGsthrough vaccination and immunization. *Global Vaccine Summit.* https://www.gavi.org/investing-gavi/resource-mobilisation-process/gavis-3rd-donor-pledging-conference-june-2020

Inc. 2U Brand. (2020, May 25). Growing a sustainable business. *Newsletter*. 2U Inc.

Kandel, R. K. (2020). Contribution of tax revenue to national revenue. *Master Thesis*. Centria University of Applied Sciences.

Kathmandu Post [KP]. (2020, July 29). As pandemic stops Nepali students from going abroad, consultancies and businesses take a hit. *Kathmandu Post National Daily*, Kantipur Publications.

Kharel, P., & Dahal, K. (2020). *Small and medium-sized enterprises in Nepal: Examining constraints on exporting*. Asian Development Bank.

Khatri, M. B. (2019). Small and medium scale enterprises and their role in employment generation in Nepal. *Tribhuvan University Journal, 33*(1). https://doi.org/10.3126/tuj.v33i1.28688

Ludeke-Freund, F., Froese, T., & Schaltegger, S. (2019). *The role of business models for sustainable consumption: A pattern approach*. Edward Elgar Publishing. https://doi.org/10.4337/9781788117814.00015

Nair, K. R., Aedo, C., Aryal, M. P., Sherpa, M., & Sharma, U. (2020, June 27). Nepal steps up remote learning during COVID-19. *Education Nepal South Asia COVID-19*.

Pant, B. (2016). Financial inclusion in Nepal: Policy review and prescriptions. *Working Paper Series*, NRB-34-WP.

Poudel, K., & Subedi, P. (2020). Impact of COVID-19 pandemic on socioeconomic and mental health aspects in Nepal. *International Journal of Social Psychiatry, 66*(8). https://doi.org/10.1177/0020764020942247

Serchan, L. (2019). *A premier education consultancy in Nepal*. PAC ASIA – Study Abroad.

Shimkhada, N. R. (2013). Problems and prospects of the cooperative sector in Nepal for promoting financial inclusion. *Enterprise Development and Micro Finance, 24*(2). https://doi.org/10.3362/1755-1986.2013.014

Singh, Sh. M. (2015). *Micro and small and medium enterprises in Nepal: Ways to grow and globalize*. https://www.youtube.com/watch?v=CmFhGR-nN2A

The Himalayan Times [THT]. (2020, May 21). Impacts of COVID-19 pandemic faced by Nepalese education. *The Himalayan Times*. International Media Network Nepal Pvt. Ltd.

Thieme, S. (2017). Educational consultants in Nepal: professionalization of services for students who want to study abroad. *Migration Infrastructures and the Production of Migrant Mobilities, 12*(2). https://doi.org/10.1080/17450101.2017.1292780

UNESCAP (2020). Micro, small and medium-sized enterprises' access to finance in Nepal. *MSME Financing Series, 3*. UNCDF.

UNESCO. (2020). *Sustainable development*. UNESCO. https://en.unesco.org/themes/education-sustainable-development/what-is-esd/sd

World Bank. (2022). *Small and medium enterprises (SMEs) finance: Improving SMEs access to finance and finding innovative solutions to unlock sources of capital*. World Bank Group. https://www.worldbank.org/en/topic/smefinance

Index

H. Dasaraju and T.T.H. Tambunan (eds.), *Role of Micro, Small and Medium Enterprises in Achieving SDGs*, Sustainable Development Goals Series, https://doi.org/10.1007/978-981-99-4829-1

GPSR Compliance

The European Union's (EU) General Product Safety Regulation (GPSR) is a set of rules that requires consumer products to be safe and our obligations to ensure this.

If you have any concerns about our products, you can contact us on ProductSafety@springernature.com

In case Publisher is established outside the EU, the EU authorized representative is:

Springer Nature Customer Service Center GmbH
Europaplatz 3
69115 Heidelberg, Germany

Batch number: 10703843

Printed by Printforce, the Netherlands